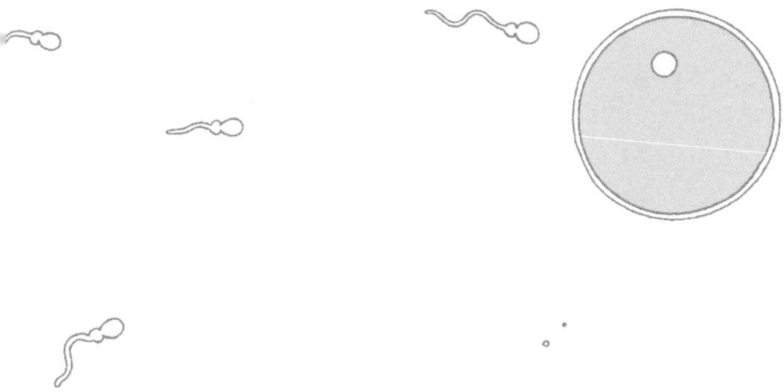

DIRTY BIOLOGY

DIRTY BIOLOGY

THE X-RATED STORY OF THE SCIENCE OF SEX

SCRIPT BY LÉO GRASSET
ART BY COLAS GRASSET

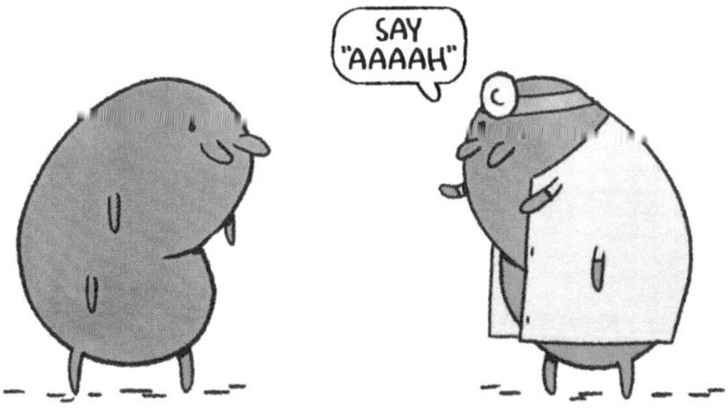
SAY "AAAAH"

GRAPHIC MUNDI

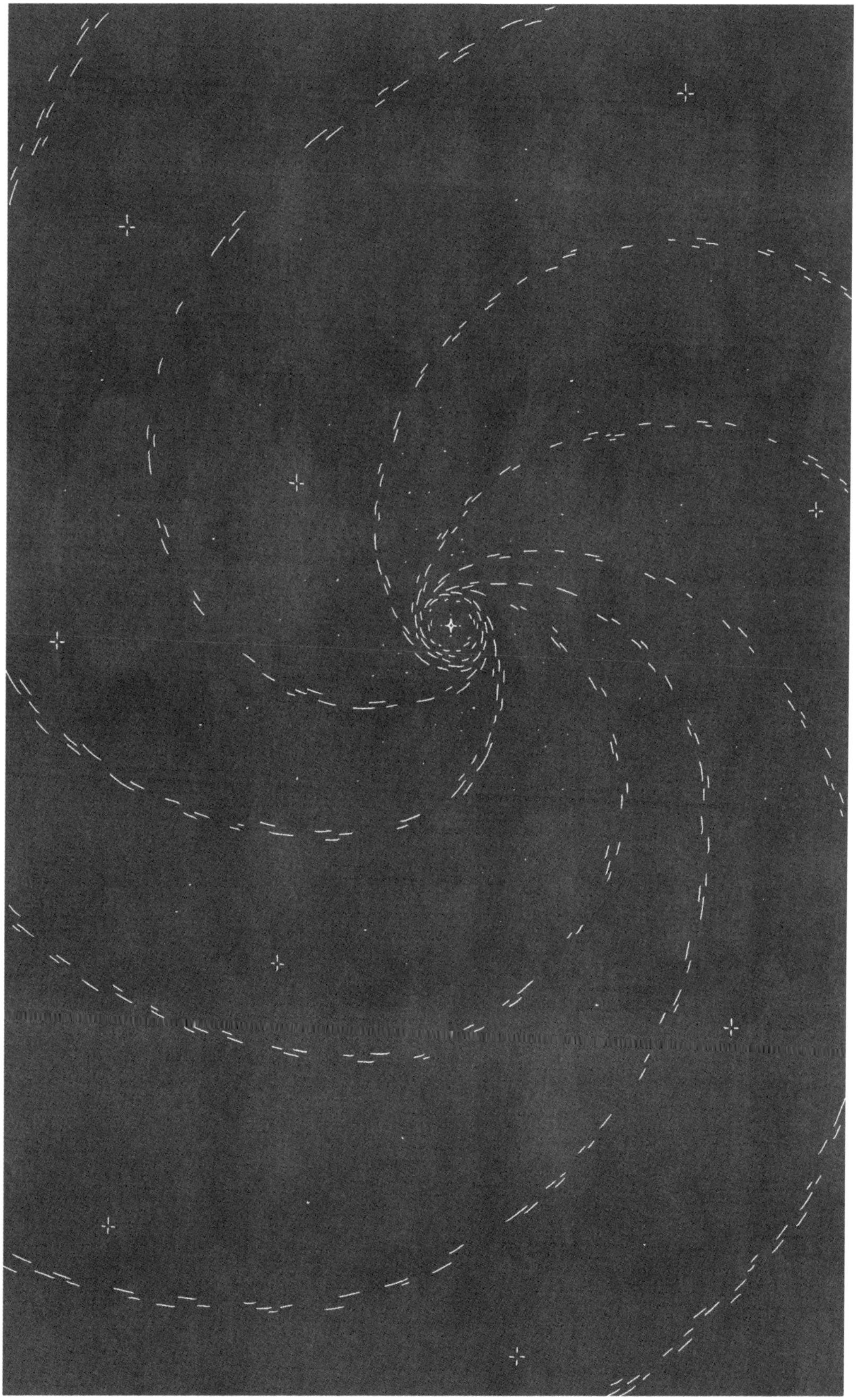

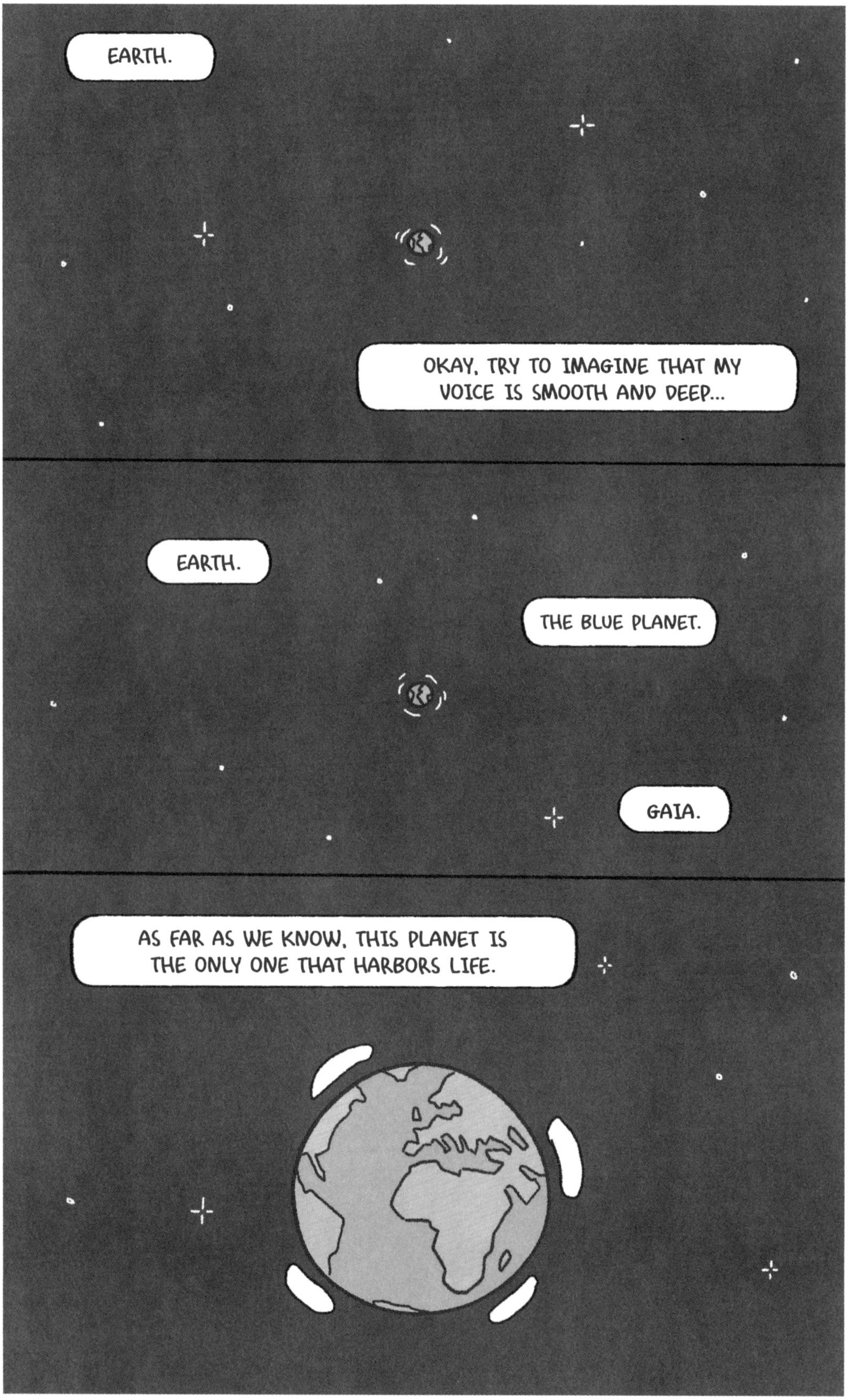
EARTH.
OKAY, TRY TO IMAGINE THAT MY VOICE IS SMOOTH AND DEEP...
EARTH.
THE BLUE PLANET.
GAIA.
AS FAR AS WE KNOW, THIS PLANET IS THE ONLY ONE THAT HARBORS LIFE.

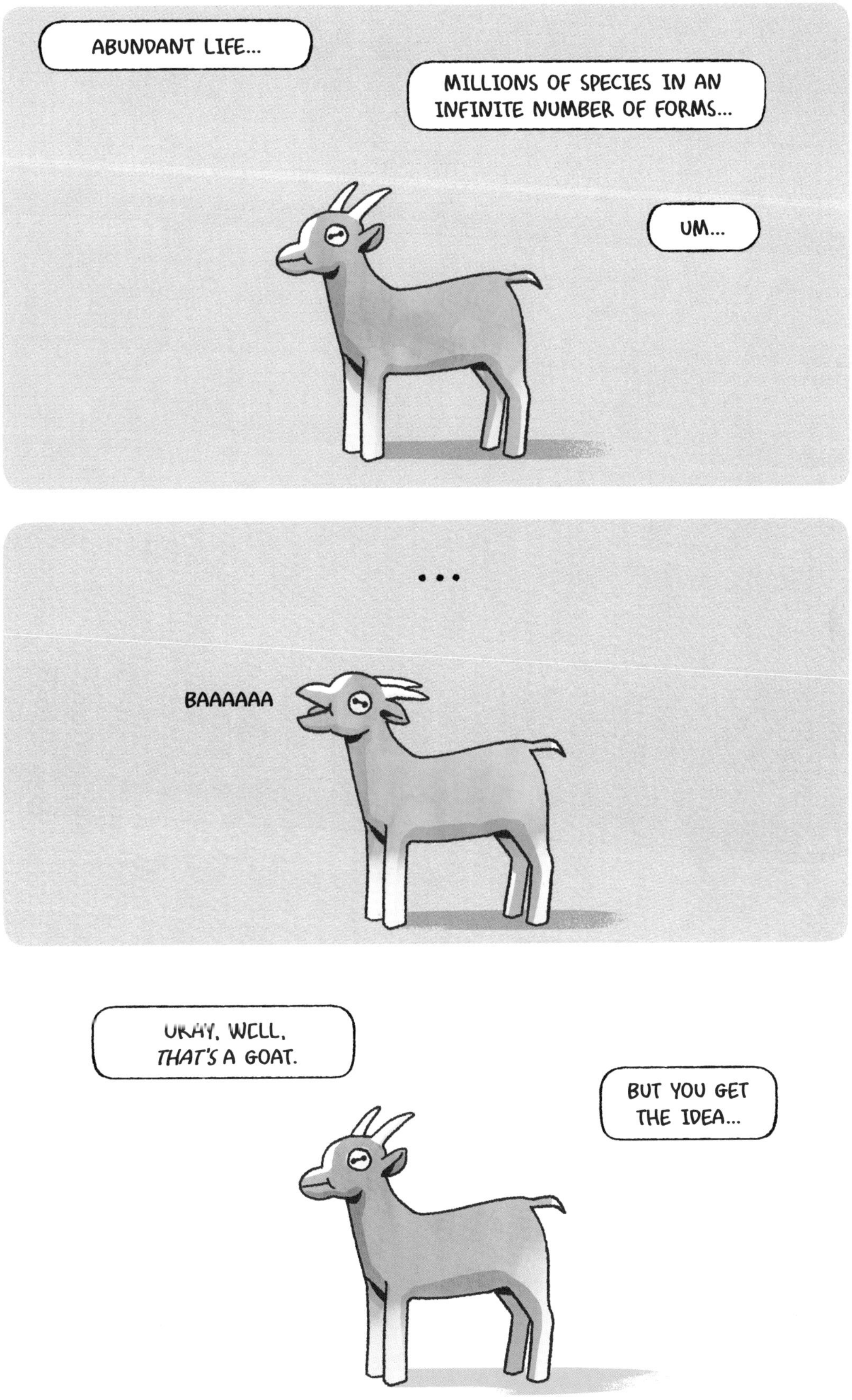
ABUNDANT LIFE...
MILLIONS OF SPECIES IN AN INFINITE NUMBER OF FORMS...
UM...
• • •
BAAAAAA
OKAY. WELL. *THAT'S* A GOAT.
BUT YOU GET THE IDEA...

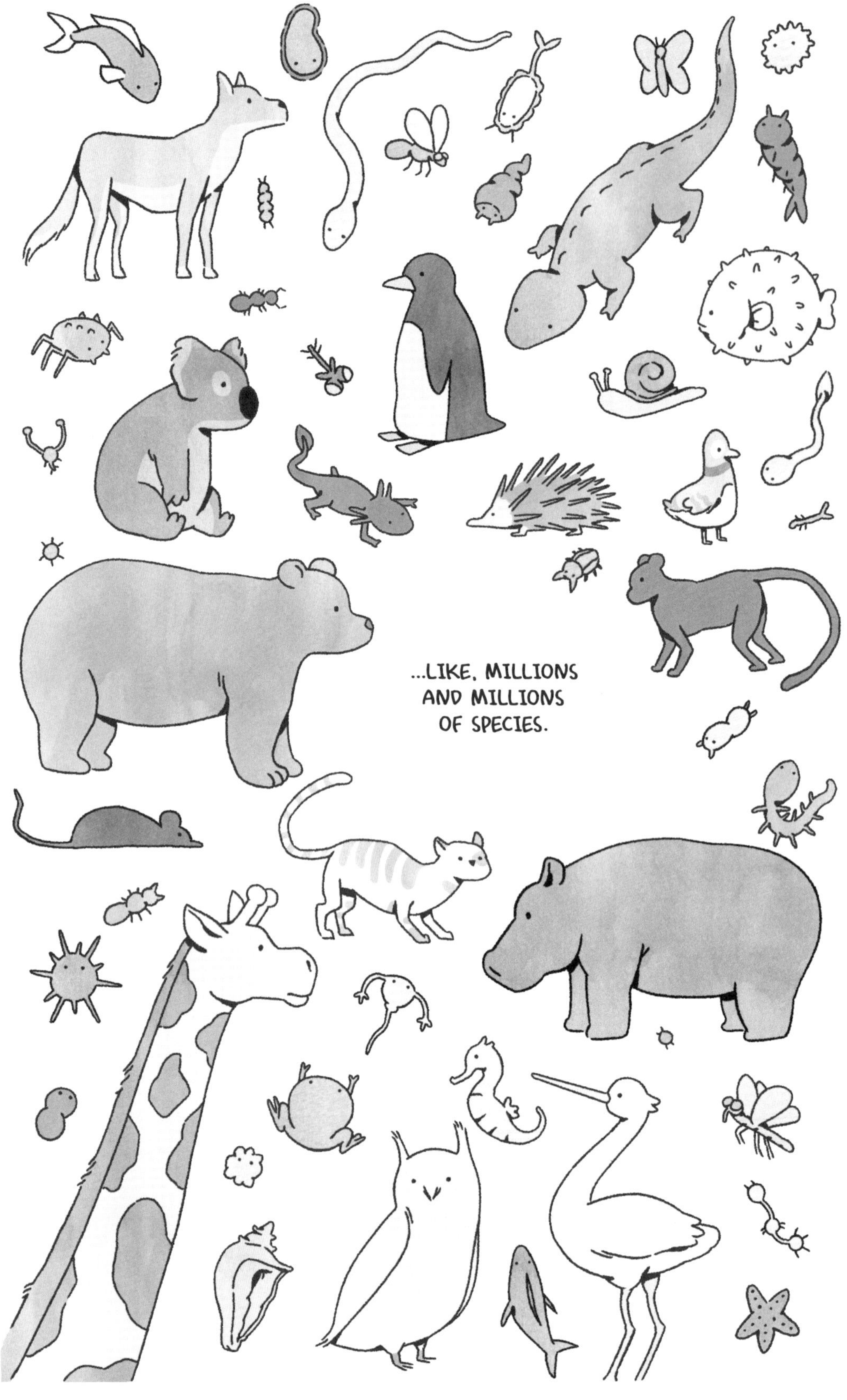
...LIKE, MILLIONS
AND MILLIONS
OF SPECIES.

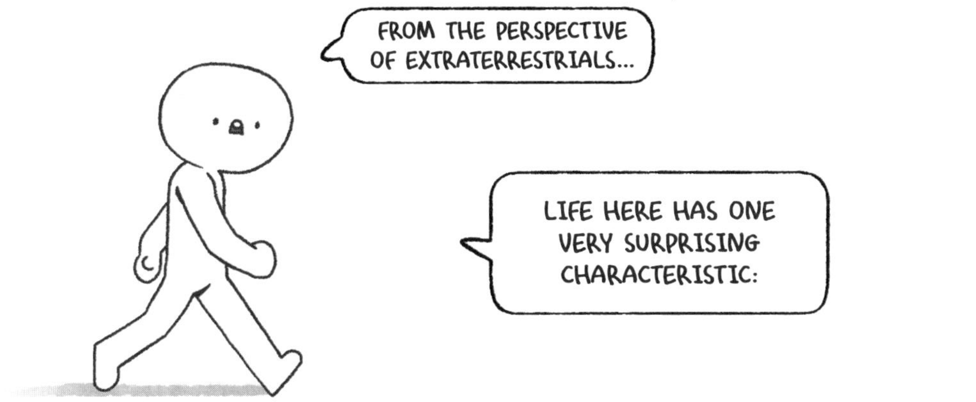
FROM THE PERSPECTIVE OF EXTRATERRESTRIALS...
LIFE HERE HAS ONE VERY SURPRISING CHARACTERISTIC:

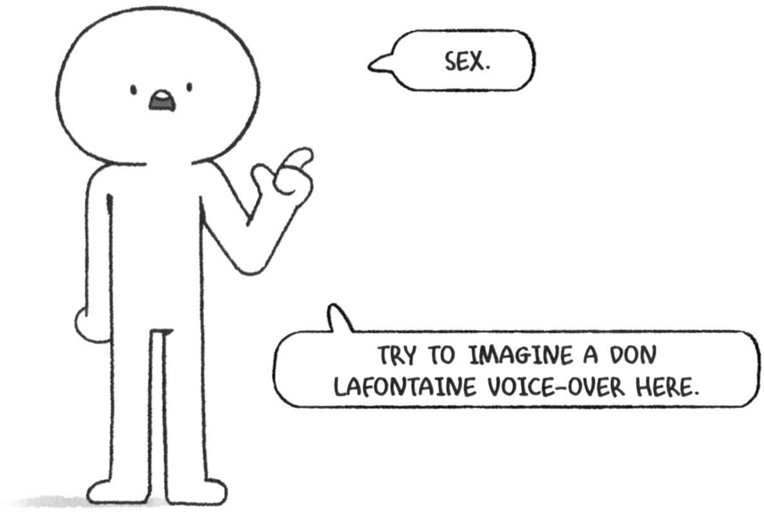
SEX.
TRY TO IMAGINE A DON LAFONTAINE VOICE-OVER HERE.

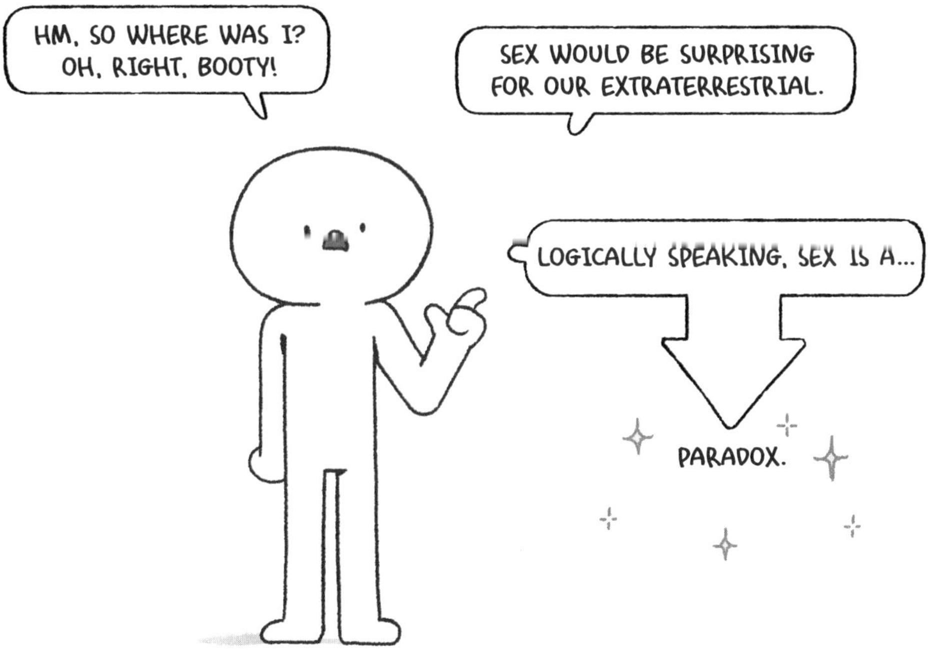
HM, SO WHERE WAS I? OH, RIGHT, BOOTY!
SEX WOULD BE SURPRISING FOR OUR EXTRATERRESTRIAL.
LOGICALLY SPEAKING, SEX IS A...
PARADOX.

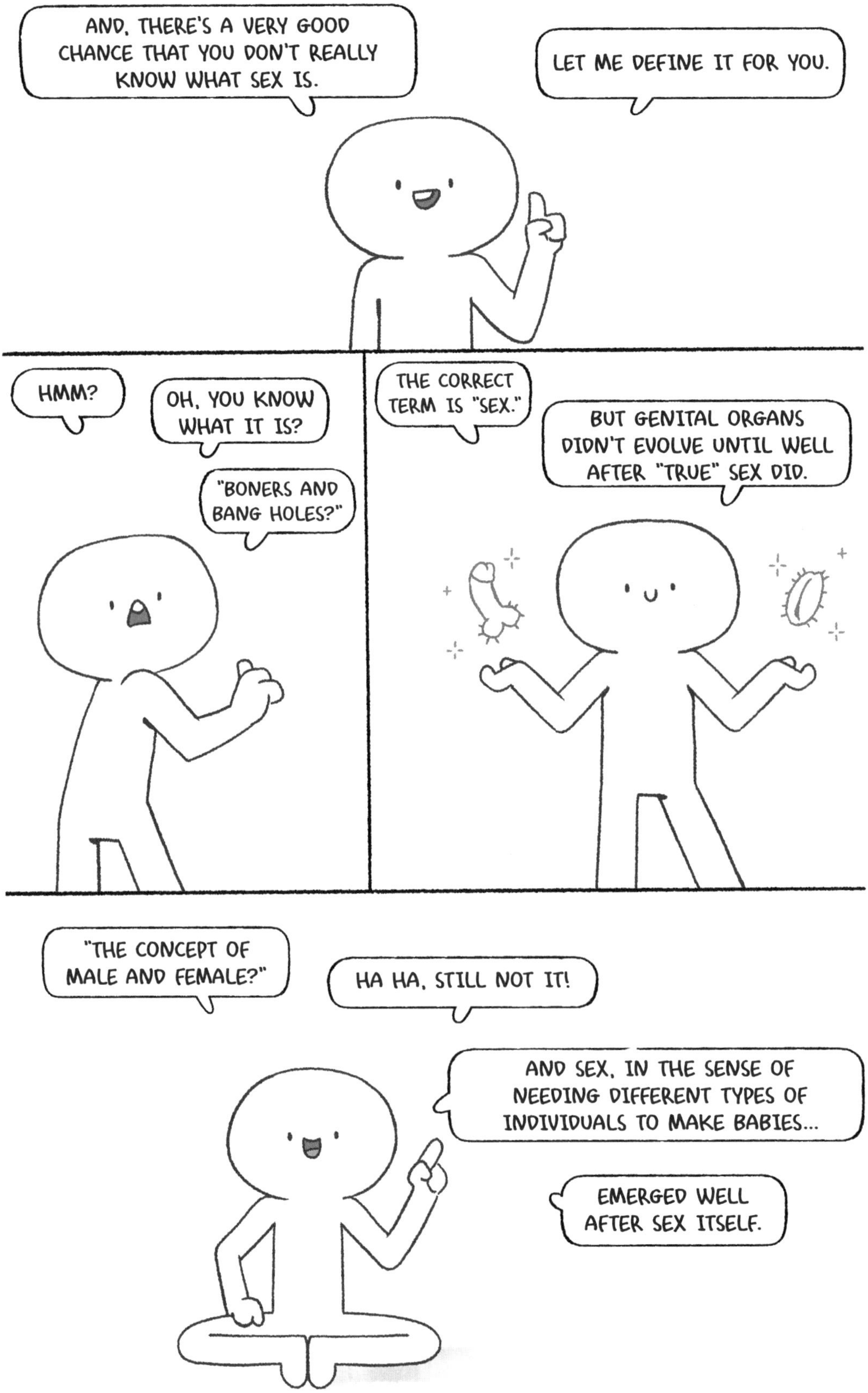

AND, THERE'S A VERY GOOD CHANCE THAT YOU DON'T REALLY KNOW WHAT SEX IS.
LET ME DEFINE IT FOR YOU.
HMM?
OH, YOU KNOW WHAT IT IS?
"BONERS AND BANG HOLES?"
THE CORRECT TERM IS "SEX."
BUT GENITAL ORGANS DIDN'T EVOLVE UNTIL WELL AFTER "TRUE" SEX DID.
"THE CONCEPT OF MALE AND FEMALE?"
HA HA, STILL NOT IT!
AND SEX, IN THE SENSE OF NEEDING DIFFERENT TYPES OF INDIVIDUALS TO MAKE BABIES...
EMERGED WELL AFTER SEX ITSELF.

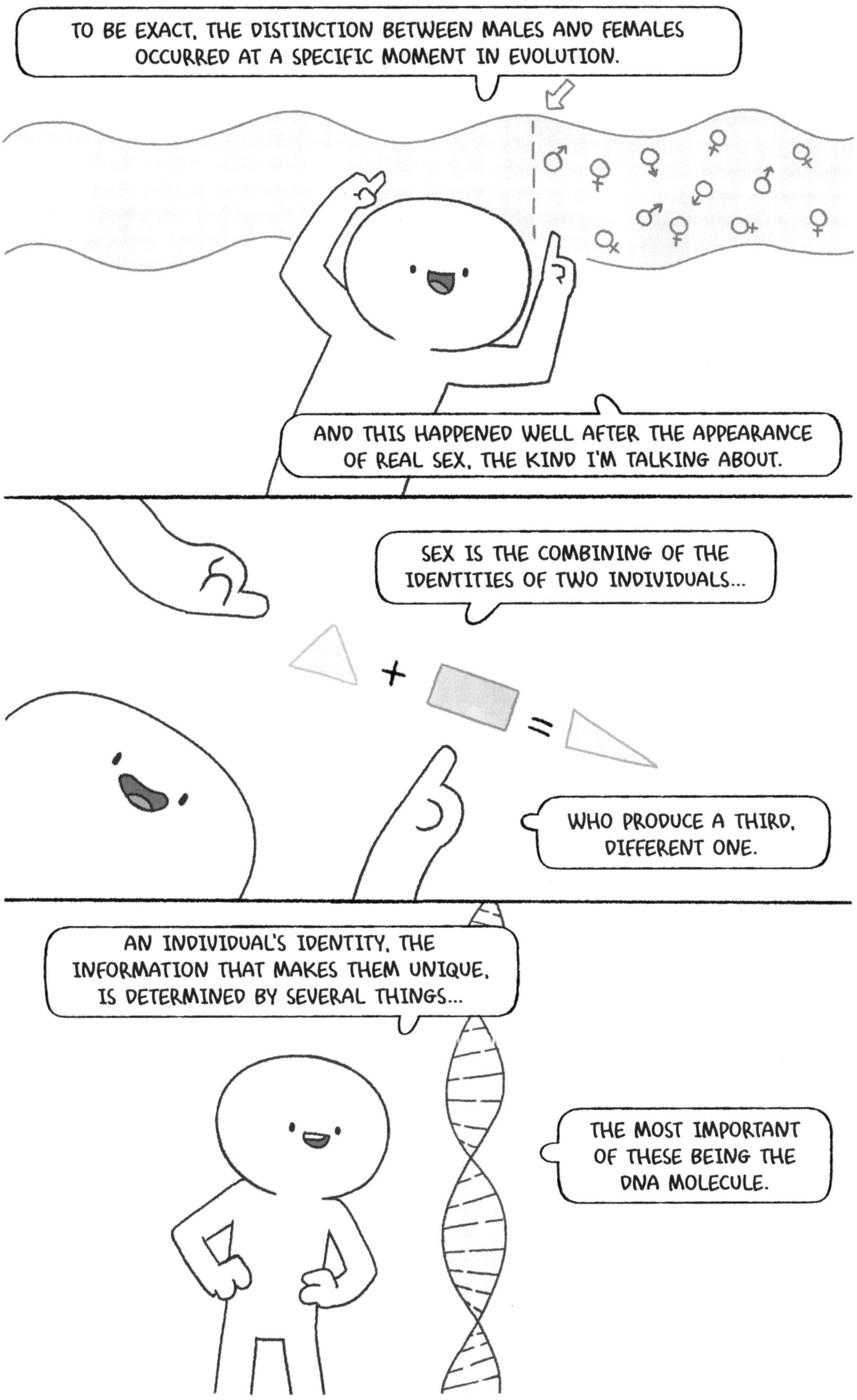
TO BE EXACT, THE DISTINCTION BETWEEN MALES AND FEMALES OCCURRED AT A SPECIFIC MOMENT IN EVOLUTION.
AND THIS HAPPENED WELL AFTER THE APPEARANCE OF REAL SEX, THE KIND I'M TALKING ABOUT.
SEX IS THE COMBINING OF THE IDENTITIES OF TWO INDIVIDUALS...
+
=
WHO PRODUCE A THIRD, DIFFERENT ONE.
AN INDIVIDUAL'S IDENTITY, THE INFORMATION THAT MAKES THEM UNIQUE, IS DETERMINED BY SEVERAL THINGS...
THE MOST IMPORTANT OF THESE BEING THE DNA MOLECULE.

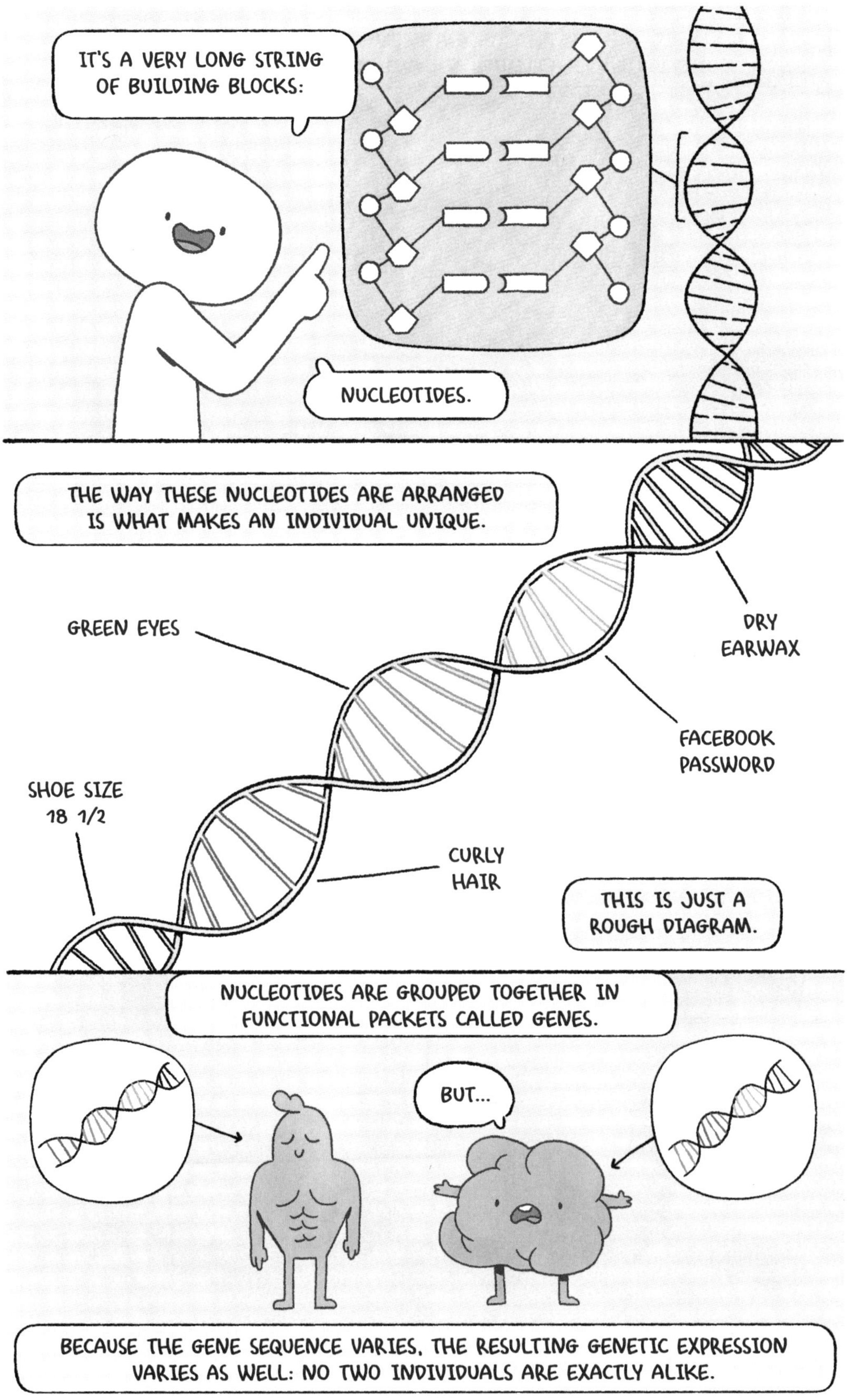
IT'S A VERY LONG STRING OF BUILDING BLOCKS:
NUCLEOTIDES.
THE WAY THESE NUCLEOTIDES ARE ARRANGED IS WHAT MAKES AN INDIVIDUAL UNIQUE.
GREEN EYES
DRY EARWAX
FACEBOOK PASSWORD
SHOE SIZE 18 1/2
CURLY HAIR
THIS IS JUST A ROUGH DIAGRAM.
NUCLEOTIDES ARE GROUPED TOGETHER IN FUNCTIONAL PACKETS CALLED GENES.
BUT...
BECAUSE THE GENE SEQUENCE VARIES, THE RESULTING GENETIC EXPRESSION VARIES AS WELL: NO TWO INDIVIDUALS ARE EXACTLY ALIKE.

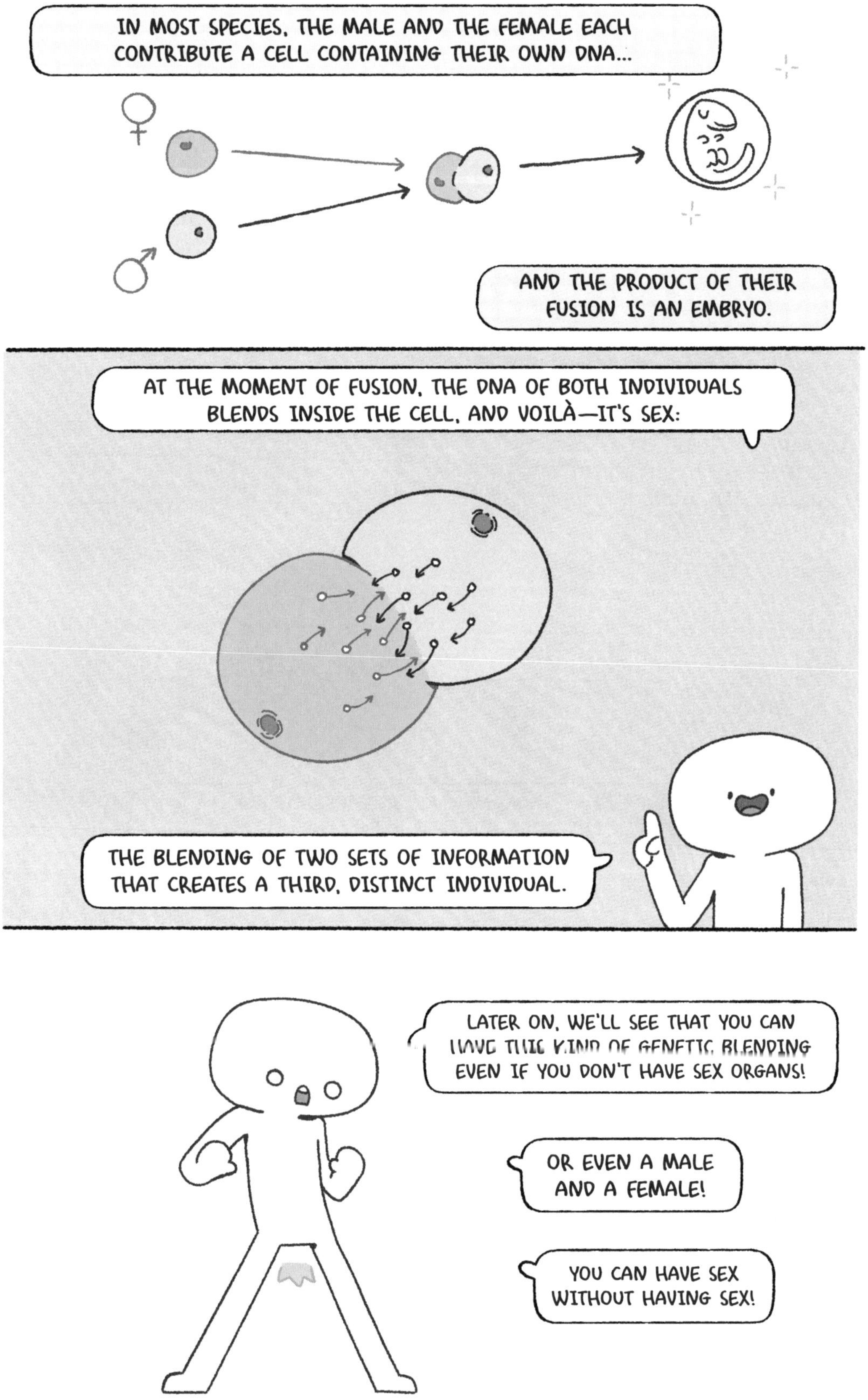
IN MOST SPECIES, THE MALE AND THE FEMALE EACH CONTRIBUTE A CELL CONTAINING THEIR OWN DNA...
AND THE PRODUCT OF THEIR FUSION IS AN EMBRYO.
AT THE MOMENT OF FUSION, THE DNA OF BOTH INDIVIDUALS BLENDS INSIDE THE CELL, AND VOILÀ—IT'S SEX:
THE BLENDING OF TWO SETS OF INFORMATION THAT CREATES A THIRD, DISTINCT INDIVIDUAL.
LATER ON, WE'LL SEE THAT YOU CAN HAVE THIS KIND OF GENETIC BLENDING EVEN IF YOU DON'T HAVE SEX ORGANS!
OR EVEN A MALE AND A FEMALE!
YOU CAN HAVE SEX WITHOUT HAVING SEX!

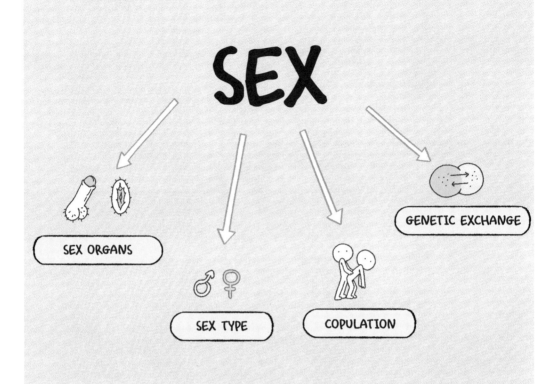

THE WORD "SEX" HAS A NUMBER OF DIFFERENT MEANINGS, REALLY.

SEX

SEX ORGANS

SEX TYPE

COPULATION

GENETIC EXCHANGE

FOR US HUMANS, ALL OF THOSE THINGS WORK TOGETHER. BUT FOR MANY OTHER SPECIES, THAT'S NOT EXACTLY THE CASE.

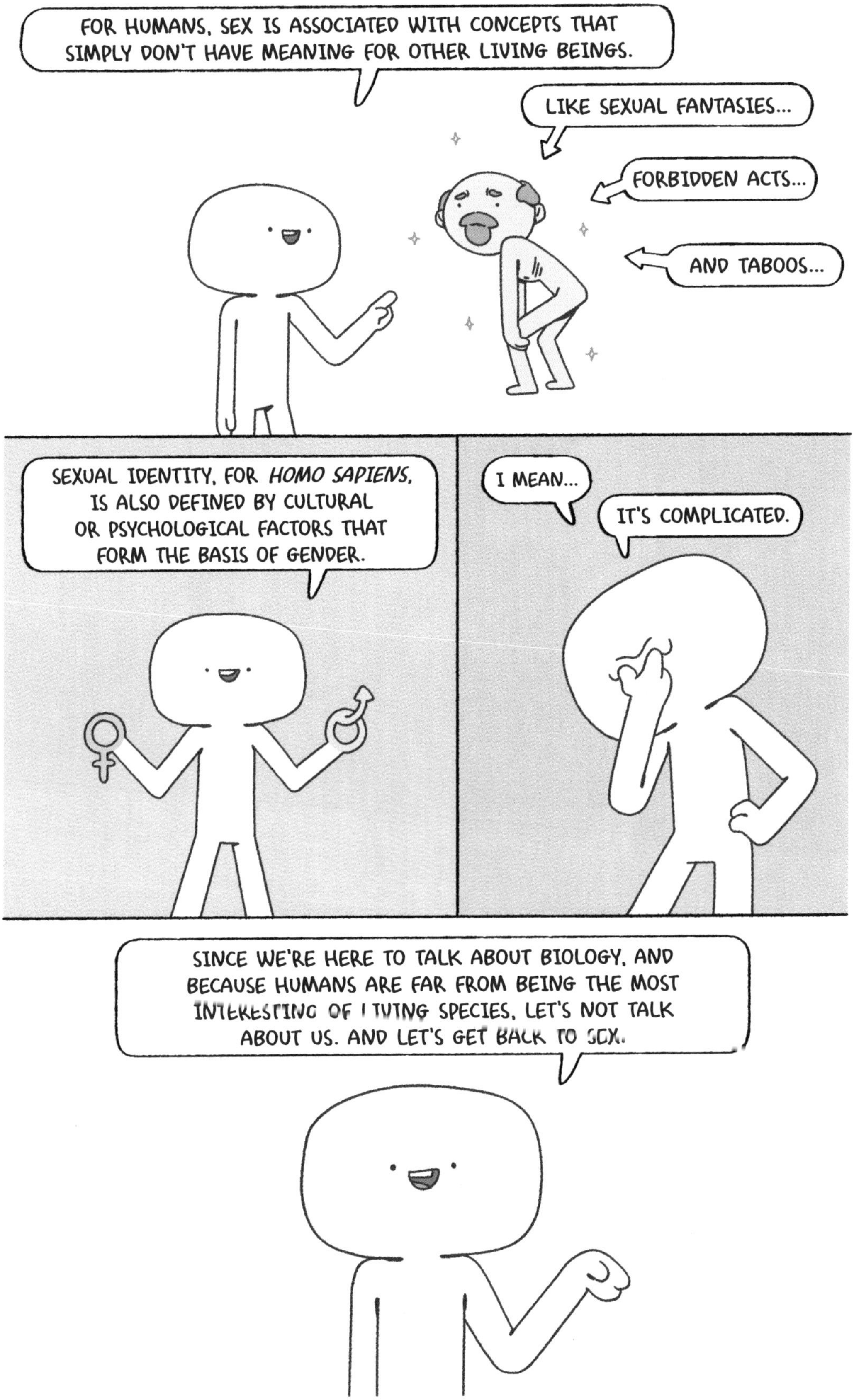
FOR HUMANS, SEX IS ASSOCIATED WITH CONCEPTS THAT SIMPLY DON'T HAVE MEANING FOR OTHER LIVING BEINGS.
LIKE SEXUAL FANTASIES...
FORBIDDEN ACTS...
AND TABOOS...
SEXUAL IDENTITY, FOR *HOMO SAPIENS*, IS ALSO DEFINED BY CULTURAL OR PSYCHOLOGICAL FACTORS THAT FORM THE BASIS OF GENDER.
I MEAN...
IT'S COMPLICATED.
SINCE WE'RE HERE TO TALK ABOUT BIOLOGY, AND BECAUSE HUMANS ARE FAR FROM BEING THE MOST INTERESTING OF LIVING SPECIES, LET'S NOT TALK ABOUT US. AND LET'S GET BACK TO SEX.

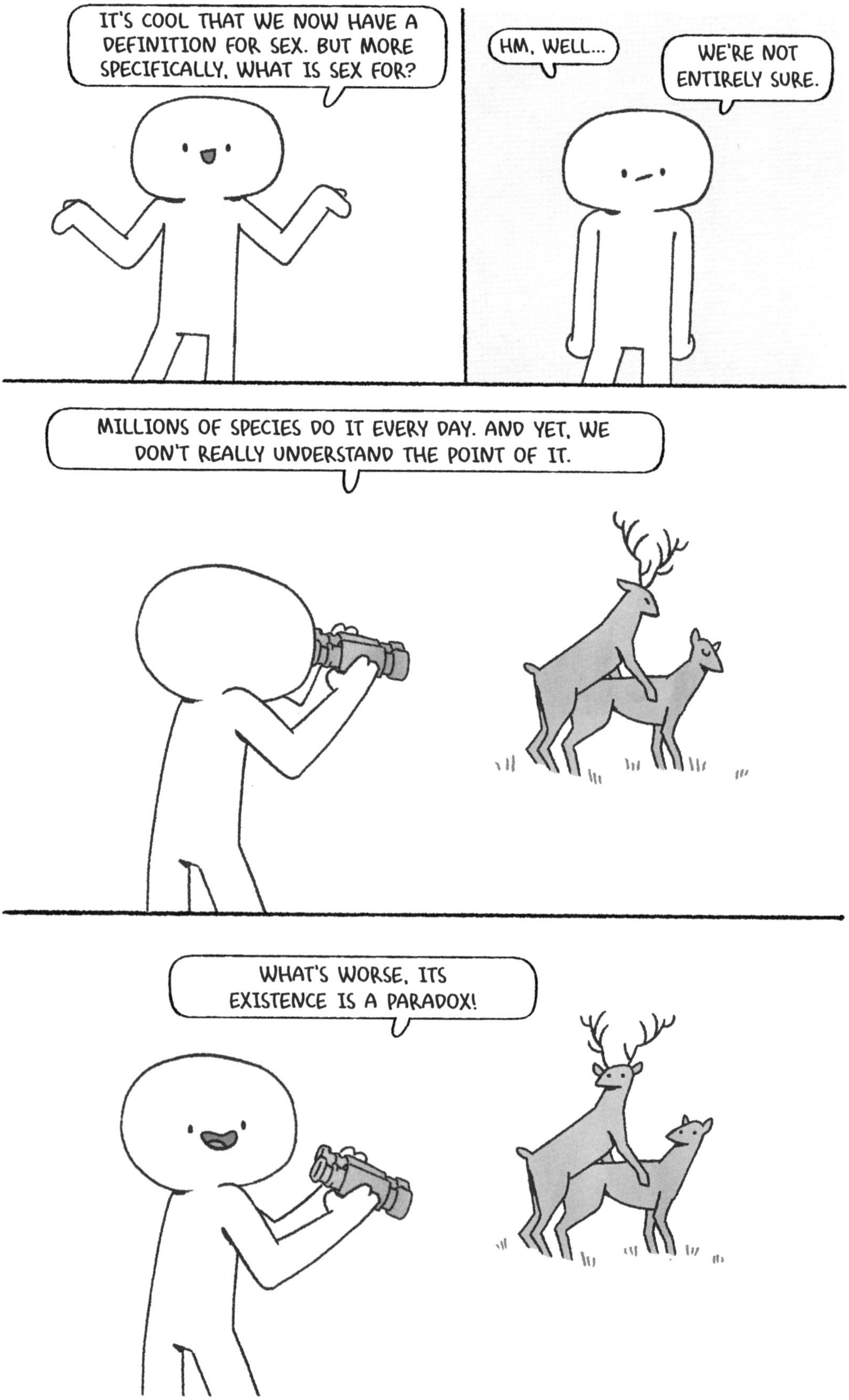
IT'S COOL THAT WE NOW HAVE A DEFINITION FOR SEX. BUT MORE SPECIFICALLY, WHAT IS SEX FOR?
HM, WELL...
WE'RE NOT ENTIRELY SURE.
MILLIONS OF SPECIES DO IT EVERY DAY. AND YET, WE DON'T REALLY UNDERSTAND THE POINT OF IT.
WHAT'S WORSE, ITS EXISTENCE IS A PARADOX!

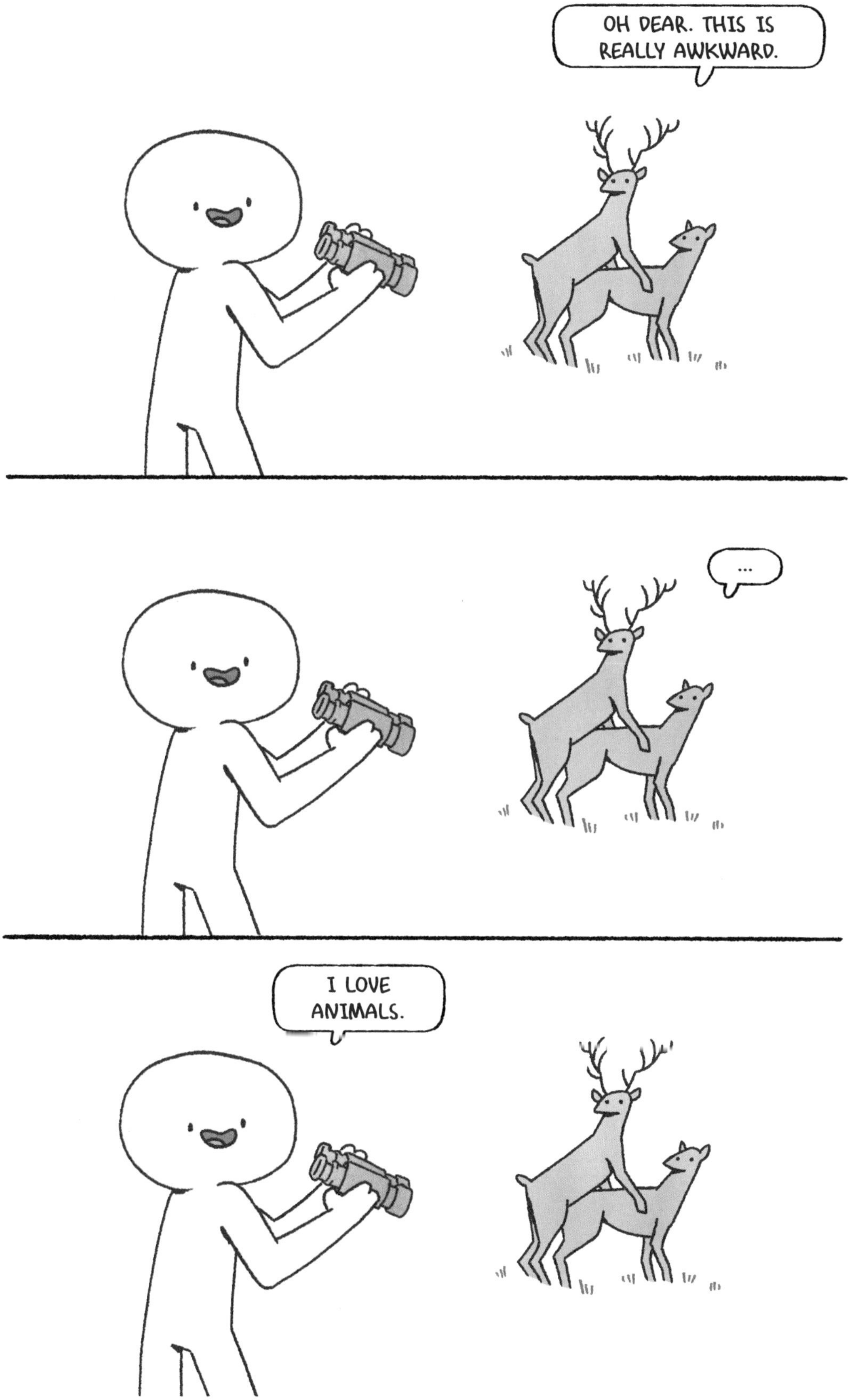
OH DEAR. THIS IS REALLY AWKWARD.
...
I LOVE ANIMALS.

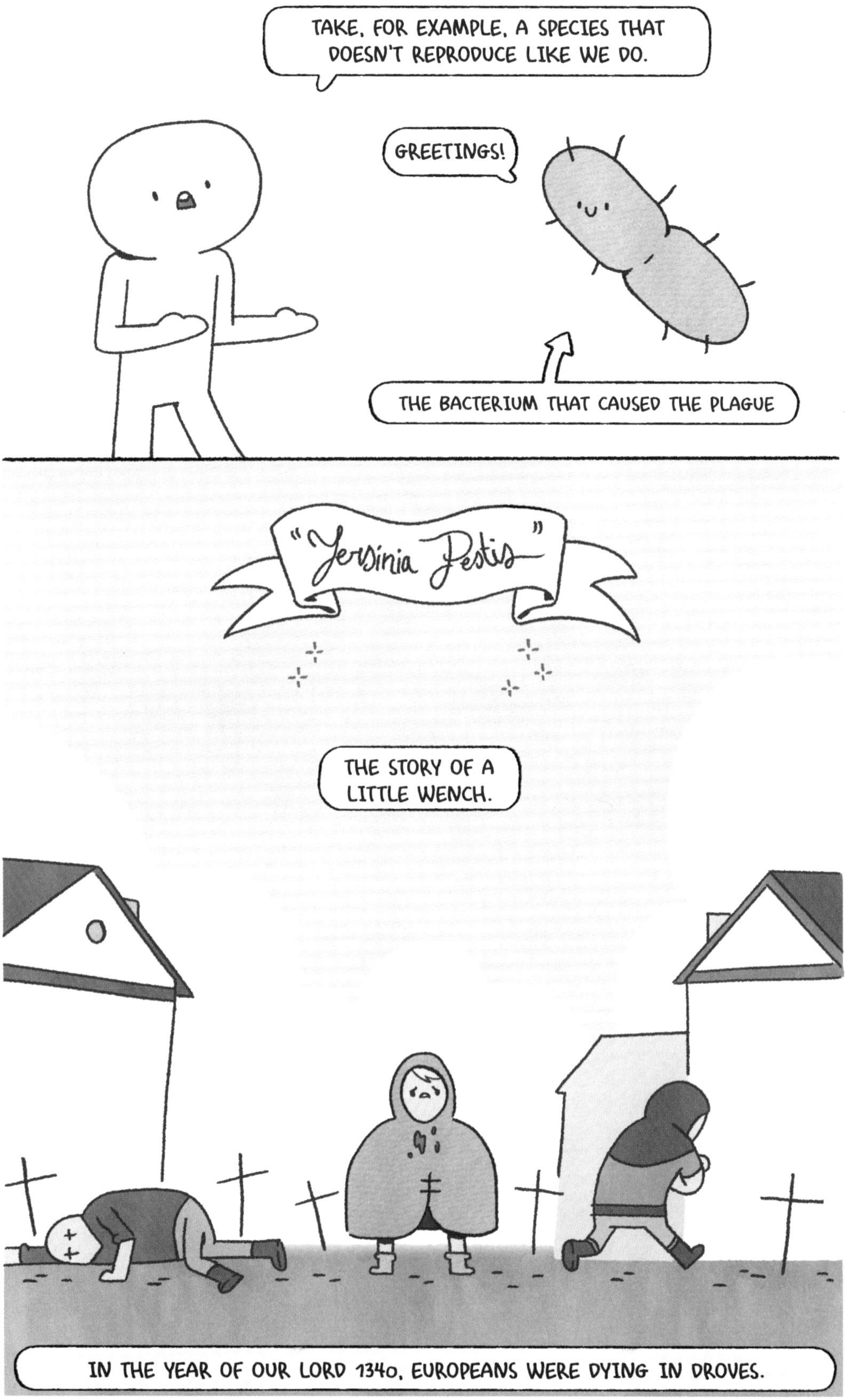
TAKE, FOR EXAMPLE, A SPECIES THAT DOESN'T REPRODUCE LIKE WE DO.
GREETINGS!
THE BACTERIUM THAT CAUSED THE PLAGUE
"Yersinia Pestis"
THE STORY OF A LITTLE WENCH.
IN THE YEAR OF OUR LORD 1340, EUROPEANS WERE DYING IN DROVES.

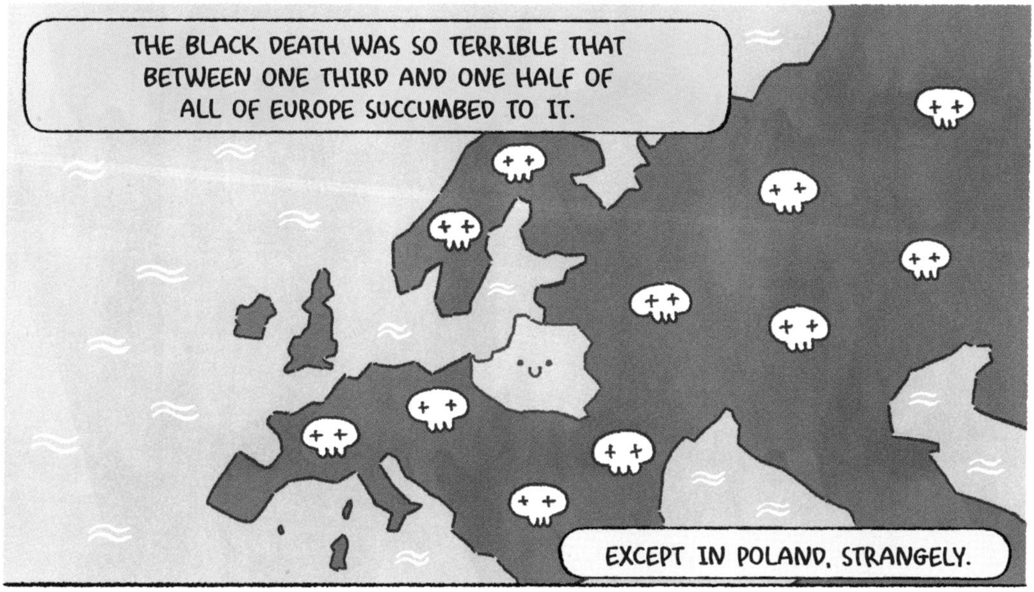
THE BLACK DEATH WAS SO TERRIBLE THAT BETWEEN ONE THIRD AND ONE HALF OF ALL OF EUROPE SUCCUMBED TO IT.
EXCEPT IN POLAND, STRANGELY.

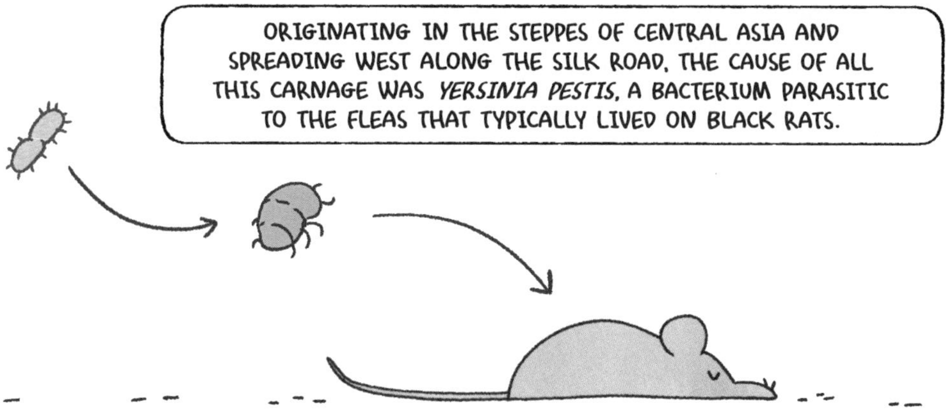
ORIGINATING IN THE STEPPES OF CENTRAL ASIA AND SPREADING WEST ALONG THE SILK ROAD, THE CAUSE OF ALL THIS CARNAGE WAS *YERSINIA PESTIS*, A BACTERIUM PARASITIC TO THE FLEAS THAT TYPICALLY LIVED ON BLACK RATS.

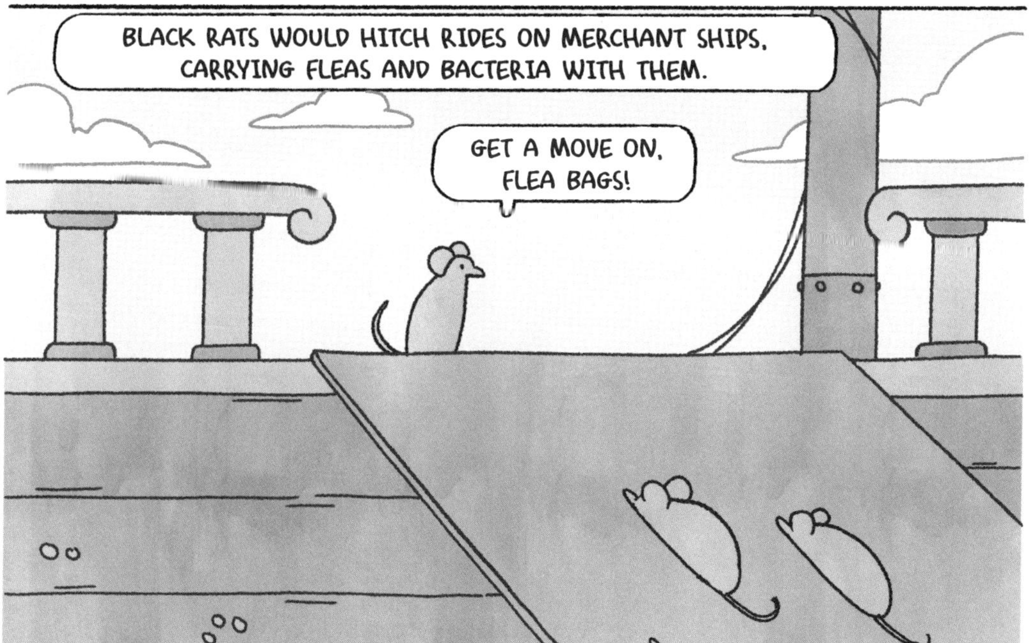
BLACK RATS WOULD HITCH RIDES ON MERCHANT SHIPS, CARRYING FLEAS AND BACTERIA WITH THEM.
GET A MOVE ON, FLEA BAGS!

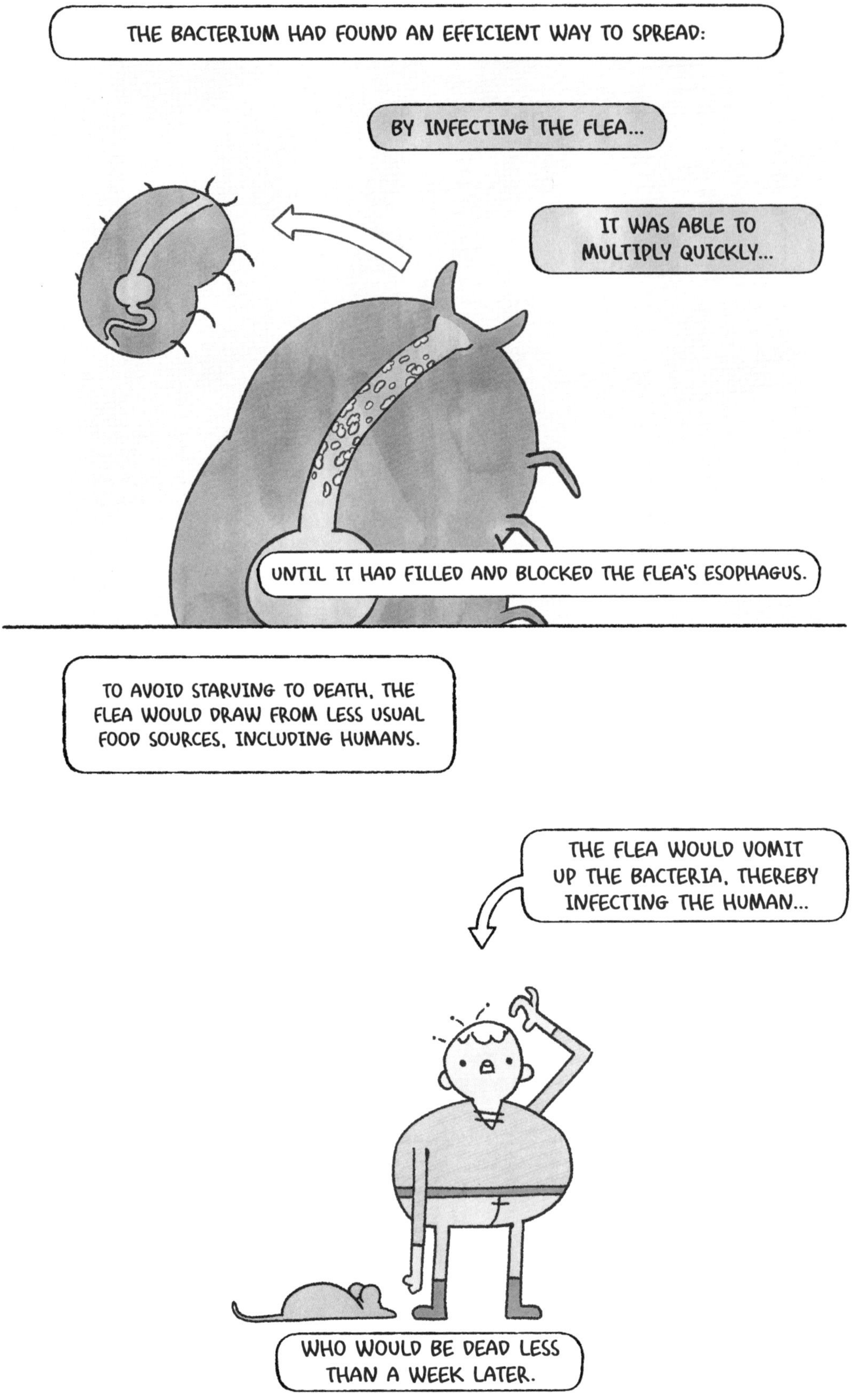
THE BACTERIUM HAD FOUND AN EFFICIENT WAY TO SPREAD:
BY INFECTING THE FLEA...
IT WAS ABLE TO MULTIPLY QUICKLY...
UNTIL IT HAD FILLED AND BLOCKED THE FLEA'S ESOPHAGUS.
TO AVOID STARVING TO DEATH, THE FLEA WOULD DRAW FROM LESS USUAL FOOD SOURCES, INCLUDING HUMANS.
THE FLEA WOULD VOMIT UP THE BACTERIA, THEREBY INFECTING THE HUMAN...
WHO WOULD BE DEAD LESS THAN A WEEK LATER.

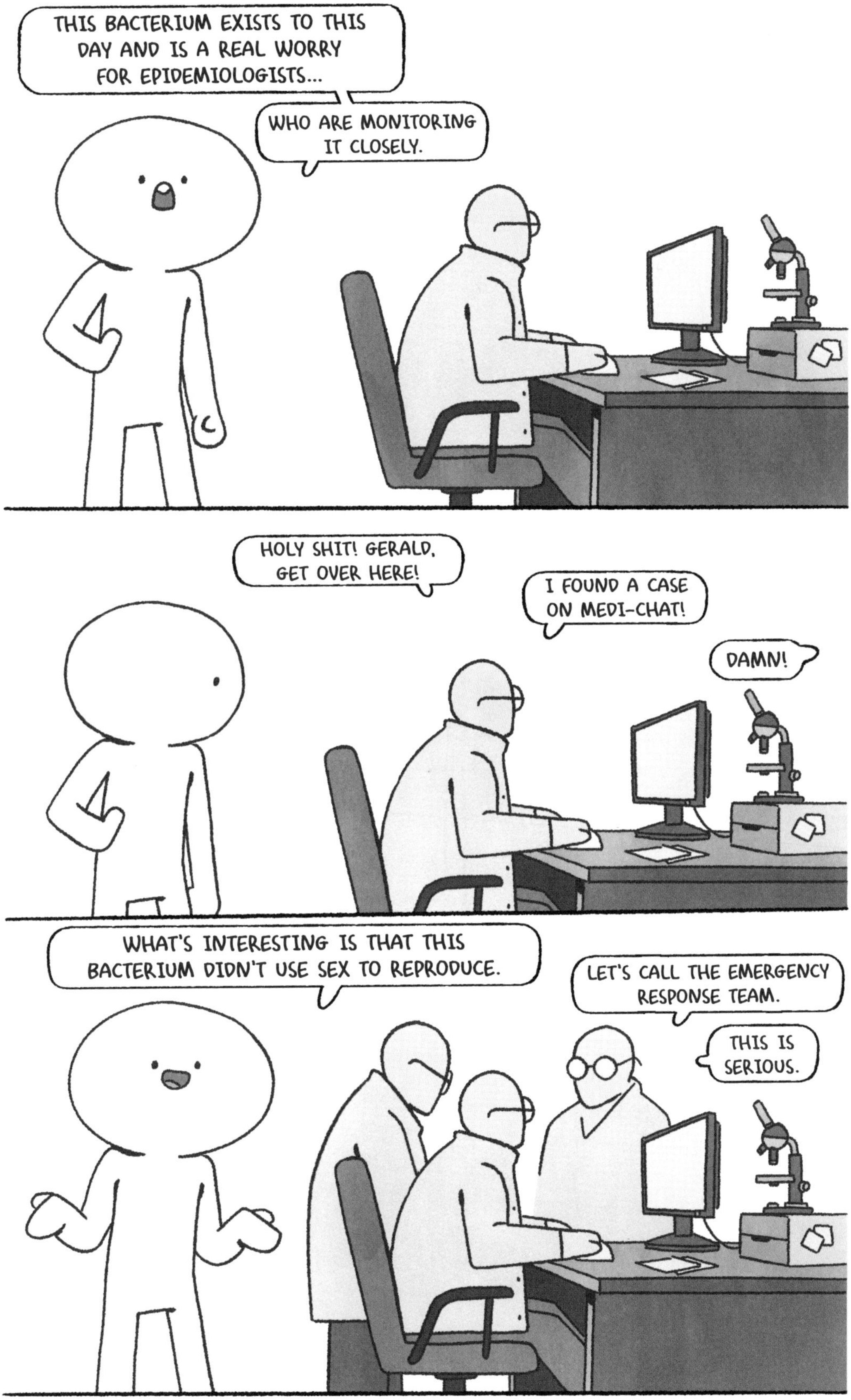
THIS BACTERIUM EXISTS TO THIS DAY AND IS A REAL WORRY FOR EPIDEMIOLOGISTS...
WHO ARE MONITORING IT CLOSELY.
HOLY SHIT! GERALD, GET OVER HERE!
I FOUND A CASE ON MEDI-CHAT!
DAMN!
WHAT'S INTERESTING IS THAT THIS BACTERIUM DIDN'T USE SEX TO REPRODUCE.
LET'S CALL THE EMERGENCY RESPONSE TEAM.
THIS IS SERIOUS.

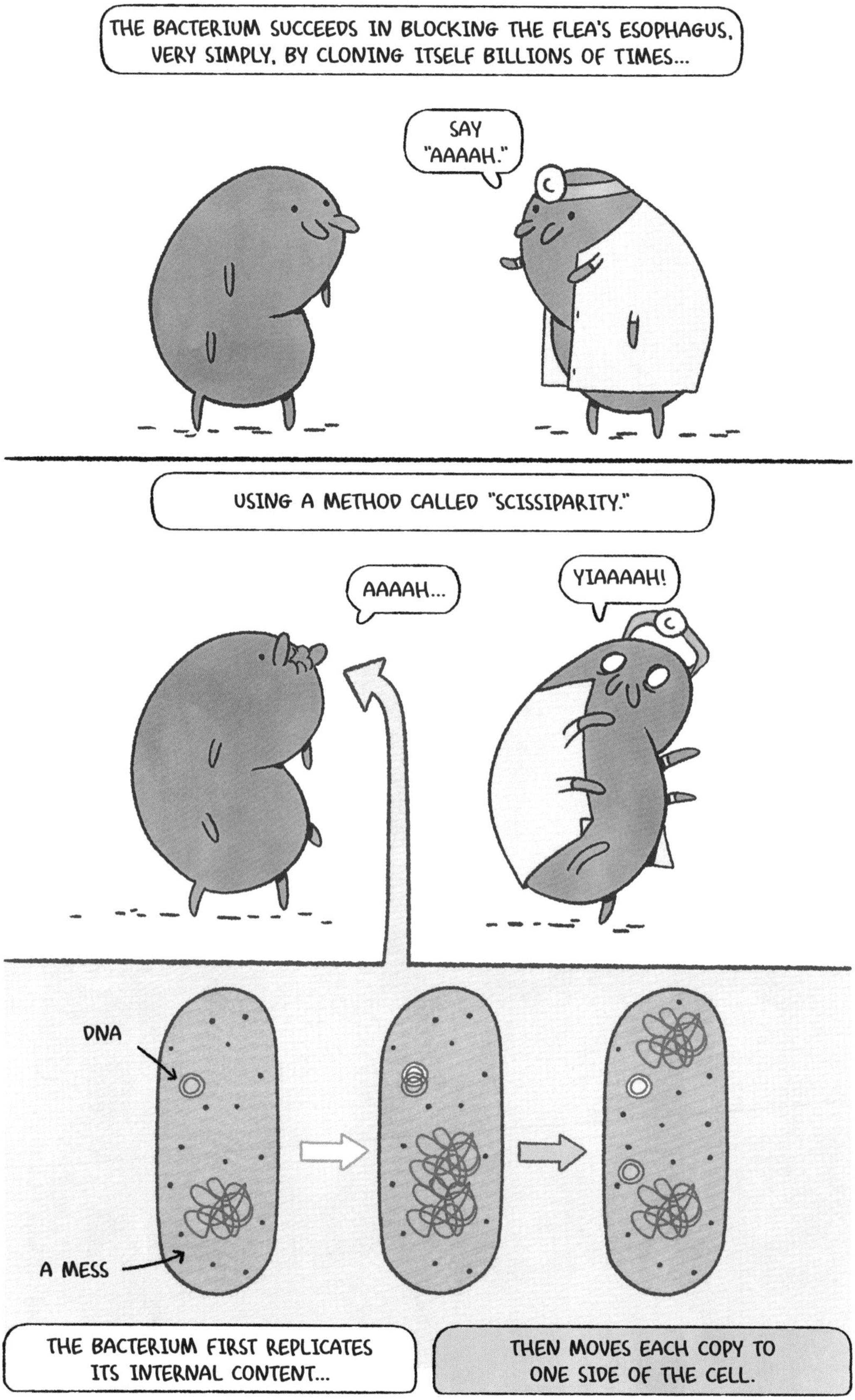
THE BACTERIUM SUCCEEDS IN BLOCKING THE FLEA'S ESOPHAGUS, VERY SIMPLY, BY CLONING ITSELF BILLIONS OF TIMES...
SAY "AAAAH."
USING A METHOD CALLED "SCISSIPARITY."
AAAAH...
YIAAAAH!
DNA
A MESS
THE BACTERIUM FIRST REPLICATES ITS INTERNAL CONTENT...
THEN MOVES EACH COPY TO ONE SIDE OF THE CELL.

IT BUILDS A WALL BETWEEN THEM, AND BAM, TWO CELLS.
EACH OF THESE CELLS WILL DIVIDE INTO TWO CELLS, THEN 8... 16... 32, ETC.
COVFEFE!
BY REPEATING AN OPERATION THAT STARTED WITH JUST ONE BACTERIUM, YOU GET BILLIONS OF IDENTICAL CELLS.
NO SEX, NO BLENDING OF GENES. JUST A COPY.
CTRL+C
CTRL+V
SEXUAL REPRODUCTION WOULD NOT HAVE RESULTED IN SUCH AN EXPLOSION IN THE NUMBER OF BACTERIA.
REPRODUCING SEXUALLY, YOU NEED TWO INDIVIDUALS TO MAKE ONE NEW ONE. IF EACH COUPLE PRODUCES A GIRL AND A BOY, THE POPULATION WILL REMAIN STABLE.

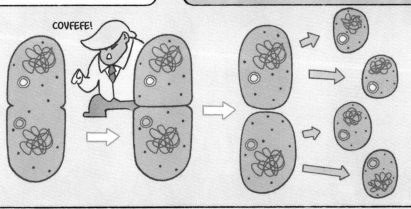

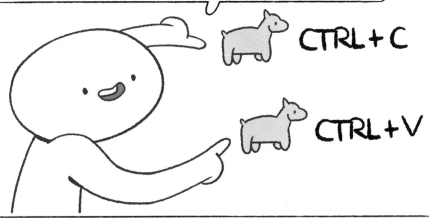

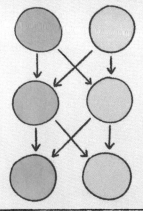

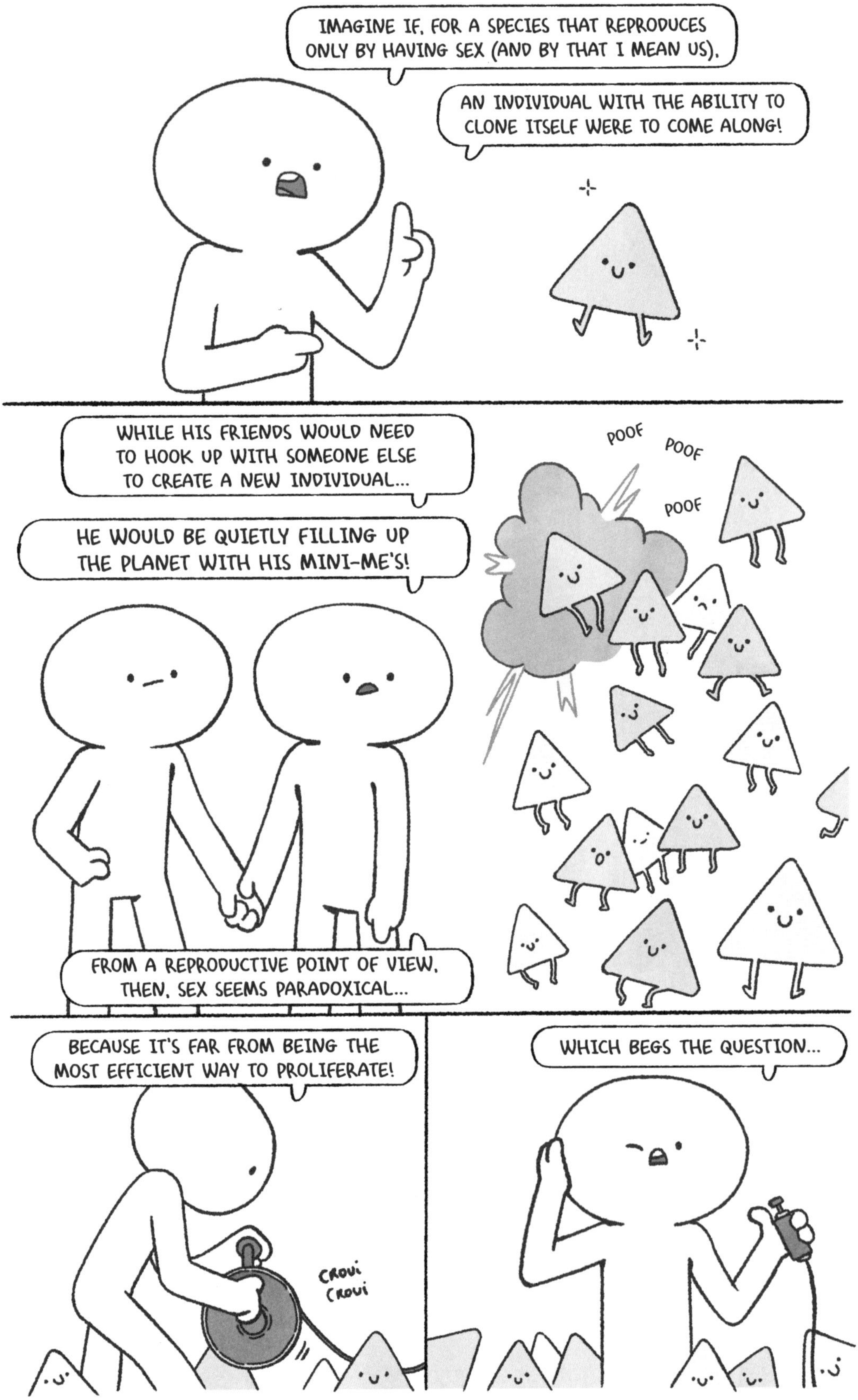
IMAGINE IF, FOR A SPECIES THAT REPRODUCES ONLY BY HAVING SEX (AND BY THAT I MEAN US),
AN INDIVIDUAL WITH THE ABILITY TO CLONE ITSELF WERE TO COME ALONG!
WHILE HIS FRIENDS WOULD NEED TO HOOK UP WITH SOMEONE ELSE TO CREATE A NEW INDIVIDUAL...
HE WOULD BE QUIETLY FILLING UP THE PLANET WITH HIS MINI-ME'S!
POOF
POOF
POOF
FROM A REPRODUCTIVE POINT OF VIEW, THEN, SEX SEEMS PARADOXICAL...
BECAUSE IT'S FAR FROM BEING THE MOST EFFICIENT WAY TO PROLIFERATE!
CROUI CROUI
WHICH BEGS THE QUESTION...

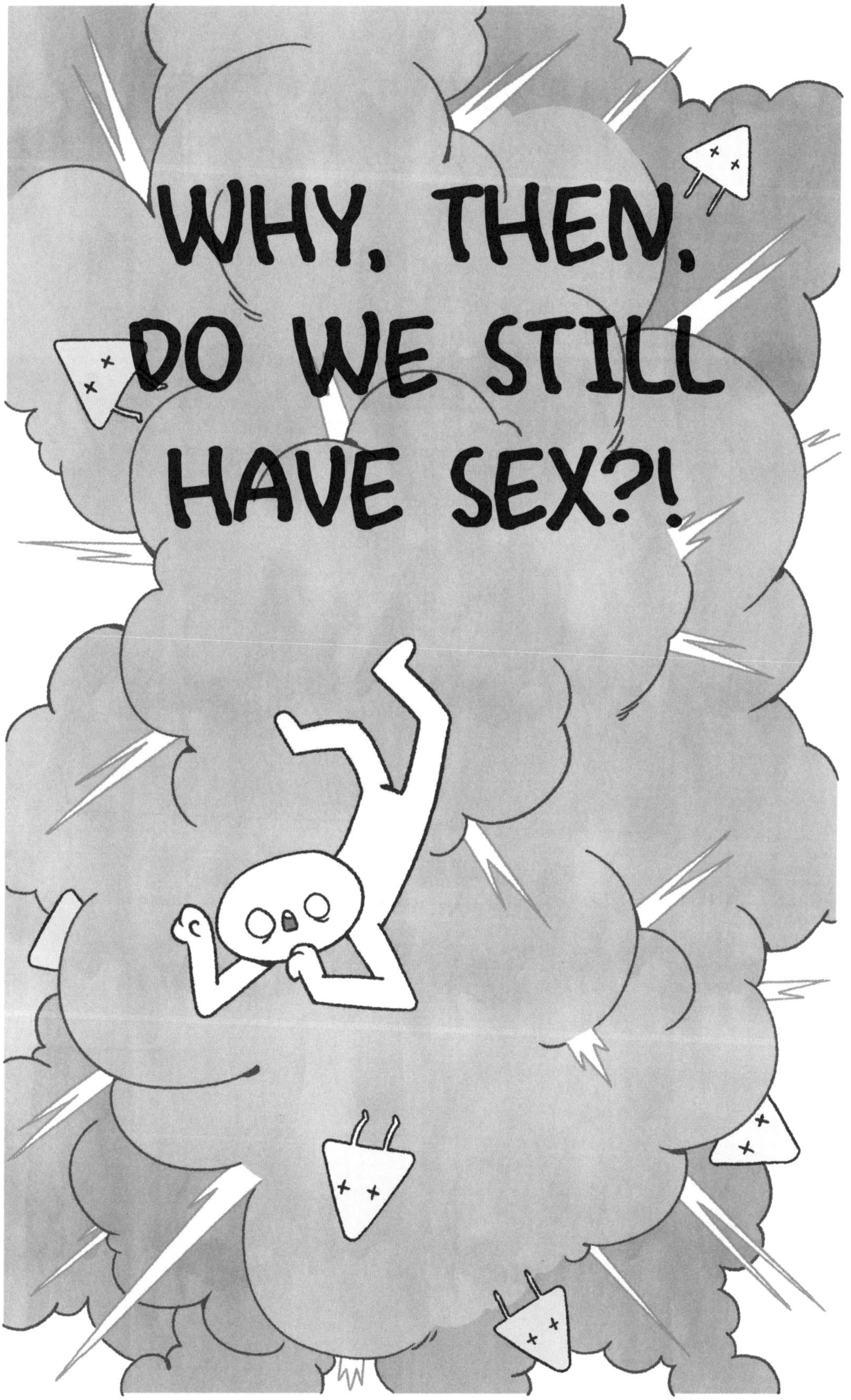
WHY, THEN, DO WE STILL HAVE SEX?!

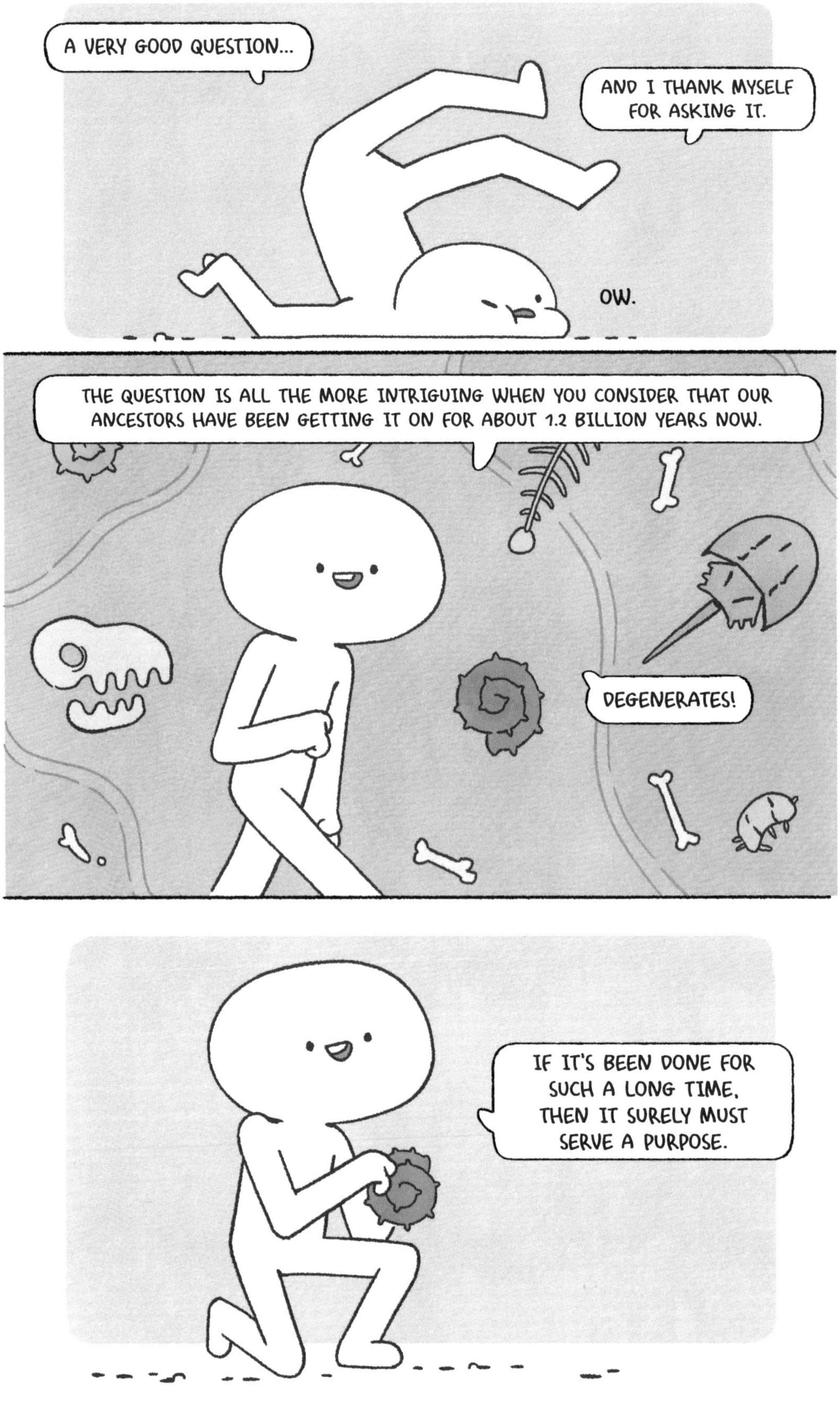

A VERY GOOD QUESTION...
AND I THANK MYSELF FOR ASKING IT.
OW.
THE QUESTION IS ALL THE MORE INTRIGUING WHEN YOU CONSIDER THAT OUR ANCESTORS HAVE BEEN GETTING IT ON FOR ABOUT 1.2 BILLION YEARS NOW.
DEGENERATES!
IF IT'S BEEN DONE FOR SUCH A LONG TIME, THEN IT SURELY MUST SERVE A PURPOSE.

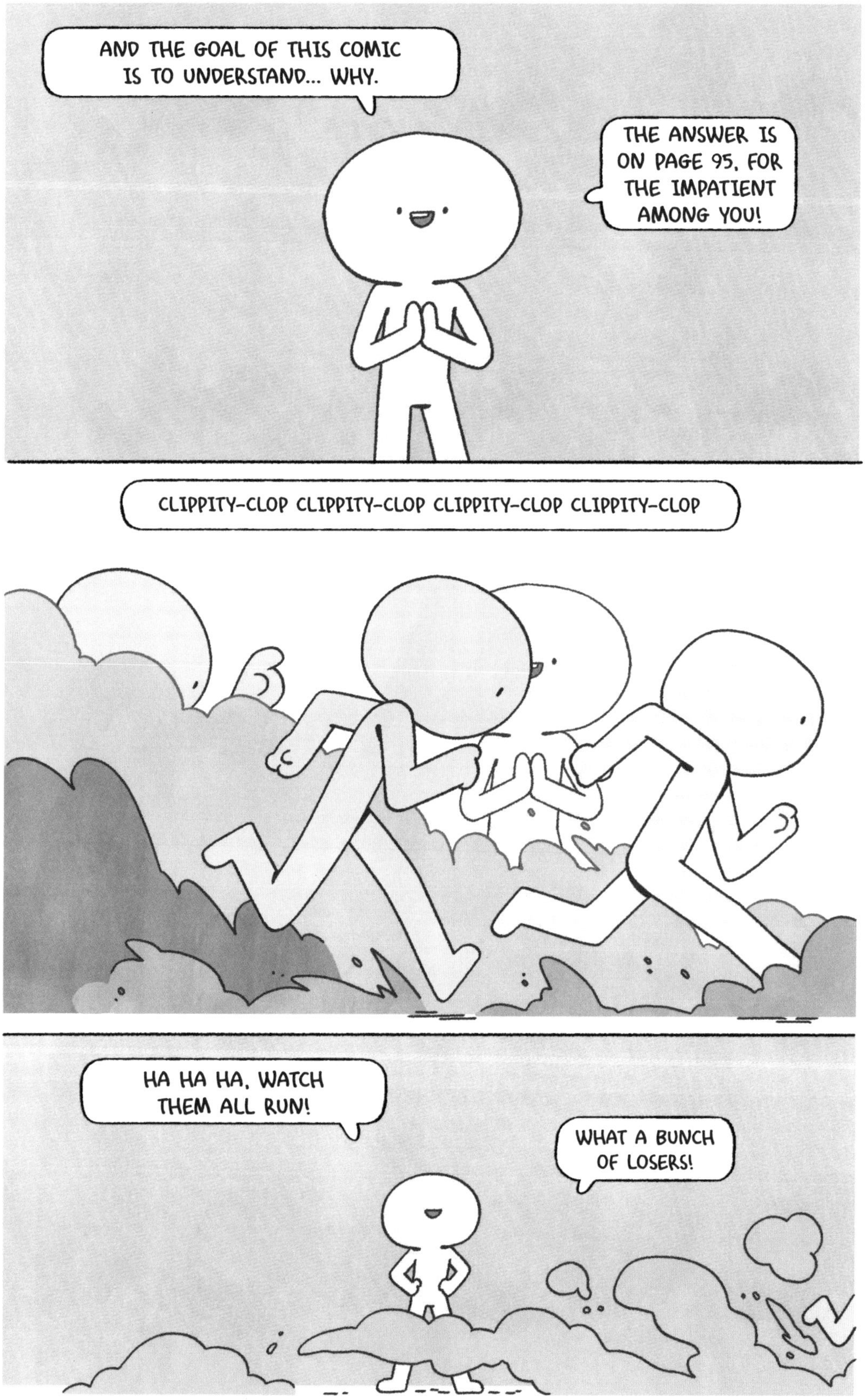
AND THE GOAL OF THIS COMIC IS TO UNDERSTAND... WHY.
THE ANSWER IS ON PAGE 95, FOR THE IMPATIENT AMONG YOU!
CLIPPITY-CLOP CLIPPITY-CLOP CLIPPITY-CLOP CLIPPITY-CLOP
HA HA HA, WATCH THEM ALL RUN!
WHAT A BUNCH OF LOSERS!

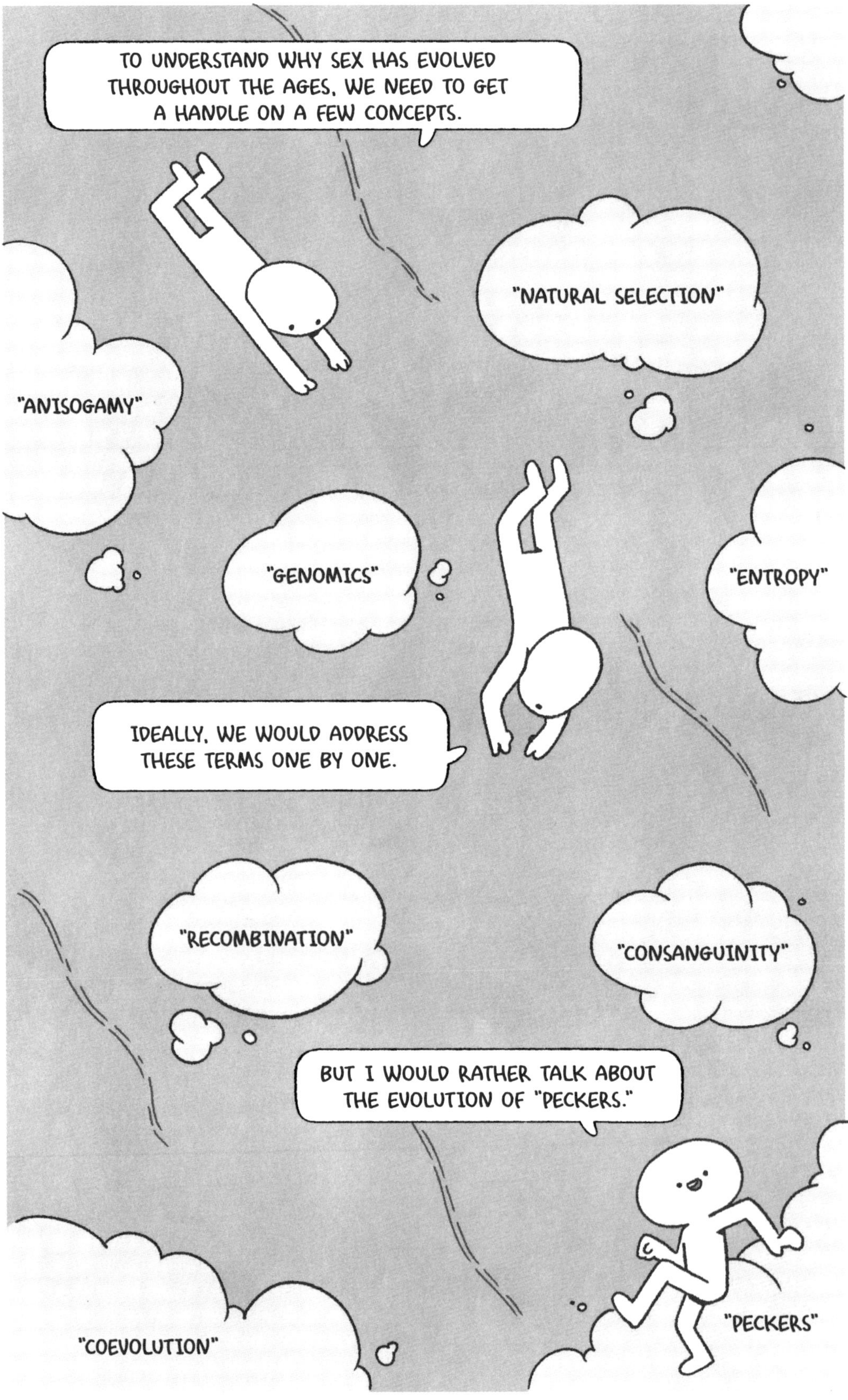
TO UNDERSTAND WHY SEX HAS EVOLVED THROUGHOUT THE AGES, WE NEED TO GET A HANDLE ON A FEW CONCEPTS.
"NATURAL SELECTION"
"ANISOGAMY"
"GENOMICS"
"ENTROPY"
IDEALLY, WE WOULD ADDRESS THESE TERMS ONE BY ONE.
"RECOMBINATION"
"CONSANGUINITY"
BUT I WOULD RATHER TALK ABOUT THE EVOLUTION OF "PECKERS."
"COEVOLUTION"
"PECKERS"

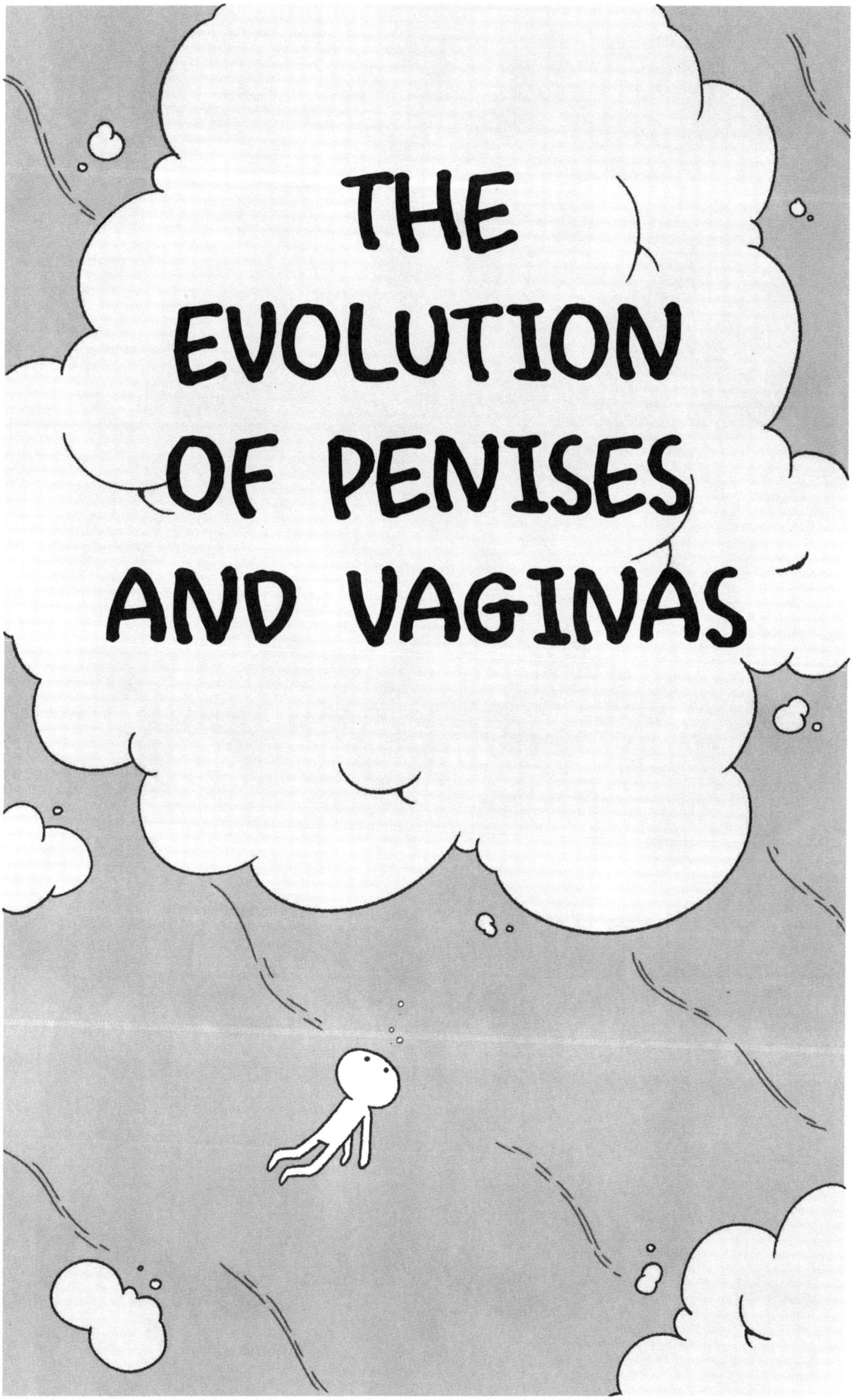
THE EVOLUTION OF PENISES AND VAGINAS

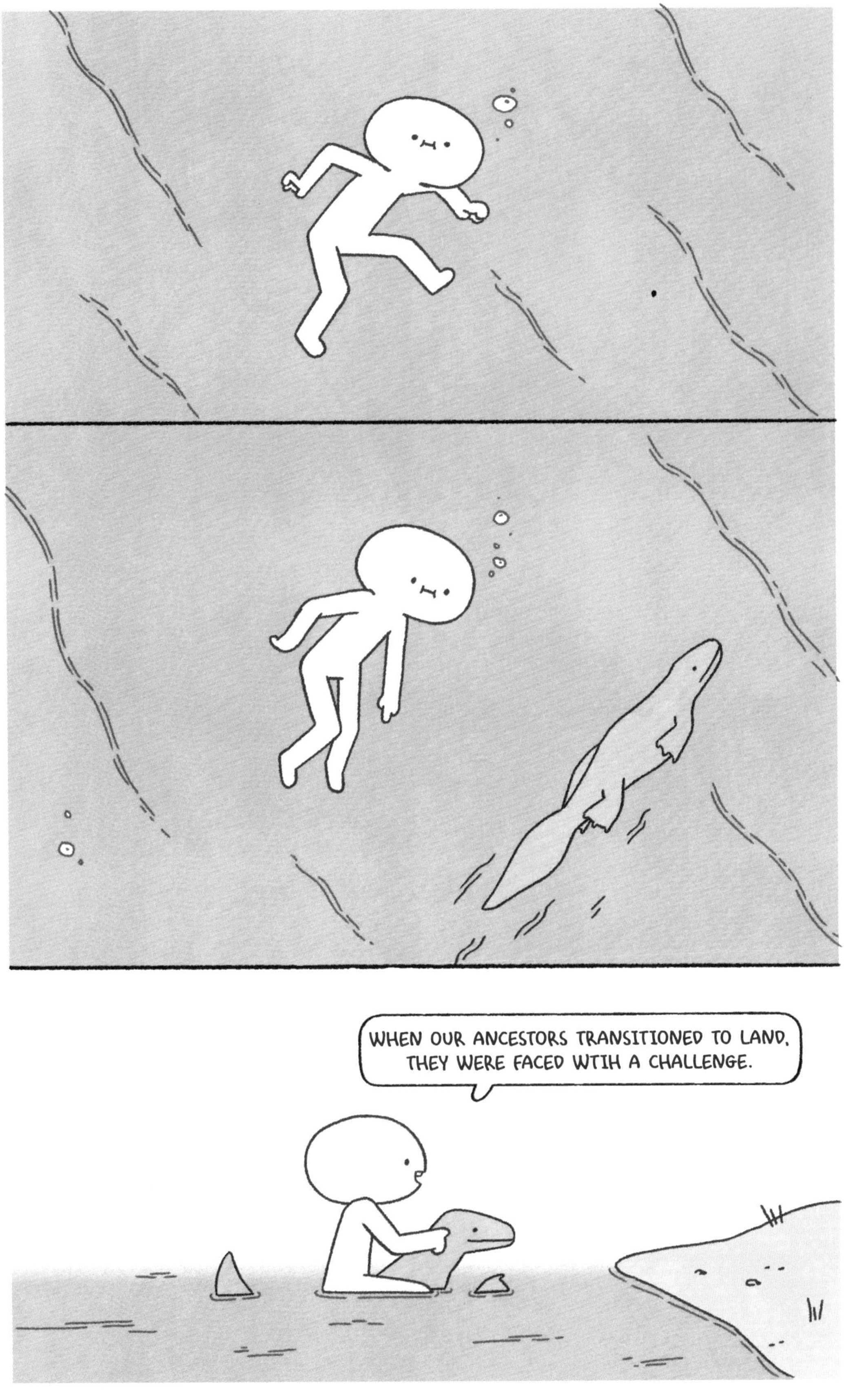
WHEN OUR ANCESTORS TRANSITIONED TO LAND, THEY WERE FACED WTIH A CHALLENGE.

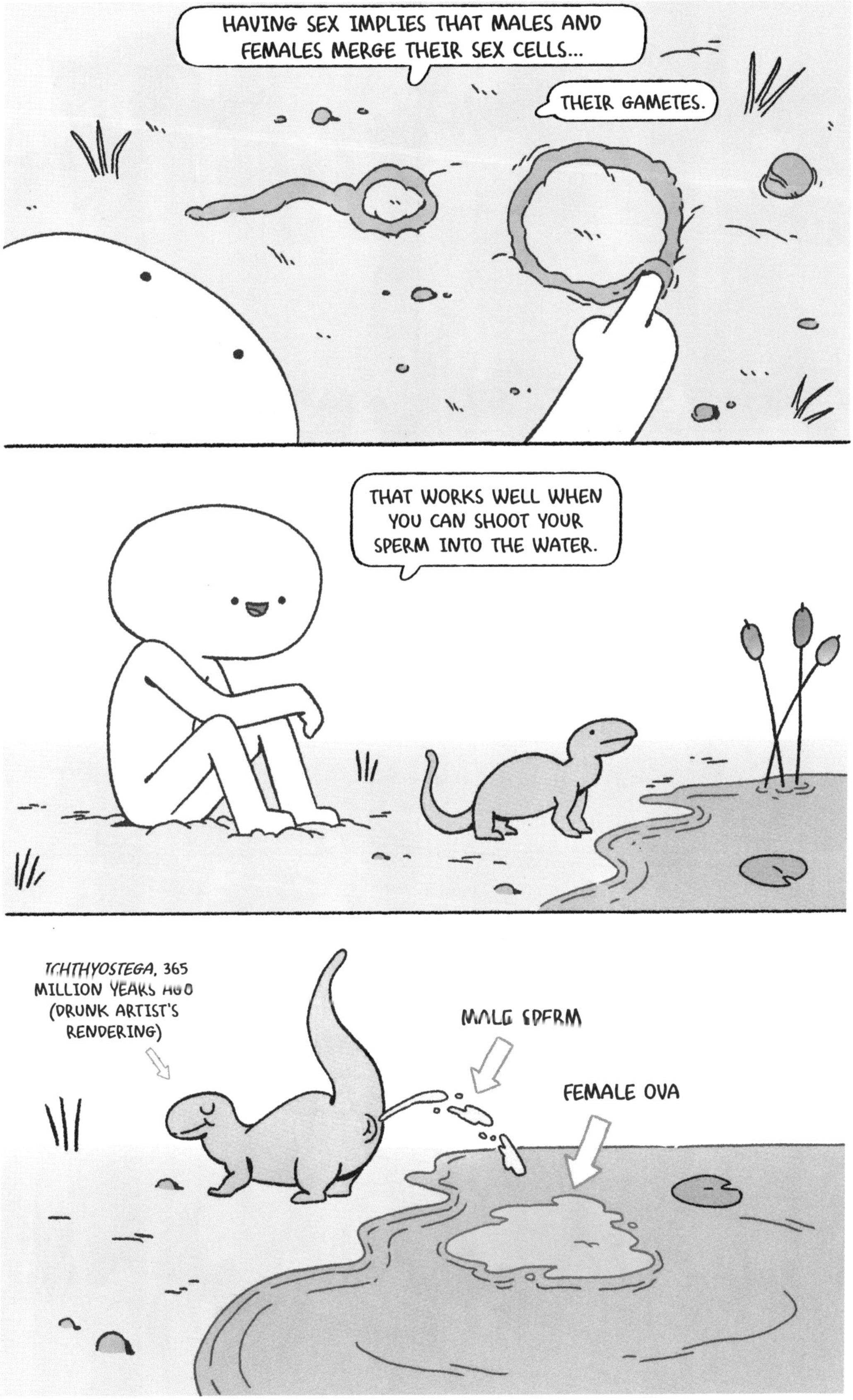
HAVING SEX IMPLIES THAT MALES AND FEMALES MERGE THEIR SEX CELLS...
THEIR GAMETES.
THAT WORKS WELL WHEN YOU CAN SHOOT YOUR SPERM INTO THE WATER.
ICHTHYOSTEGA, 365 MILLION YEARS AGO (DRUNK ARTIST'S RENDERING)
MALE SPERM
FEMALE OVA

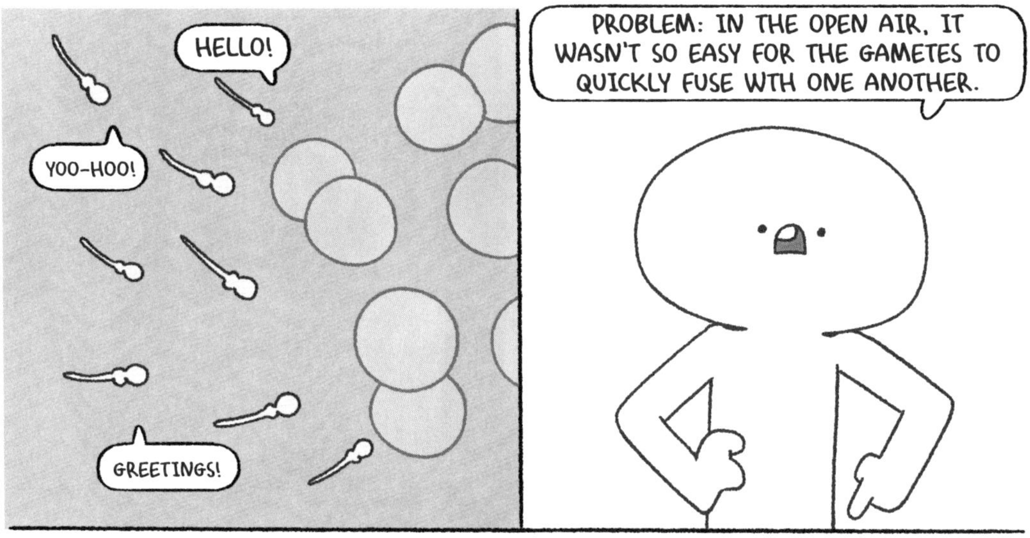
HELLO!
YOO-HOO!
GREETINGS!
PROBLEM: IN THE OPEN AIR, IT WASN'T SO EASY FOR THE GAMETES TO QUICKLY FUSE WTH ONE ANOTHER.

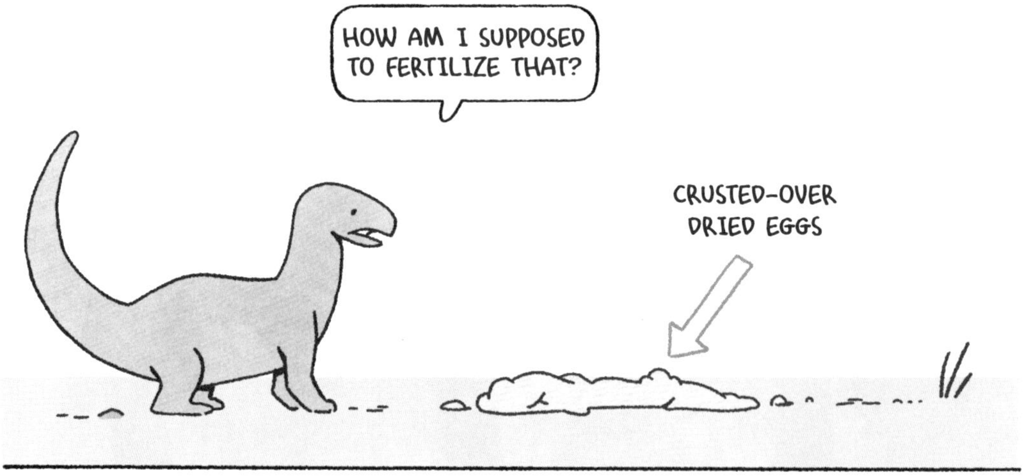
HOW AM I SUPPOSED TO FERTILIZE THAT?
CRUSTED-OVER DRIED EGGS

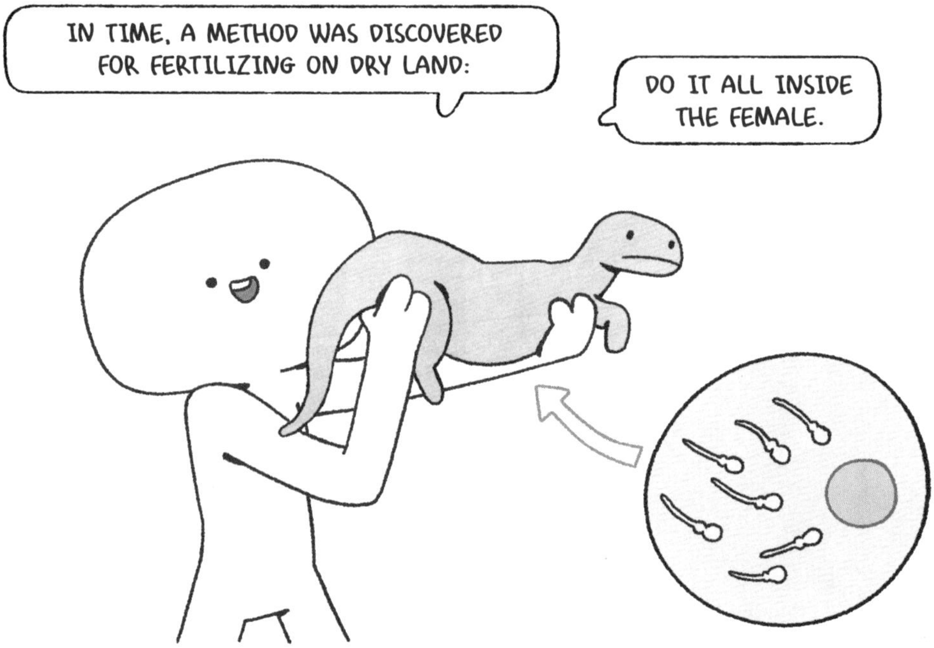
IN TIME, A METHOD WAS DISCOVERED FOR FERTILIZING ON DRY LAND:
DO IT ALL INSIDE THE FEMALE.

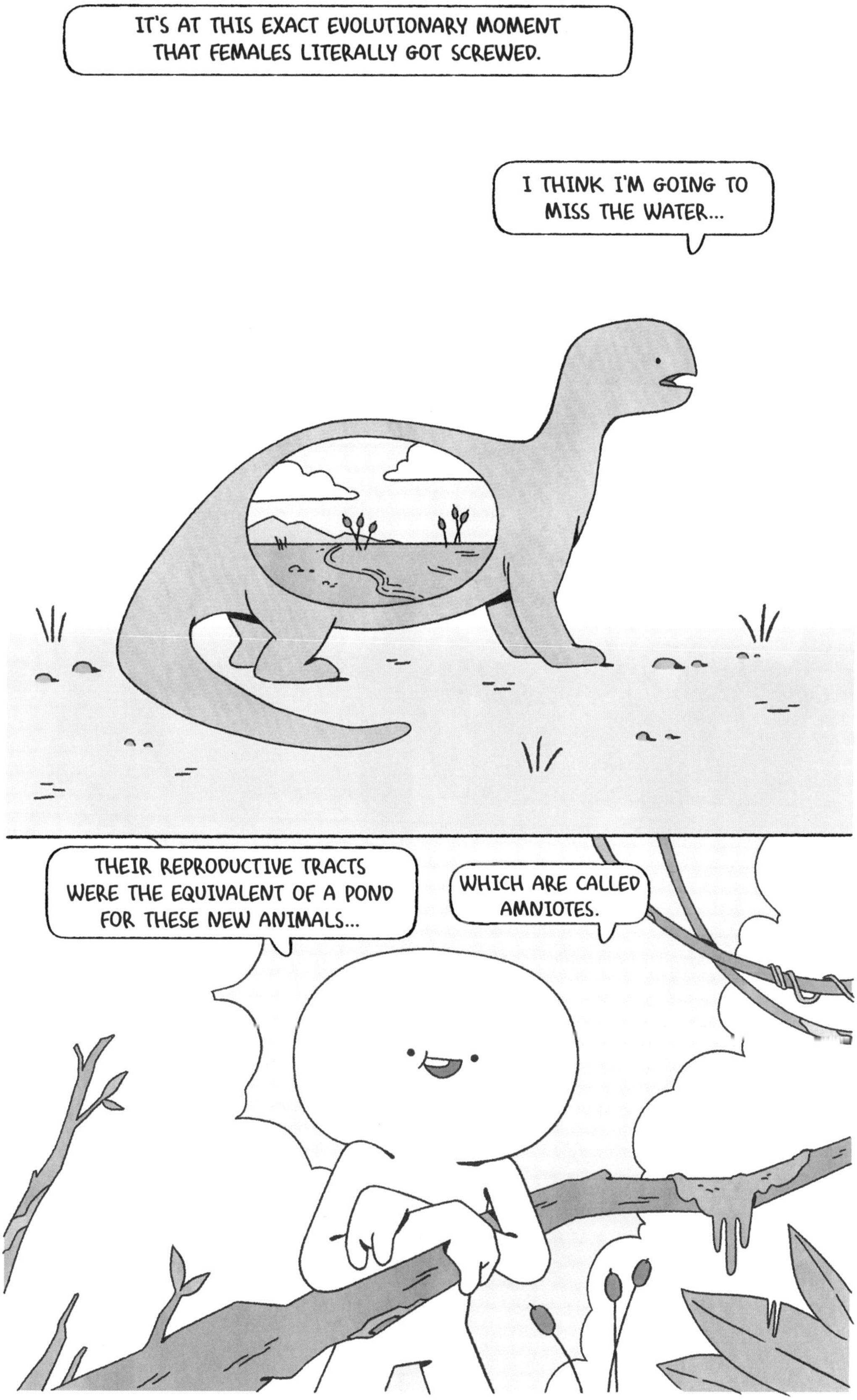
IT'S AT THIS EXACT EVOLUTIONARY MOMENT THAT FEMALES LITERALLY GOT SCREWED.
I THINK I'M GOING TO MISS THE WATER...
THEIR REPRODUCTIVE TRACTS WERE THE EQUIVALENT OF A POND FOR THESE NEW ANIMALS...
WHICH ARE CALLED AMNIOTES.

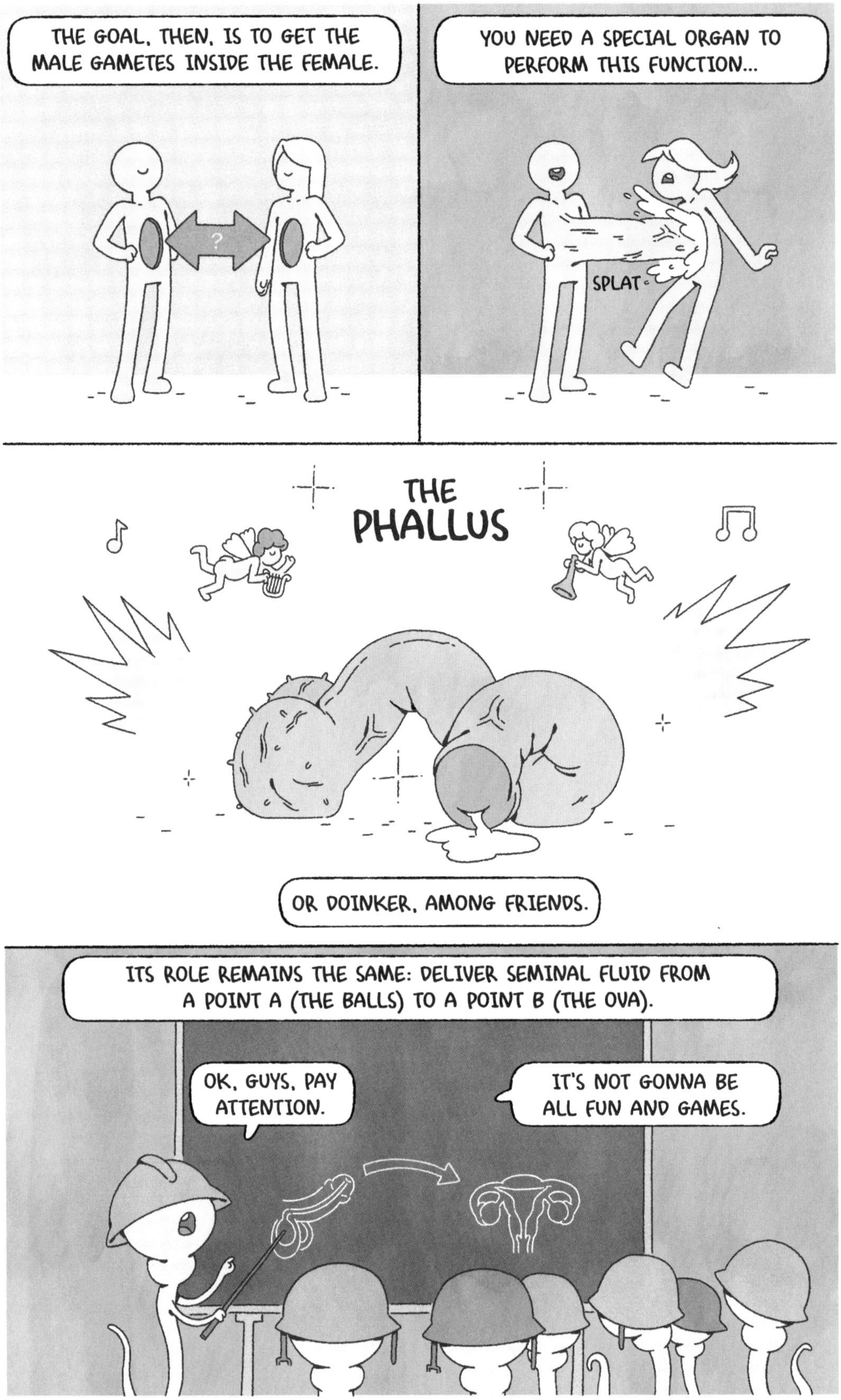
THE GOAL, THEN, IS TO GET THE MALE GAMETES INSIDE THE FEMALE.
?
YOU NEED A SPECIAL ORGAN TO PERFORM THIS FUNCTION...
SPLAT
THE PHALLUS
OR DOINKER, AMONG FRIENDS.
ITS ROLE REMAINS THE SAME: DELIVER SEMINAL FLUID FROM A POINT A (THE BALLS) TO A POINT B (THE OVA).
OK, GUYS, PAY ATTENTION.
IT'S NOT GONNA BE ALL FUN AND GAMES.

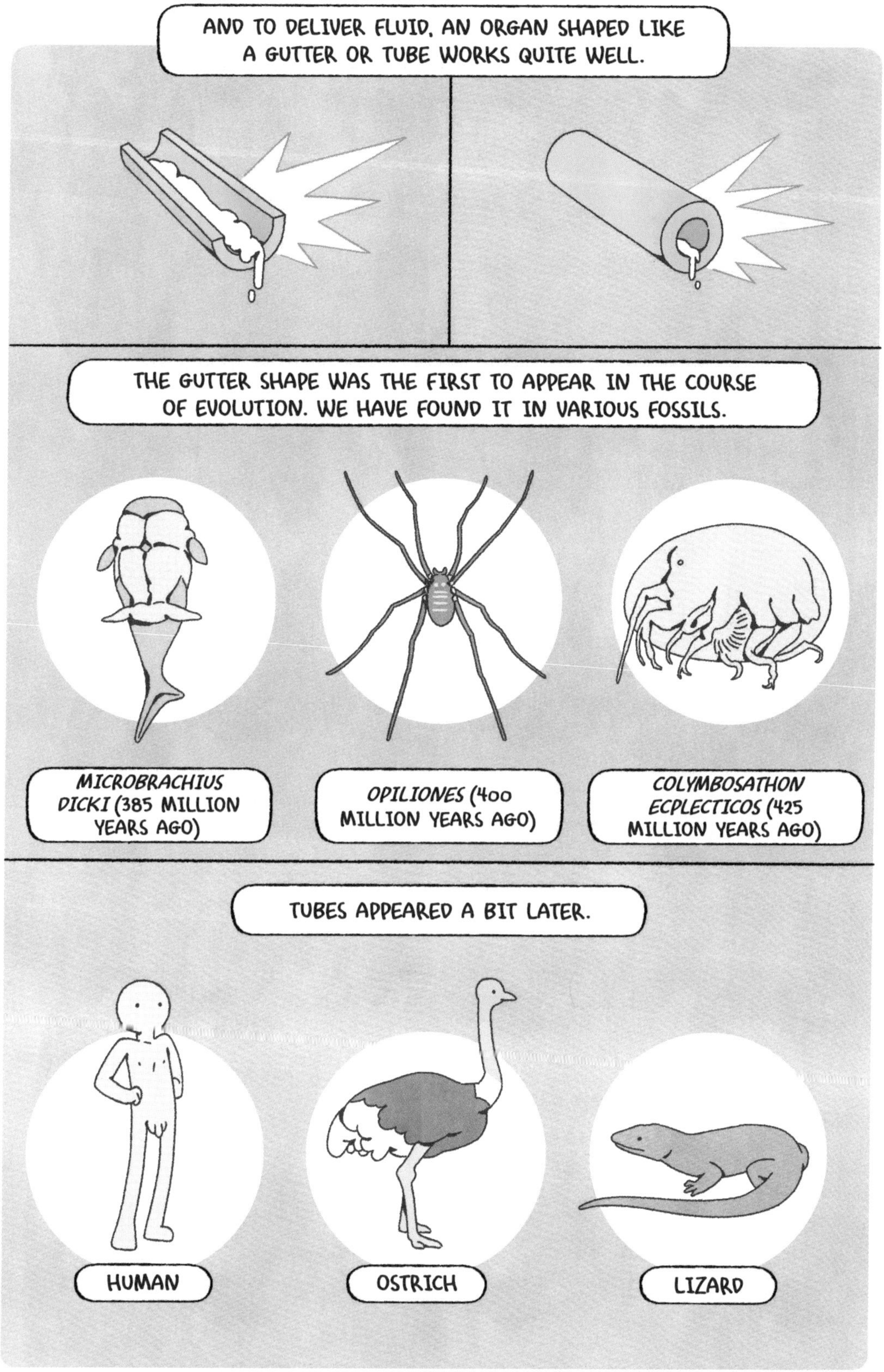
AND TO DELIVER FLUID, AN ORGAN SHAPED LIKE A GUTTER OR TUBE WORKS QUITE WELL.
THE GUTTER SHAPE WAS THE FIRST TO APPEAR IN THE COURSE OF EVOLUTION. WE HAVE FOUND IT IN VARIOUS FOSSILS.
MICROBRACHIUS DICKI (385 MILLION YEARS AGO)
OPILIONES (400 MILLION YEARS AGO)
COLYMBOSATHON ECPLECTICOS (425 MILLION YEARS AGO)
TUBES APPEARED A BIT LATER.
HUMAN
OSTRICH
LIZARD

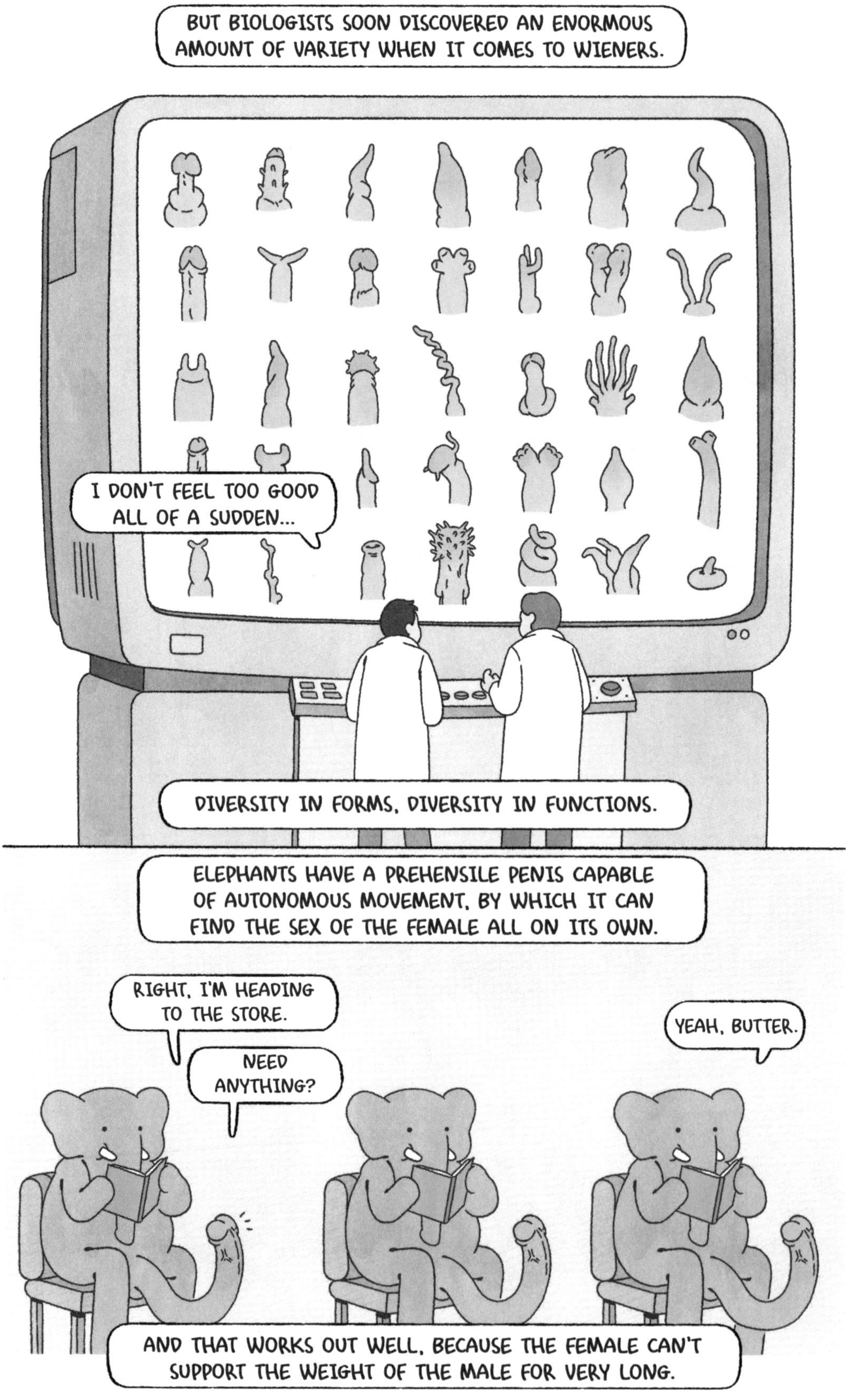
BUT BIOLOGISTS SOON DISCOVERED AN ENORMOUS AMOUNT OF VARIETY WHEN IT COMES TO WIENERS.
I DON'T FEEL TOO GOOD ALL OF A SUDDEN...
DIVERSITY IN FORMS, DIVERSITY IN FUNCTIONS.
ELEPHANTS HAVE A PREHENSILE PENIS CAPABLE OF AUTONOMOUS MOVEMENT, BY WHICH IT CAN FIND THE SEX OF THE FEMALE ALL ON ITS OWN.
RIGHT, I'M HEADING TO THE STORE.
NEED ANYTHING?
YEAH, BUTTER.
AND THAT WORKS OUT WELL, BECAUSE THE FEMALE CAN'T SUPPORT THE WEIGHT OF THE MALE FOR VERY LONG.

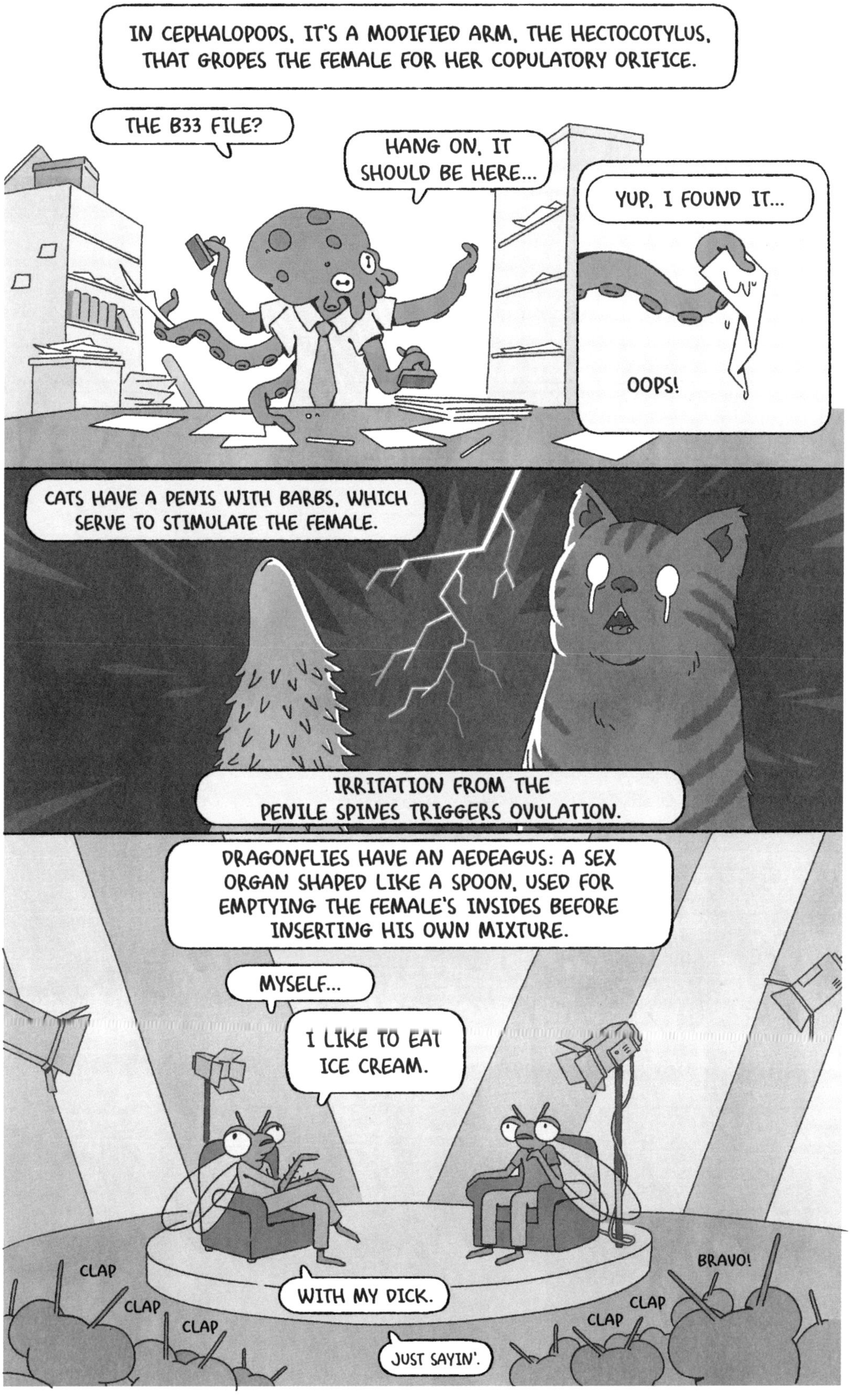
IN CEPHALOPODS, IT'S A MODIFIED ARM, THE HECTOCOTYLUS, THAT GROPES THE FEMALE FOR HER COPULATORY ORIFICE.
THE B33 FILE?
HANG ON, IT SHOULD BE HERE...
YUP, I FOUND IT...
OOPS!
CATS HAVE A PENIS WITH BARBS, WHICH SERVE TO STIMULATE THE FEMALE.
IRRITATION FROM THE PENILE SPINES TRIGGERS OVULATION.
DRAGONFLIES HAVE AN AEDEAGUS: A SEX ORGAN SHAPED LIKE A SPOON, USED FOR EMPTYING THE FEMALE'S INSIDES BEFORE INSERTING HIS OWN MIXTURE.
MYSELF...
I LIKE TO EAT ICE CREAM.
WITH MY DICK.
CLAP
CLAP
CLAP
CLAP
CLAP
BRAVO!
JUST SAYIN'.

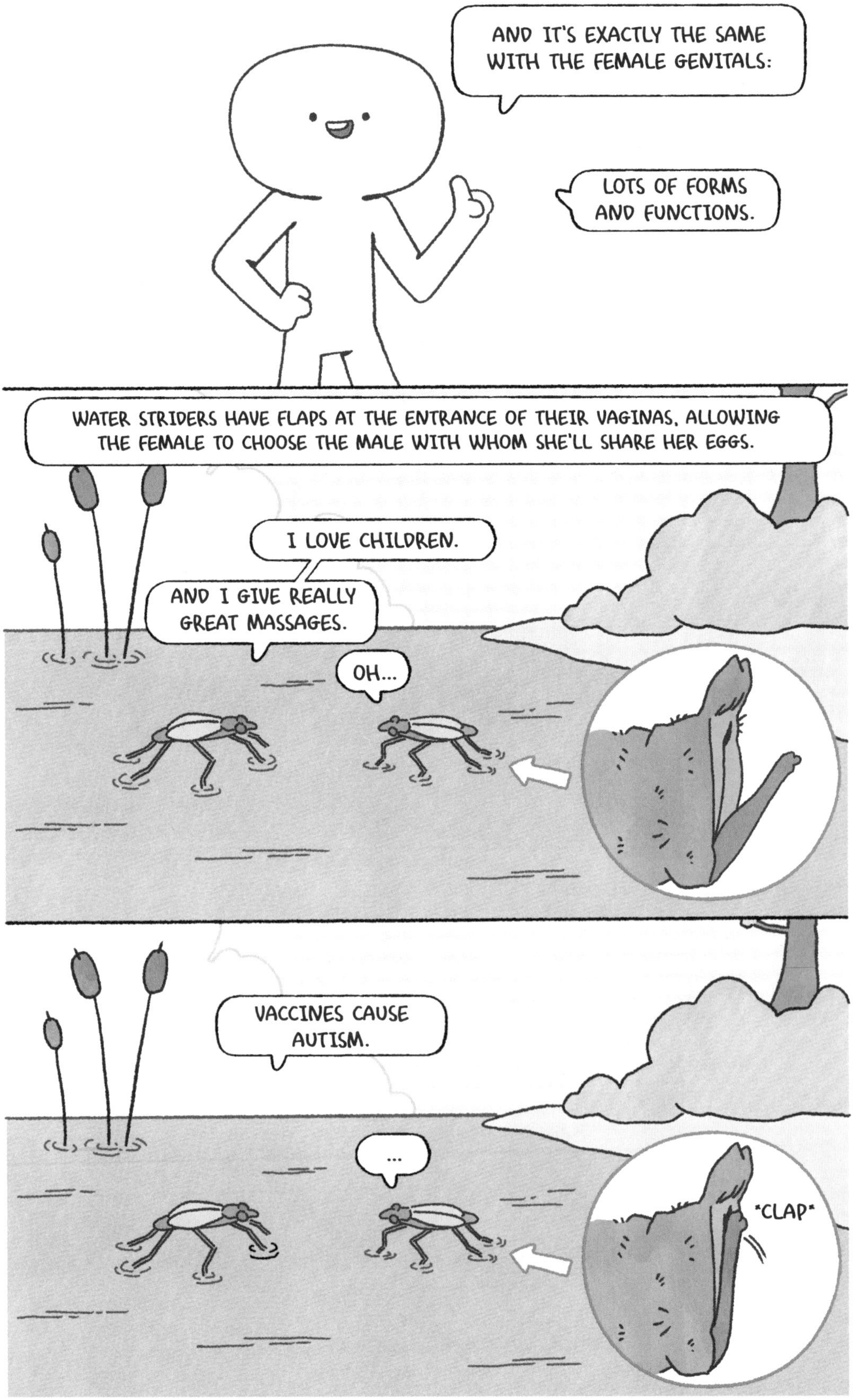
AND IT'S EXACTLY THE SAME WITH THE FEMALE GENITALS:
LOTS OF FORMS AND FUNCTIONS.
WATER STRIDERS HAVE FLAPS AT THE ENTRANCE OF THEIR VAGINAS, ALLOWING THE FEMALE TO CHOOSE THE MALE WITH WHOM SHE'LL SHARE HER EGGS.
I LOVE CHILDREN.
AND I GIVE REALLY GREAT MASSAGES.
OH...
VACCINES CAUSE AUTISM.
...
CLAP

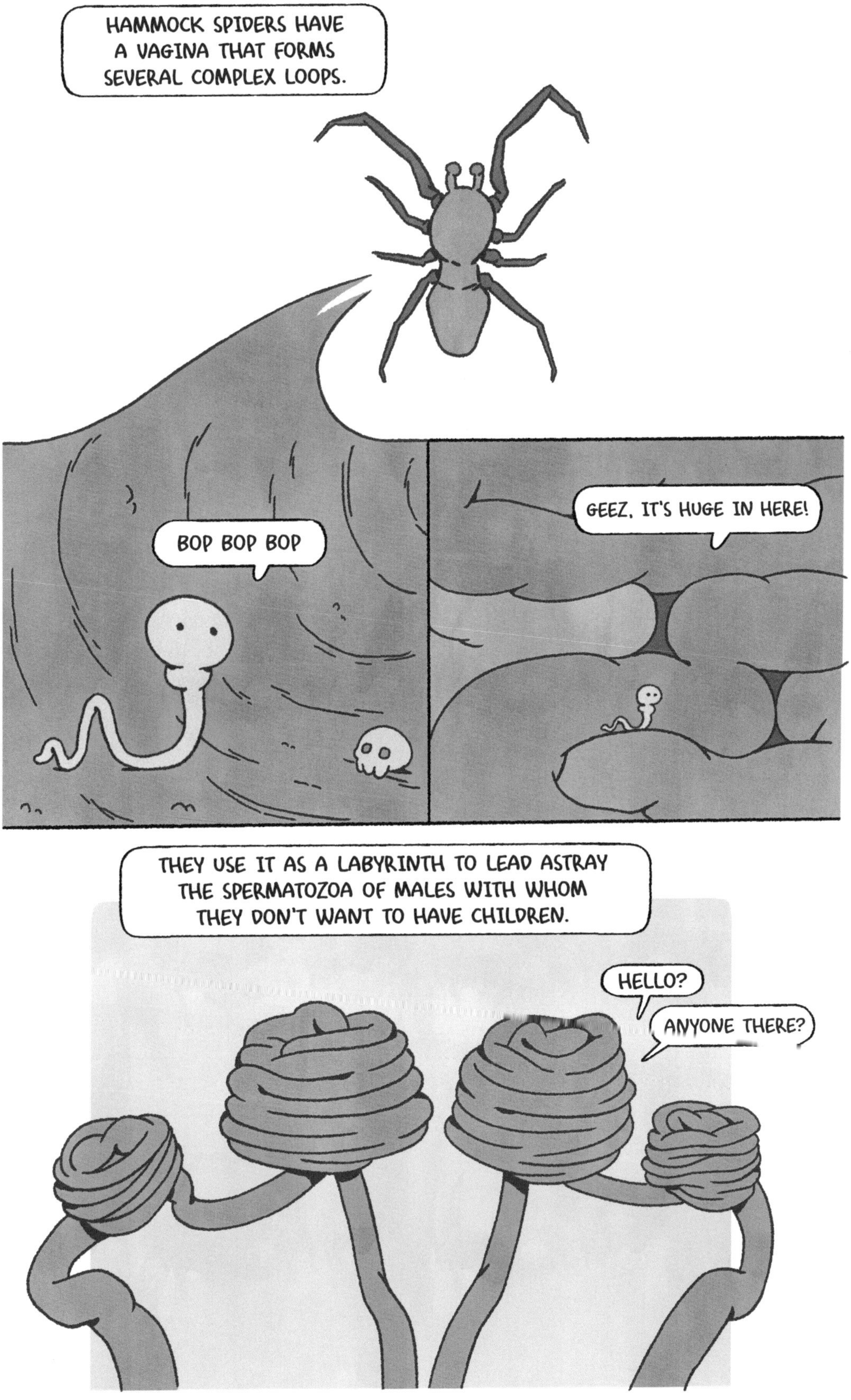
HAMMOCK SPIDERS HAVE A VAGINA THAT FORMS SEVERAL COMPLEX LOOPS.
BOP BOP BOP
GEEZ, IT'S HUGE IN HERE!
THEY USE IT AS A LABYRINTH TO LEAD ASTRAY THE SPERMATOZOA OF MALES WITH WHOM THEY DON'T WANT TO HAVE CHILDREN.
HELLO?
ANYONE THERE?

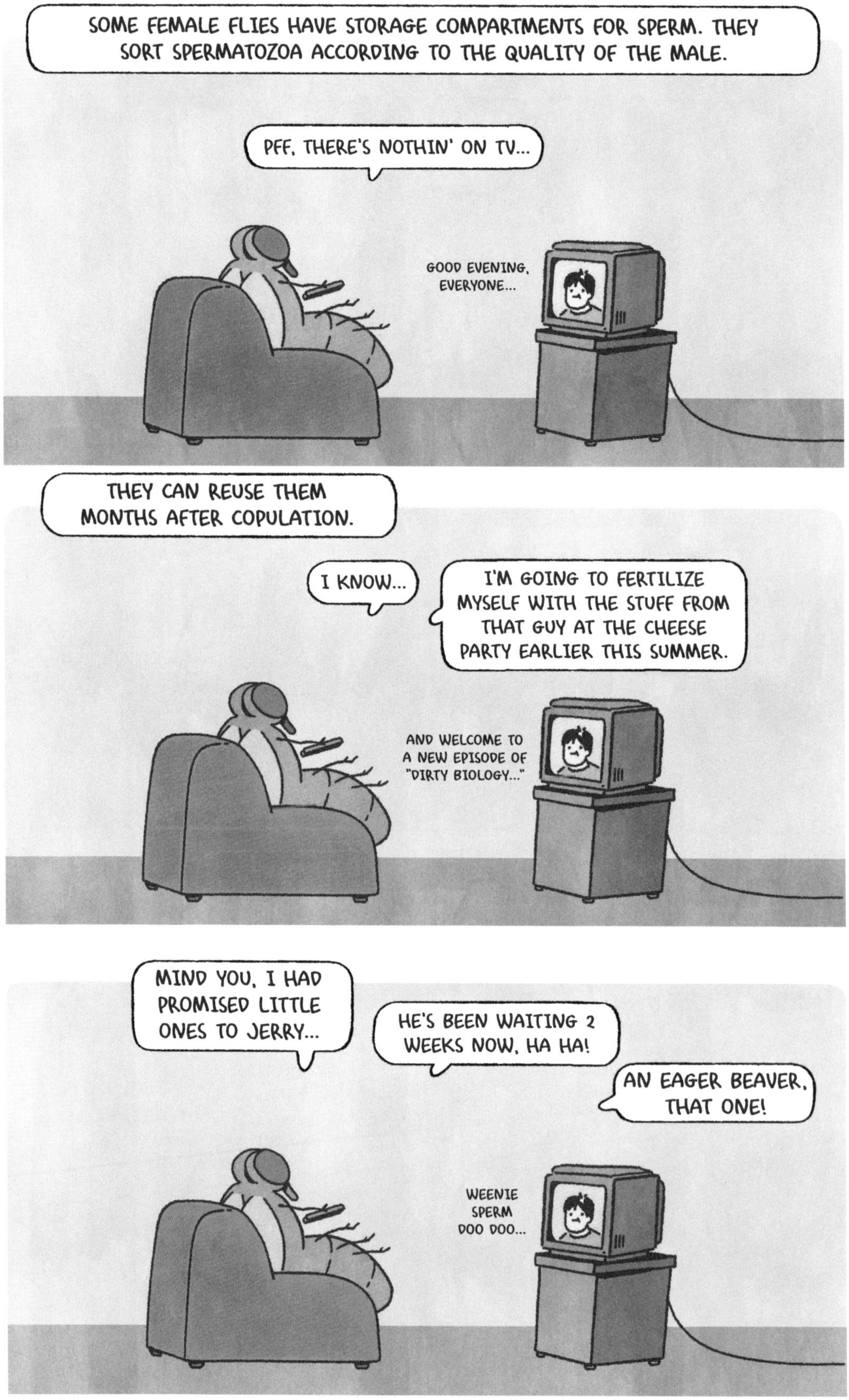
SOME FEMALE FLIES HAVE STORAGE COMPARTMENTS FOR SPERM. THEY SORT SPERMATOZOA ACCORDING TO THE QUALITY OF THE MALE.
PFF. THERE'S NOTHIN' ON TV...
GOOD EVENING, EVERYONE...
THEY CAN REUSE THEM MONTHS AFTER COPULATION.
I KNOW...
I'M GOING TO FERTILIZE MYSELF WITH THE STUFF FROM THAT GUY AT THE CHEESE PARTY EARLIER THIS SUMMER.
AND WELCOME TO A NEW EPISODE OF "DIRTY BIOLOGY..."
MIND YOU, I HAD PROMISED LITTLE ONES TO JERRY...
HE'S BEEN WAITING 2 WEEKS NOW, HA HA!
AN EAGER BEAVER, THAT ONE!
WEENIE SPERM DOO DOO...

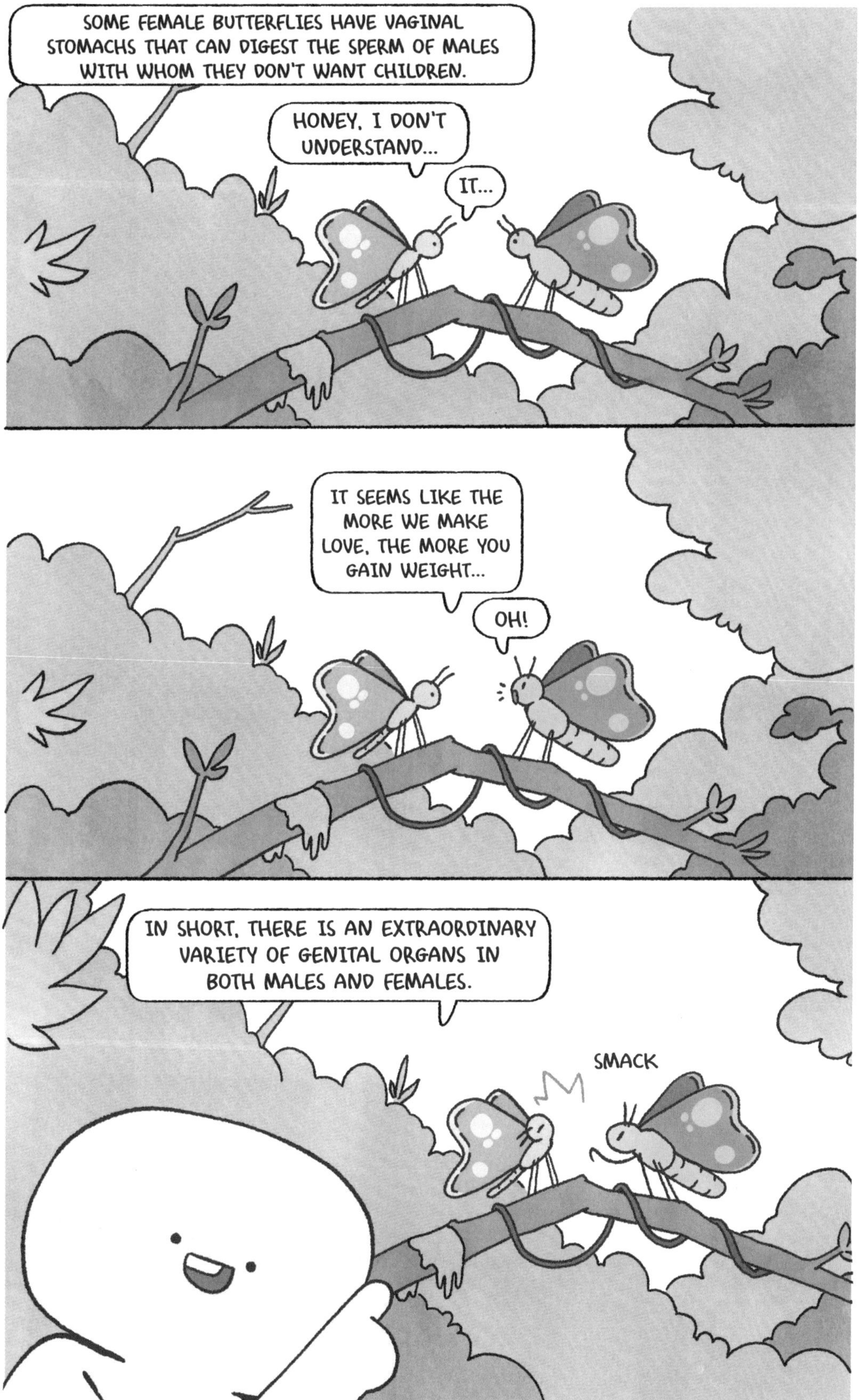
SOME FEMALE BUTTERFLIES HAVE VAGINAL STOMACHS THAT CAN DIGEST THE SPERM OF MALES WITH WHOM THEY DON'T WANT CHILDREN.
HONEY, I DON'T UNDERSTAND...
IT...
IT SEEMS LIKE THE MORE WE MAKE LOVE, THE MORE YOU GAIN WEIGHT...
OH!
IN SHORT, THERE IS AN EXTRAORDINARY VARIETY OF GENITAL ORGANS IN BOTH MALES AND FEMALES.
SMACK

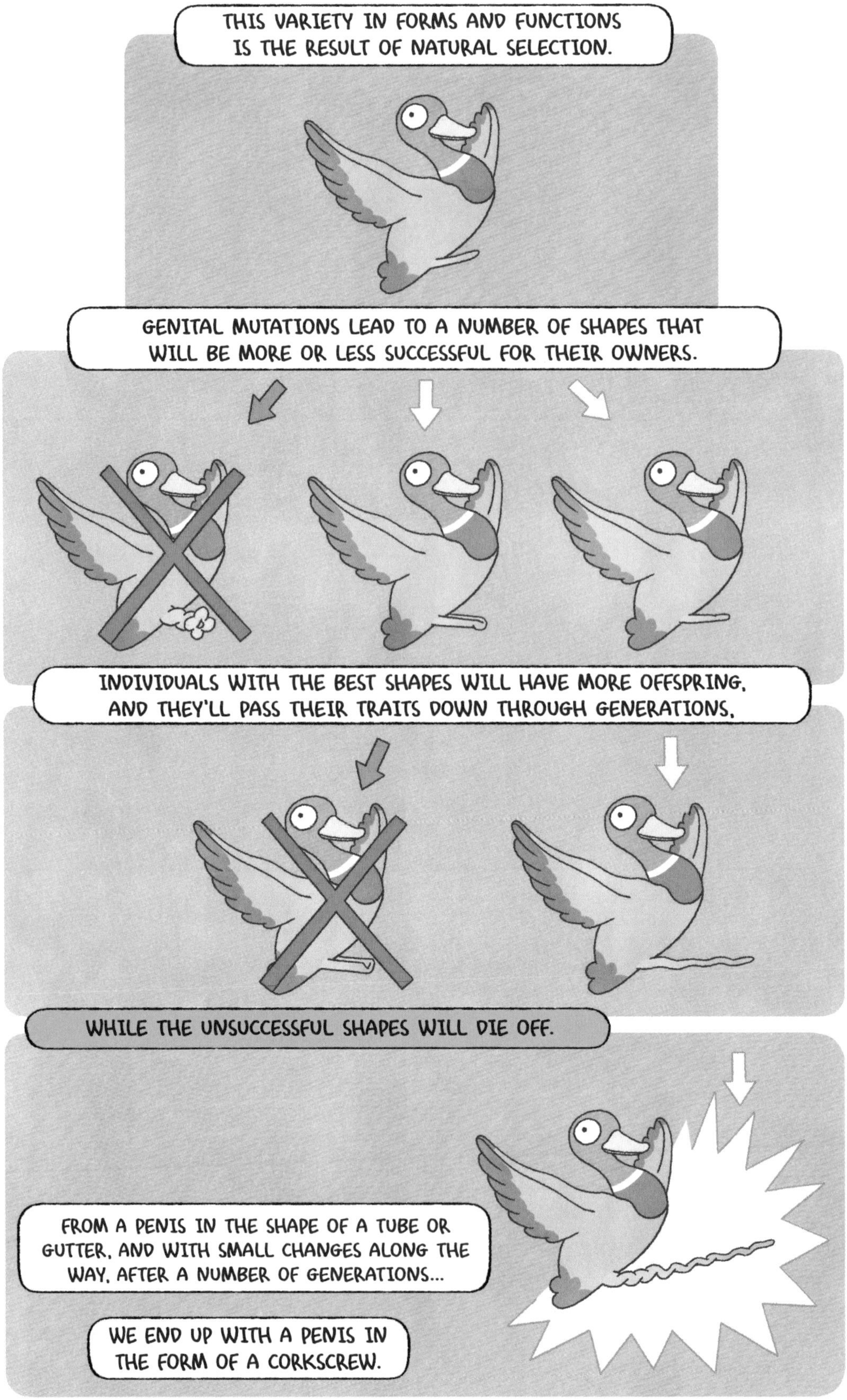
THIS VARIETY IN FORMS AND FUNCTIONS IS THE RESULT OF NATURAL SELECTION.
GENITAL MUTATIONS LEAD TO A NUMBER OF SHAPES THAT WILL BE MORE OR LESS SUCCESSFUL FOR THEIR OWNERS.
INDIVIDUALS WITH THE BEST SHAPES WILL HAVE MORE OFFSPRING, AND THEY'LL PASS THEIR TRAITS DOWN THROUGH GENERATIONS,
WHILE THE UNSUCCESSFUL SHAPES WILL DIE OFF.
FROM A PENIS IN THE SHAPE OF A TUBE OR GUTTER, AND WITH SMALL CHANGES ALONG THE WAY, AFTER A NUMBER OF GENERATIONS...
WE END UP WITH A PENIS IN THE FORM OF A CORKSCREW.

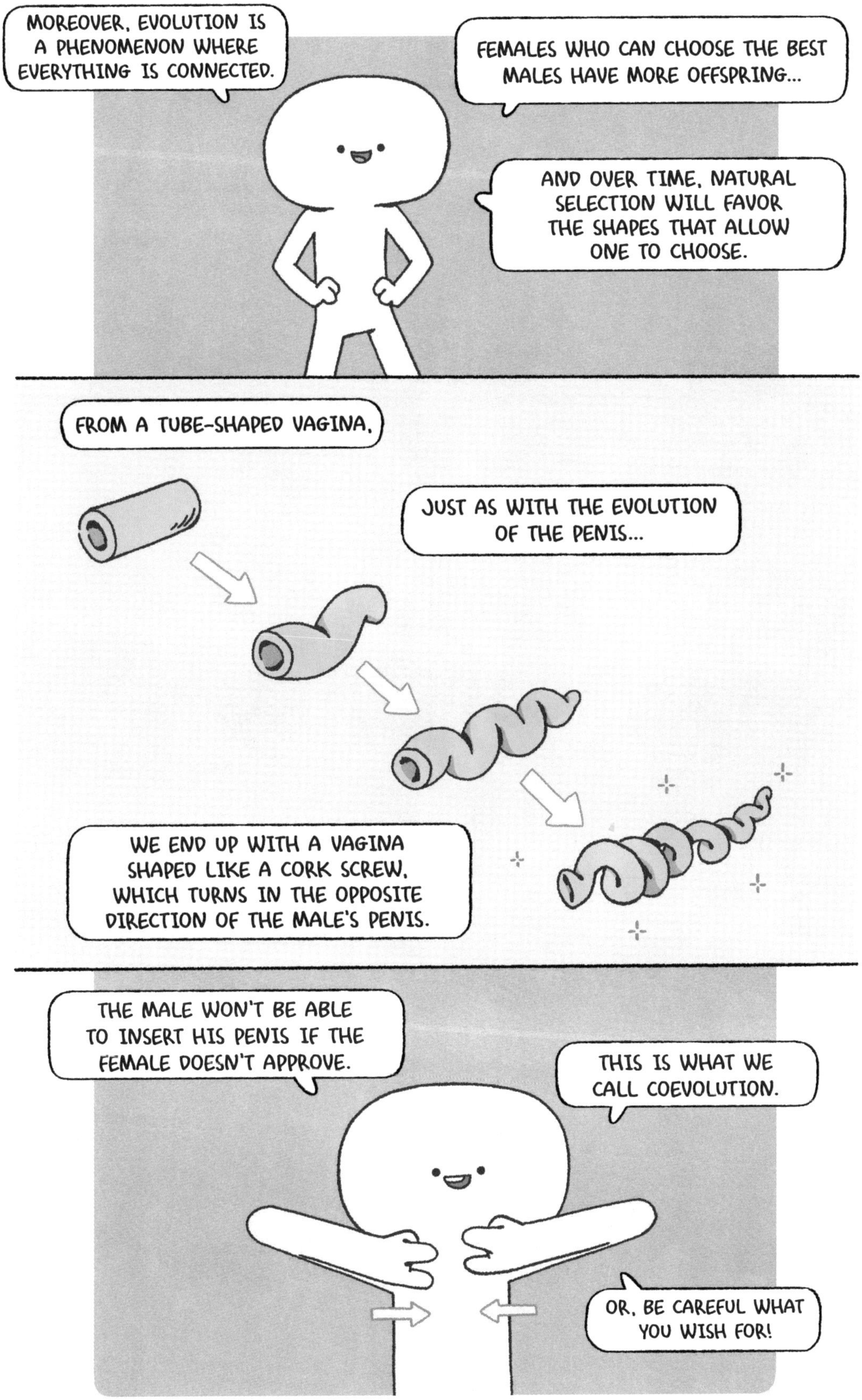
MOREOVER, EVOLUTION IS A PHENOMENON WHERE EVERYTHING IS CONNECTED.
FEMALES WHO CAN CHOOSE THE BEST MALES HAVE MORE OFFSPRING...
AND OVER TIME, NATURAL SELECTION WILL FAVOR THE SHAPES THAT ALLOW ONE TO CHOOSE.
FROM A TUBE-SHAPED VAGINA,
JUST AS WITH THE EVOLUTION OF THE PENIS...
WE END UP WITH A VAGINA SHAPED LIKE A CORK SCREW, WHICH TURNS IN THE OPPOSITE DIRECTION OF THE MALE'S PENIS.
THE MALE WON'T BE ABLE TO INSERT HIS PENIS IF THE FEMALE DOESN'T APPROVE.
THIS IS WHAT WE CALL COEVOLUTION.
OR, BE CAREFUL WHAT YOU WISH FOR!

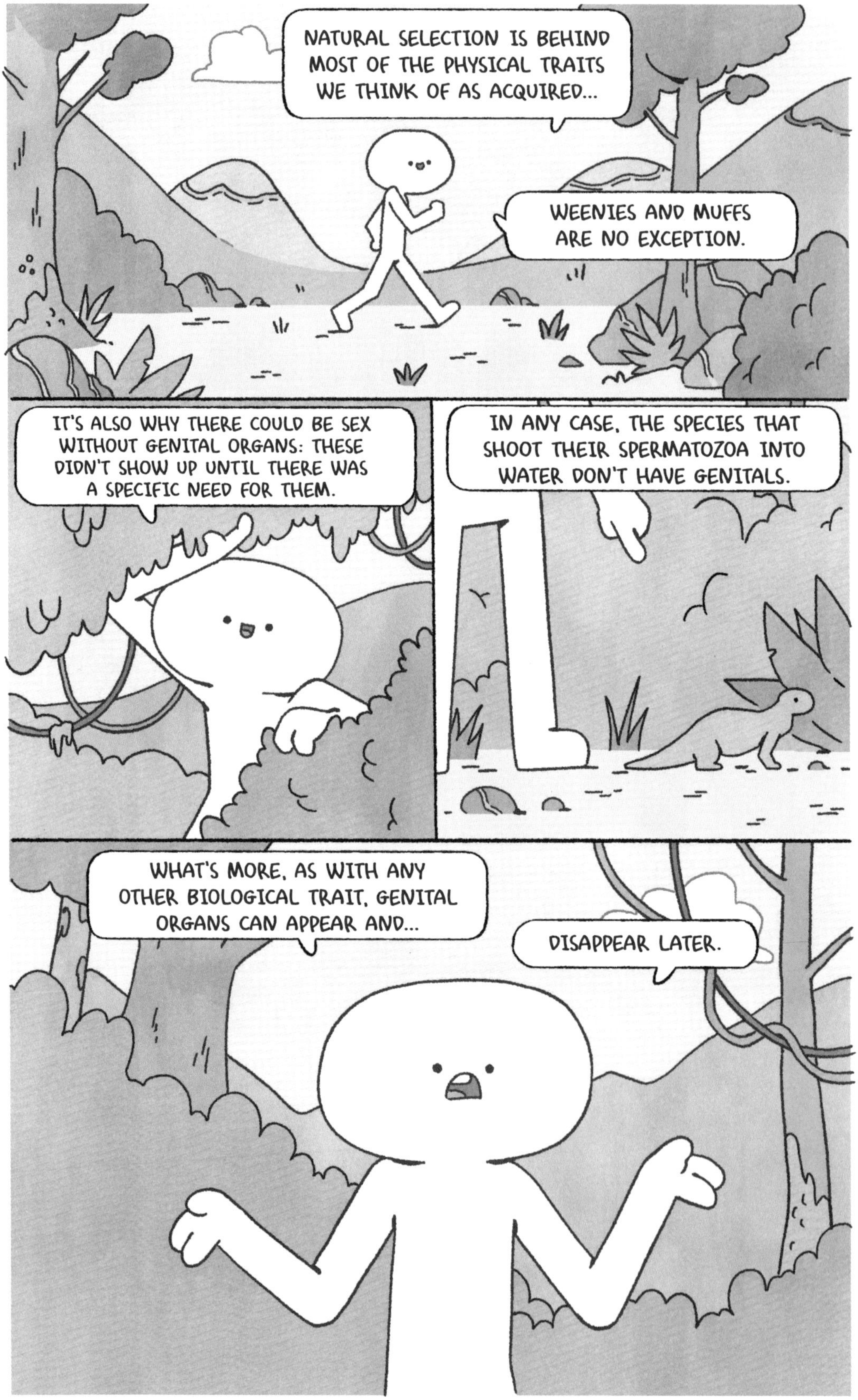
NATURAL SELECTION IS BEHIND MOST OF THE PHYSICAL TRAITS WE THINK OF AS ACQUIRED...
WEENIES AND MUFFS ARE NO EXCEPTION.
IT'S ALSO WHY THERE COULD BE SEX WITHOUT GENITAL ORGANS: THESE DIDN'T SHOW UP UNTIL THERE WAS A SPECIFIC NEED FOR THEM.
IN ANY CASE, THE SPECIES THAT SHOOT THEIR SPERMATOZOA INTO WATER DON'T HAVE GENITALS.
WHAT'S MORE, AS WITH ANY OTHER BIOLOGICAL TRAIT, GENITAL ORGANS CAN APPEAR AND...
DISAPPEAR LATER.

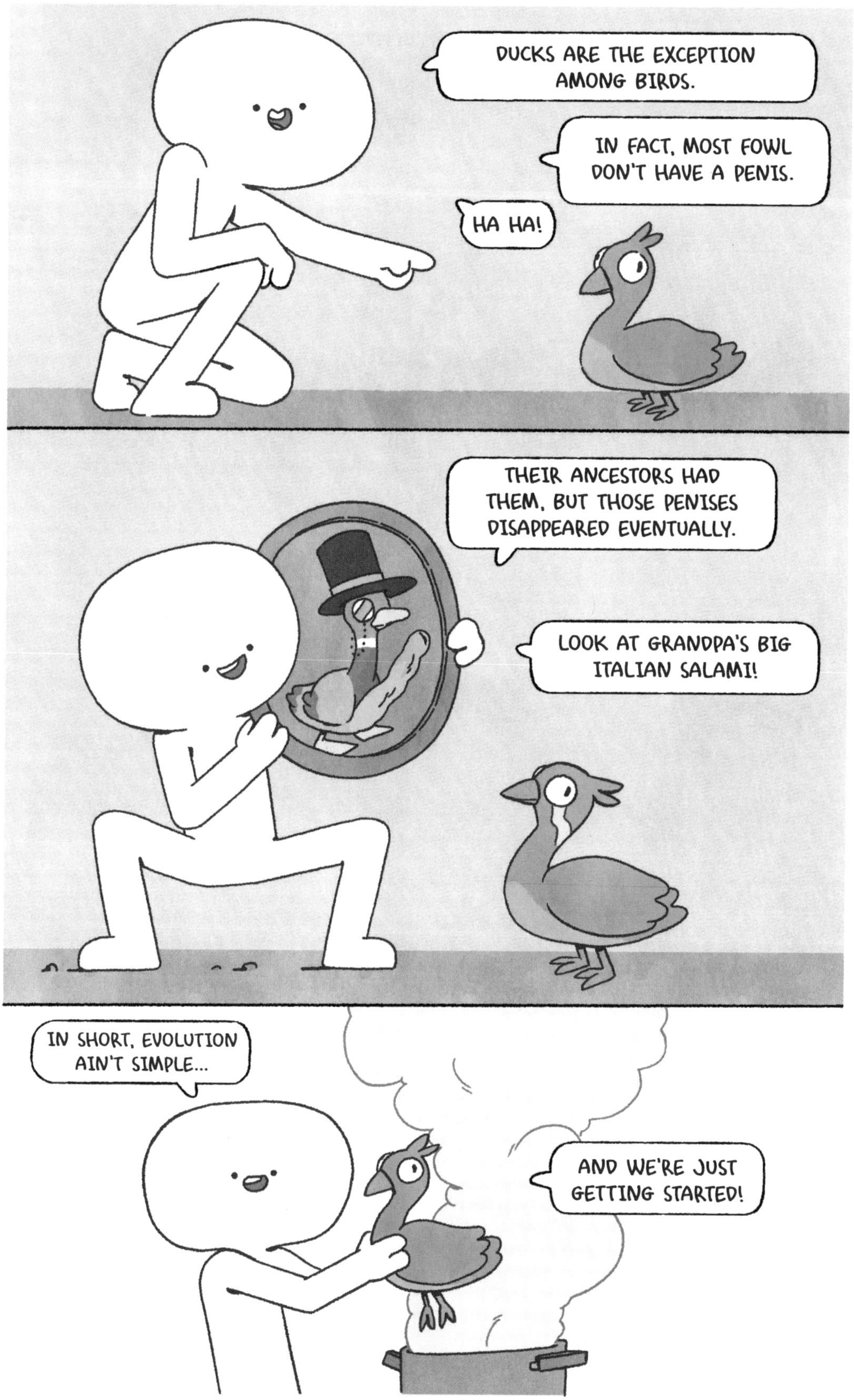
DUCKS ARE THE EXCEPTION AMONG BIRDS.
IN FACT, MOST FOWL DON'T HAVE A PENIS.
HA HA!
THEIR ANCESTORS HAD THEM, BUT THOSE PENISES DISAPPEARED EVENTUALLY.
LOOK AT GRANDPA'S BIG ITALIAN SALAMI!
IN SHORT, EVOLUTION AIN'T SIMPLE...
AND WE'RE JUST GETTING STARTED!

WHY TWO SEXES?

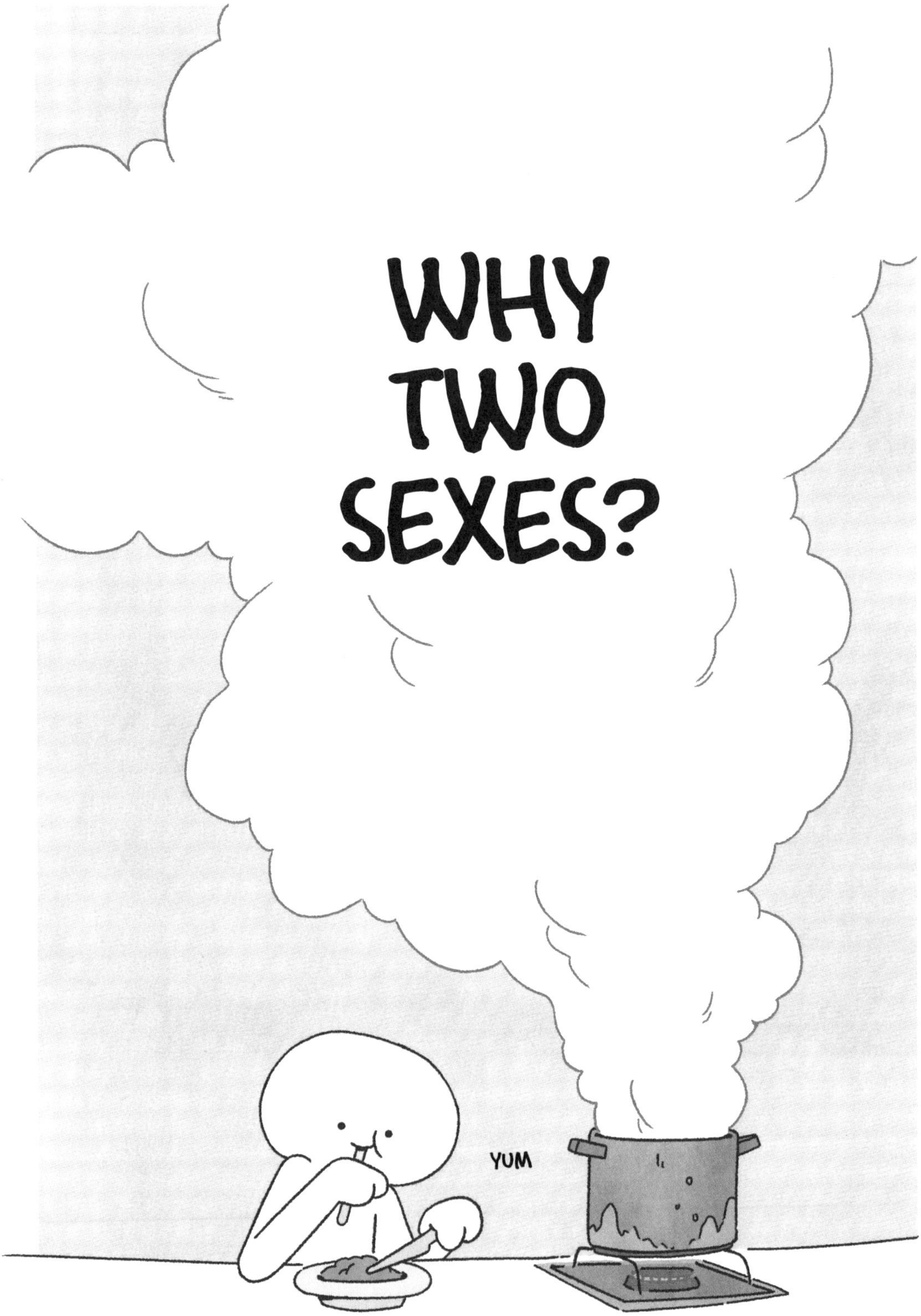
YUM

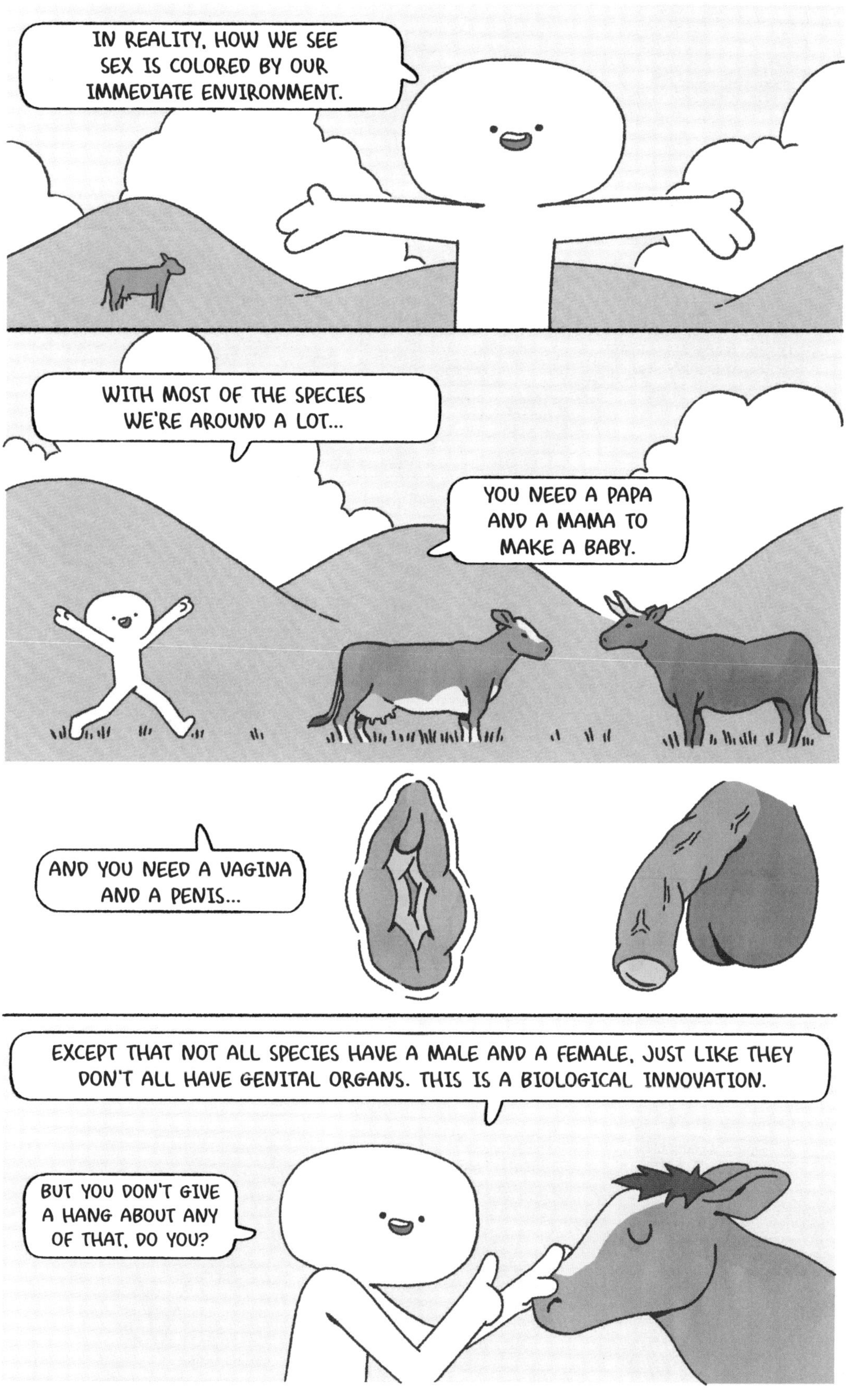
IN REALITY, HOW WE SEE SEX IS COLORED BY OUR IMMEDIATE ENVIRONMENT.
WITH MOST OF THE SPECIES WE'RE AROUND A LOT...
YOU NEED A PAPA AND A MAMA TO MAKE A BABY.
AND YOU NEED A VAGINA AND A PENIS...
EXCEPT THAT NOT ALL SPECIES HAVE A MALE AND A FEMALE, JUST LIKE THEY DON'T ALL HAVE GENITAL ORGANS. THIS IS A BIOLOGICAL INNOVATION.
BUT YOU DON'T GIVE A HANG ABOUT ANY OF THAT, DO YOU?

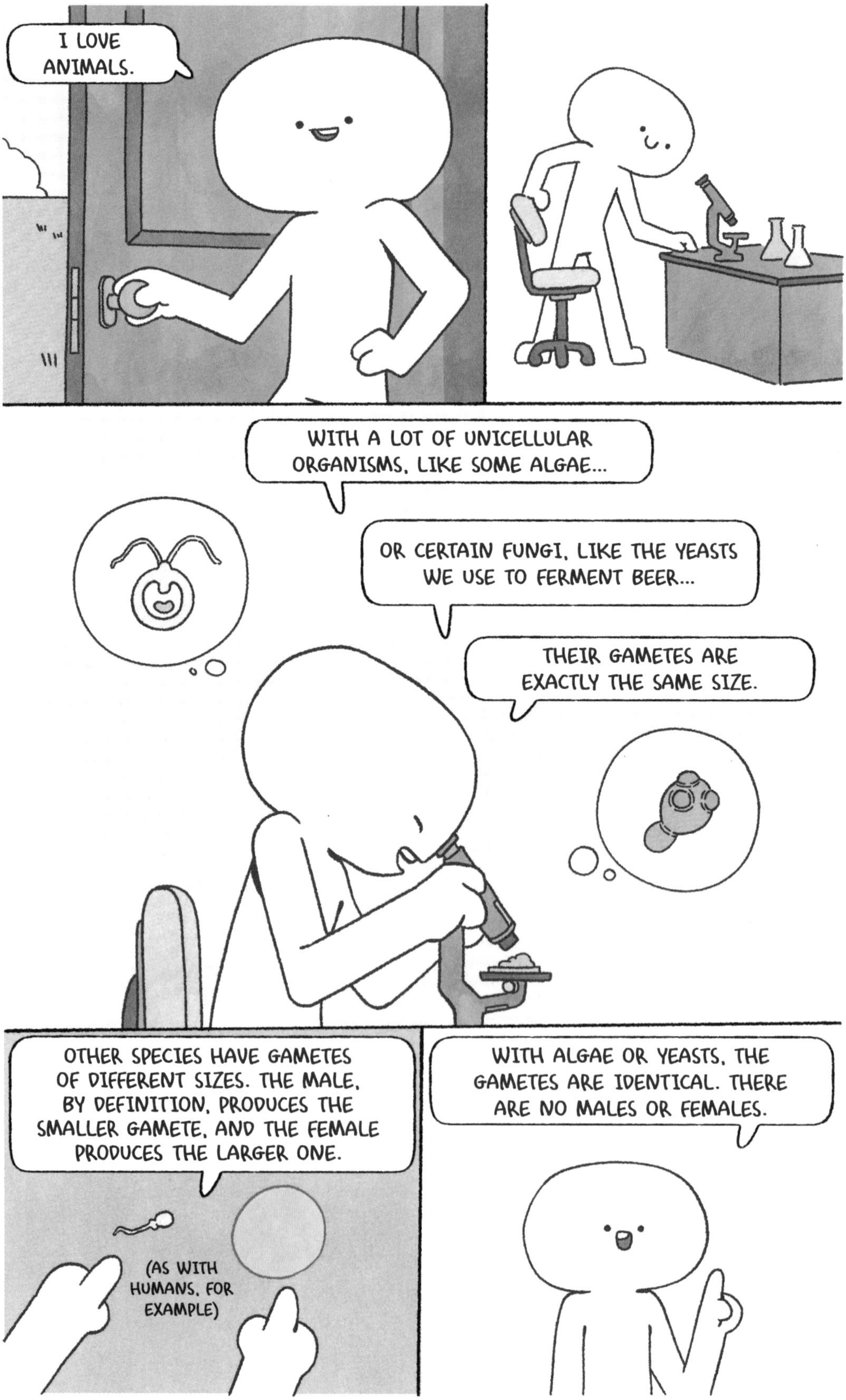
I LOVE ANIMALS.
WITH A LOT OF UNICELLULAR ORGANISMS, LIKE SOME ALGAE...
OR CERTAIN FUNGI, LIKE THE YEASTS WE USE TO FERMENT BEER...
THEIR GAMETES ARE EXACTLY THE SAME SIZE.
OTHER SPECIES HAVE GAMETES OF DIFFERENT SIZES. THE MALE, BY DEFINITION, PRODUCES THE SMALLER GAMETE, AND THE FEMALE PRODUCES THE LARGER ONE.
(AS WITH HUMANS, FOR EXAMPLE)
WITH ALGAE OR YEASTS, THE GAMETES ARE IDENTICAL. THERE ARE NO MALES OR FEMALES.

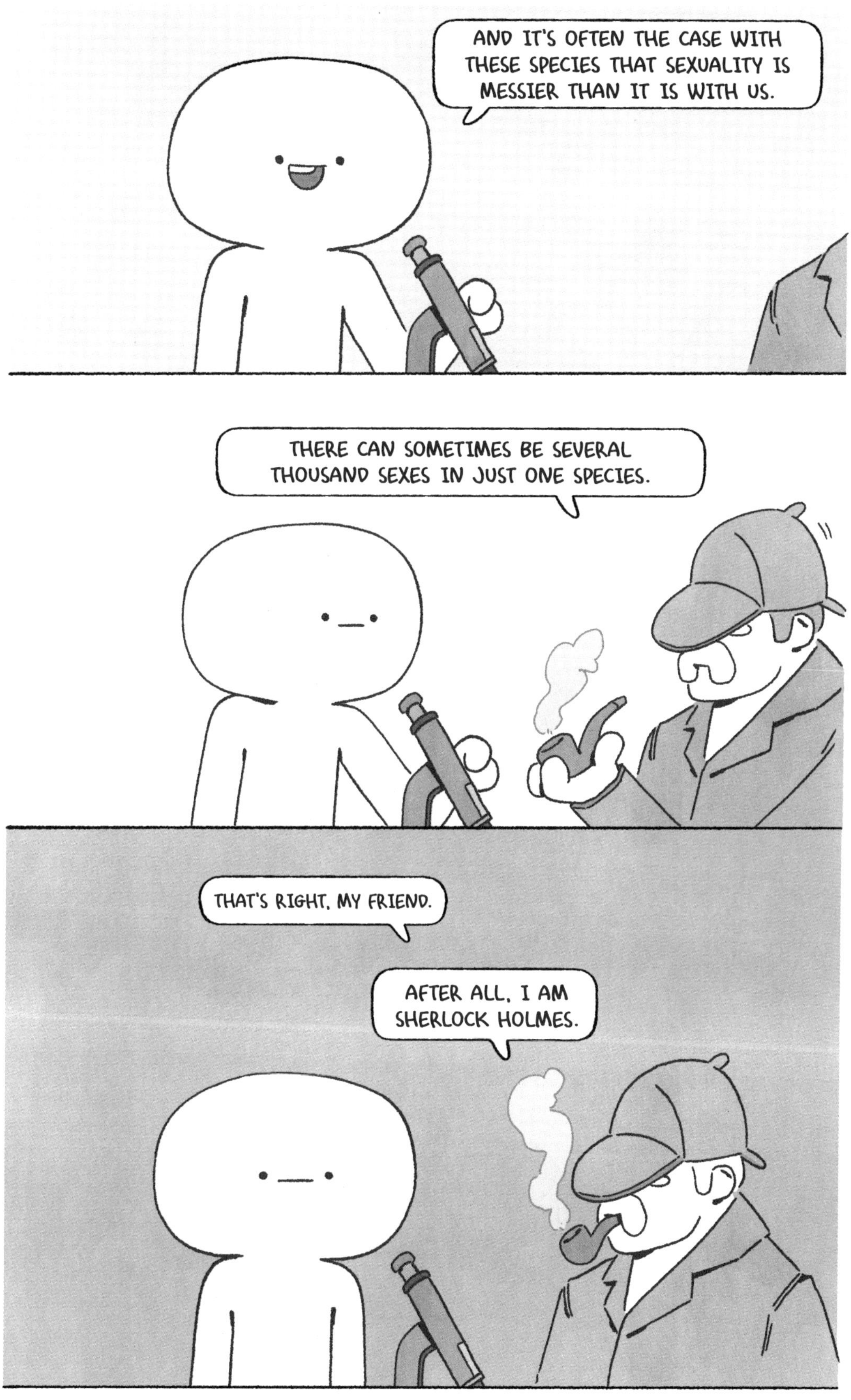
AND IT'S OFTEN THE CASE WITH THESE SPECIES THAT SEXUALITY IS MESSIER THAN IT IS WITH US.
THERE CAN SOMETIMES BE SEVERAL THOUSAND SEXES IN JUST ONE SPECIES.
THAT'S RIGHT, MY FRIEND.
AFTER ALL, I AM SHERLOCK HOLMES.

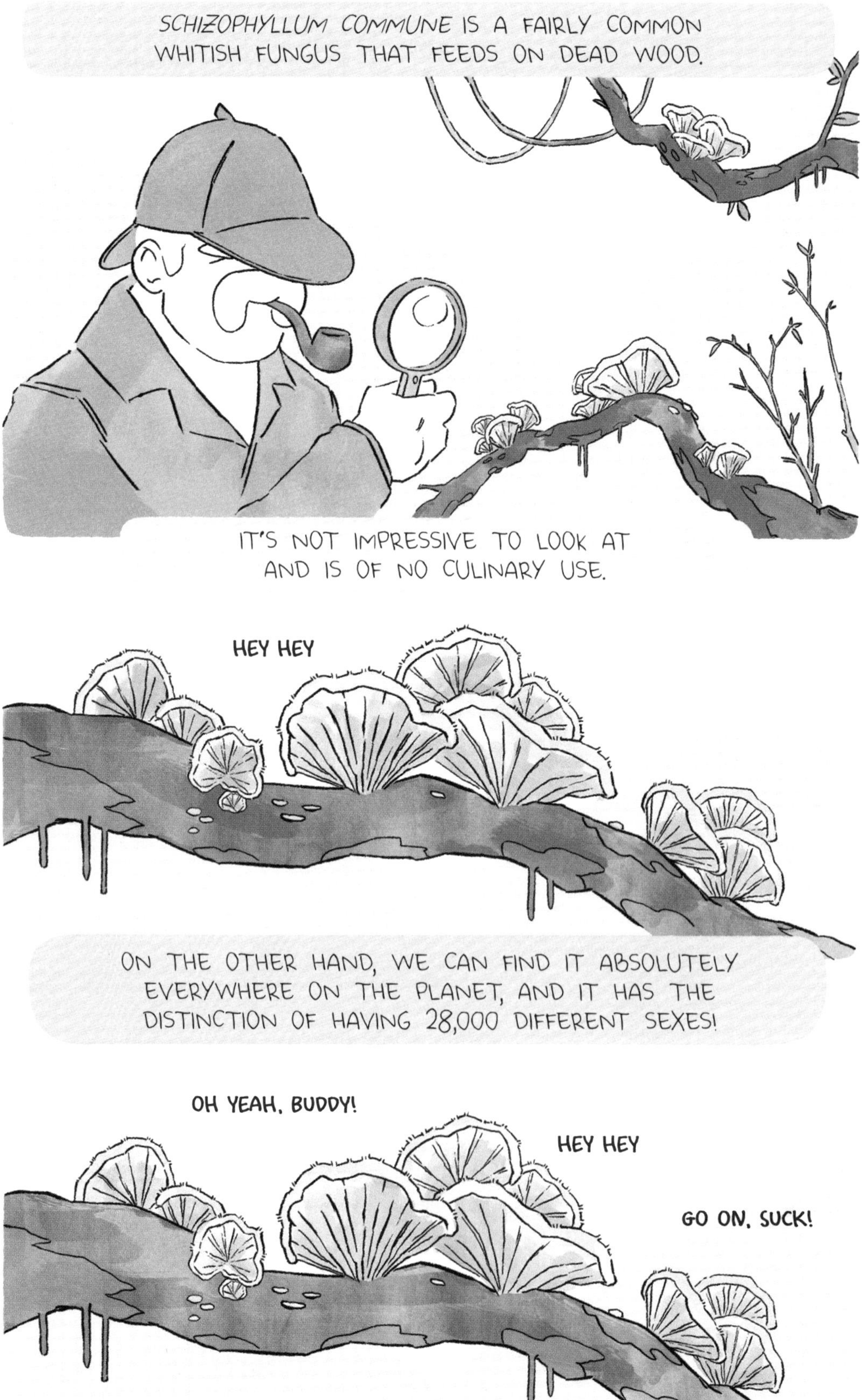
SCHIZOPHYLLUM COMMUNE IS A FAIRLY COMMON WHITISH FUNGUS THAT FEEDS ON DEAD WOOD.
IT'S NOT IMPRESSIVE TO LOOK AT AND IS OF NO CULINARY USE.
HEY HEY
ON THE OTHER HAND, WE CAN FIND IT ABSOLUTELY EVERYWHERE ON THE PLANET, AND IT HAS THE DISTINCTION OF HAVING 28,000 DIFFERENT SEXES!
OH YEAH, BUDDY!
HEY HEY
GO ON, SUCK!

HUMANS, LIKE ALL ANIMALS, HAVE
TWO SEXES FOR REPRODUCTION.

KNOB
GASH

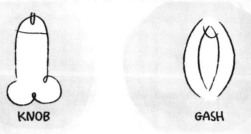

MALES' GAMETES ARE INCOMPATIBLE WITH THOSE OF
OTHER MALES. THE SAME GOES FOR FEMALES.

CLEVER...

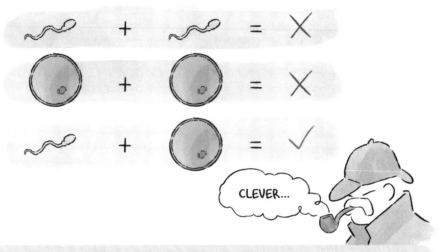

IT'S THE SAME WITH FUNGI, EXCEPT THAT THE
GAMETES ARE OF THE SAME SIZE AND SHAPE.

TYPE 1
TYPE 2

I THOUGHT SO...

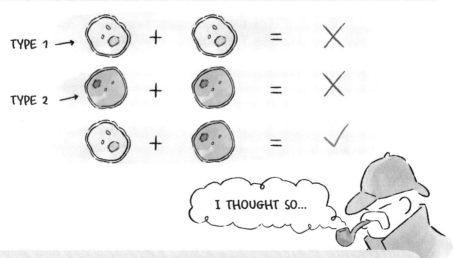

THAT'S WHY THESE ARE CALLED SEX TYPES. IT'S PRETTY
MUCH THE SAME AS WITH HUMANS: CELLS OF ANY GIVEN
TYPE CANNOT FUSE WITH OTHER CELLS OF THE SAME TYPE.

IN *SACCHAROMYCES CEREVISIAE*, YEASTS (THE UNICELLULAR FUNGUS USED FOR NATURAL FERMENTATION IN BEER, WINE, AND BREAD)...

THERE ARE TWO SEX TYPES:

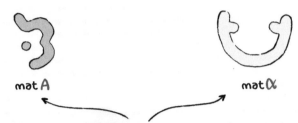

mat A
mat α

THEIR NAME COMES FROM THE "MAT" GENE, WHICH DETERMINES SEX TYPE.

IN OTHER SPECIES, THERE CAN BE MORE THAN 2 VERSIONS OF THIS GENE, AND THUS SEVERAL SEX TYPES.

FOR EXAMPLE, A COPROPHILOUS FUNGUS:

mat A
mat α
mat β

OR AN AMANITA:

mat A
mat δ
mat α
mat β
mat γ

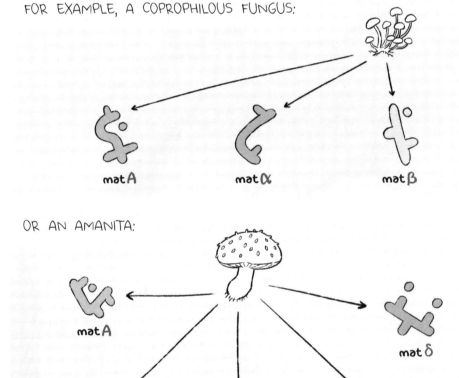

(GAMETES SHOWN HERE ARE, AGAIN, A DRUNK ARTIST'S RENDERING. SORRY.)

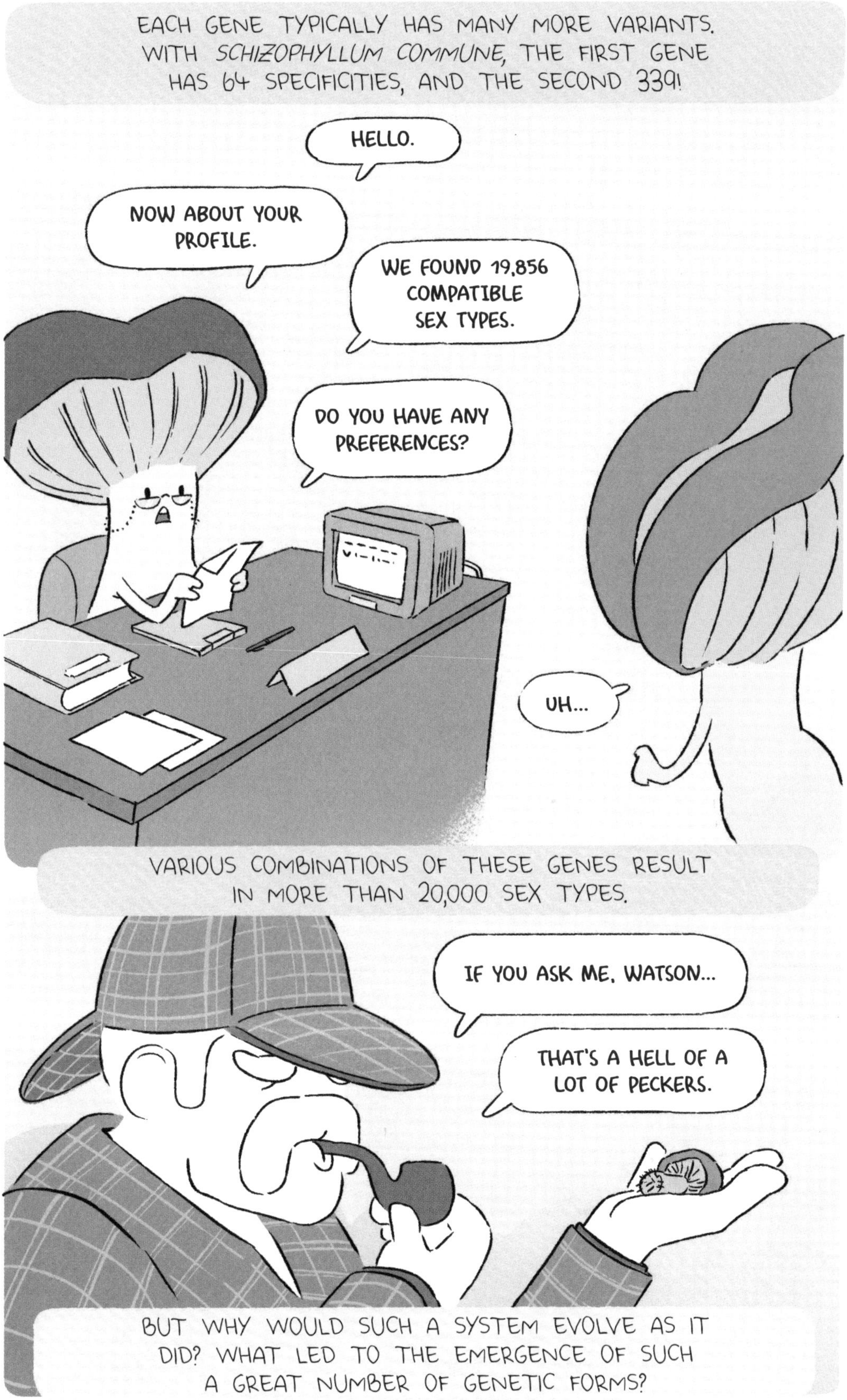
EACH GENE TYPICALLY HAS MANY MORE VARIANTS. WITH *SCHIZOPHYLLUM COMMUNE*, THE FIRST GENE HAS 64 SPECIFICITIES, AND THE SECOND 339!
HELLO.
NOW ABOUT YOUR PROFILE.
WE FOUND 19,856 COMPATIBLE SEX TYPES.
DO YOU HAVE ANY PREFERENCES?
UH...
VARIOUS COMBINATIONS OF THESE GENES RESULT IN MORE THAN 20,000 SEX TYPES.
IF YOU ASK ME, WATSON...
THAT'S A HELL OF A LOT OF PECKERS.
BUT WHY WOULD SUCH A SYSTEM EVOLVE AS IT DID? WHAT LED TO THE EMERGENCE OF SUCH A GREAT NUMBER OF GENETIC FORMS?

ONE POSSIBLE EXPLANATION IS THAT, WITH ONLY TWO SEX TYPES, THE SPORES EJECTED OUT INTO NATURE WOULD ENCOUNTER POTENTIAL PARTNERS ONLY 50% OF THE TIME.

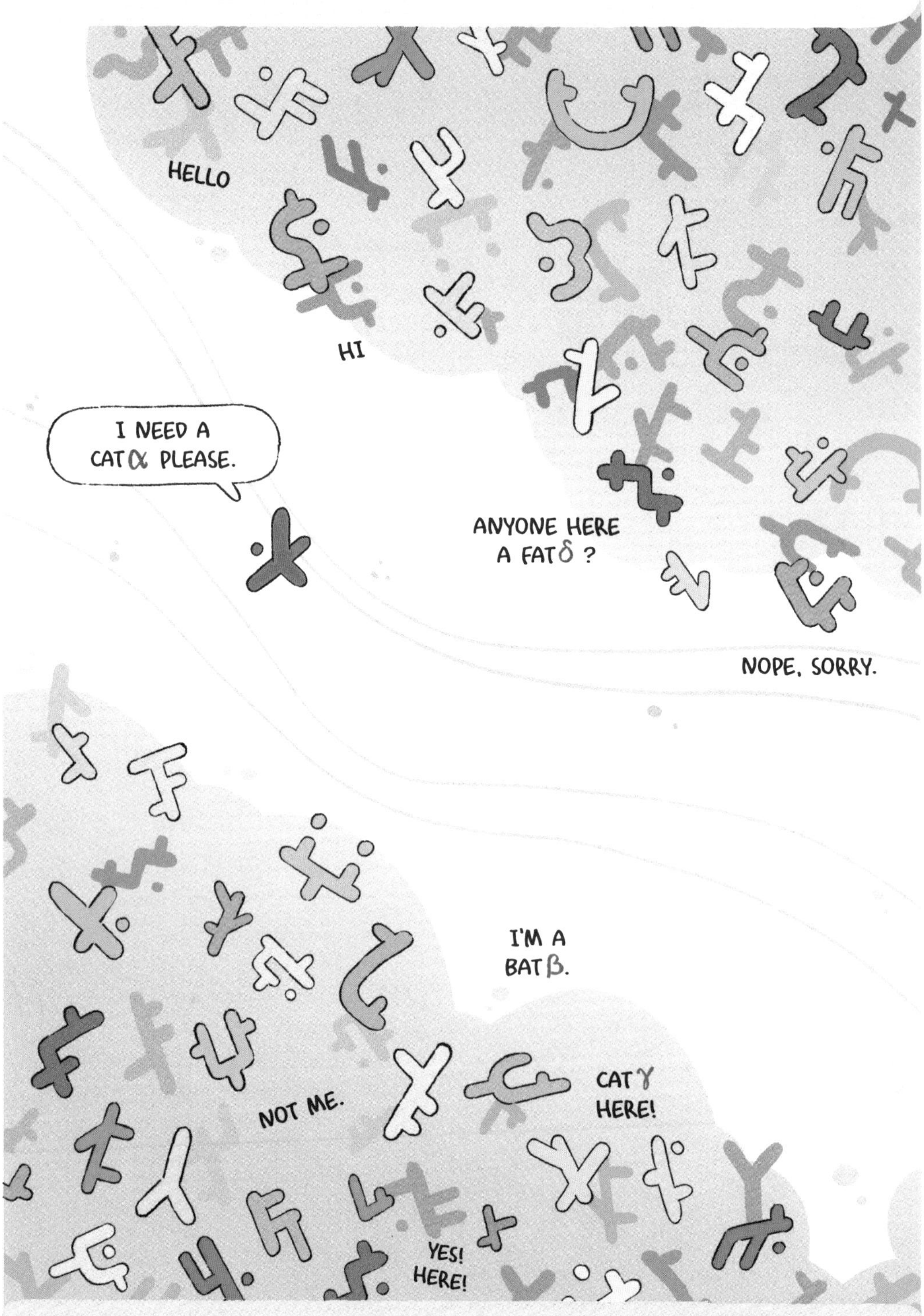
HELLO
HI
I NEED A CATα PLEASE.
ANYONE HERE A FATδ ?
NOPE. SORRY.
I'M A BATβ.
NOT ME.
CATγ HERE!
YES! HERE!

WHEN YOU HAVE 20,000 POTENTIAL PARTNERS, YOU'LL FIND SOMEONE WHO IS COMPATIBLE ALMOST ALL THE TIME.

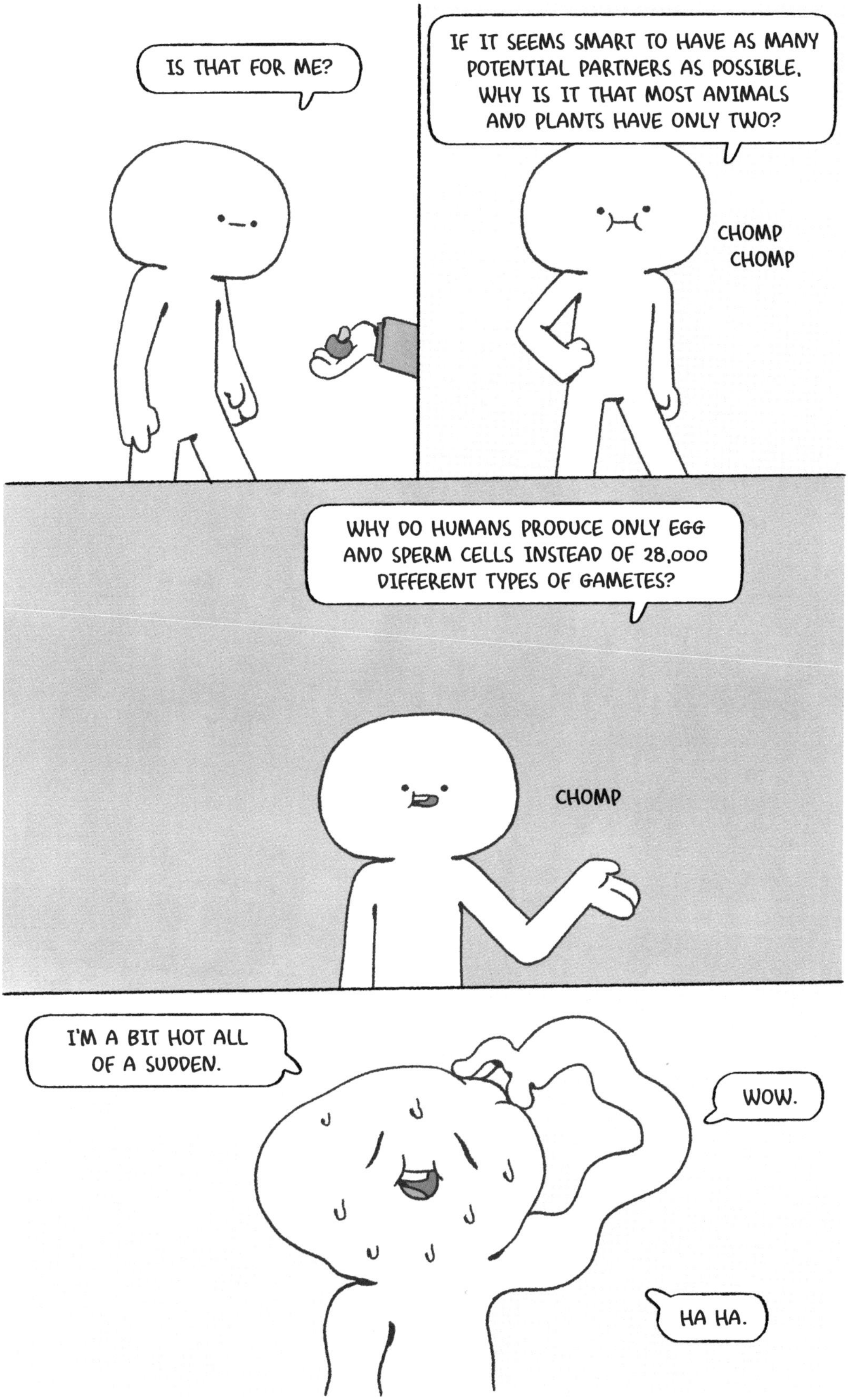
IS THAT FOR ME?
IF IT SEEMS SMART TO HAVE AS MANY POTENTIAL PARTNERS AS POSSIBLE, WHY IS IT THAT MOST ANIMALS AND PLANTS HAVE ONLY TWO?
CHOMP CHOMP
WHY DO HUMANS PRODUCE ONLY EGG AND SPERM CELLS INSTEAD OF 28,000 DIFFERENT TYPES OF GAMETES?
CHOMP
I'M A BIT HOT ALL OF A SUDDEN.
WOW.
HA HA.

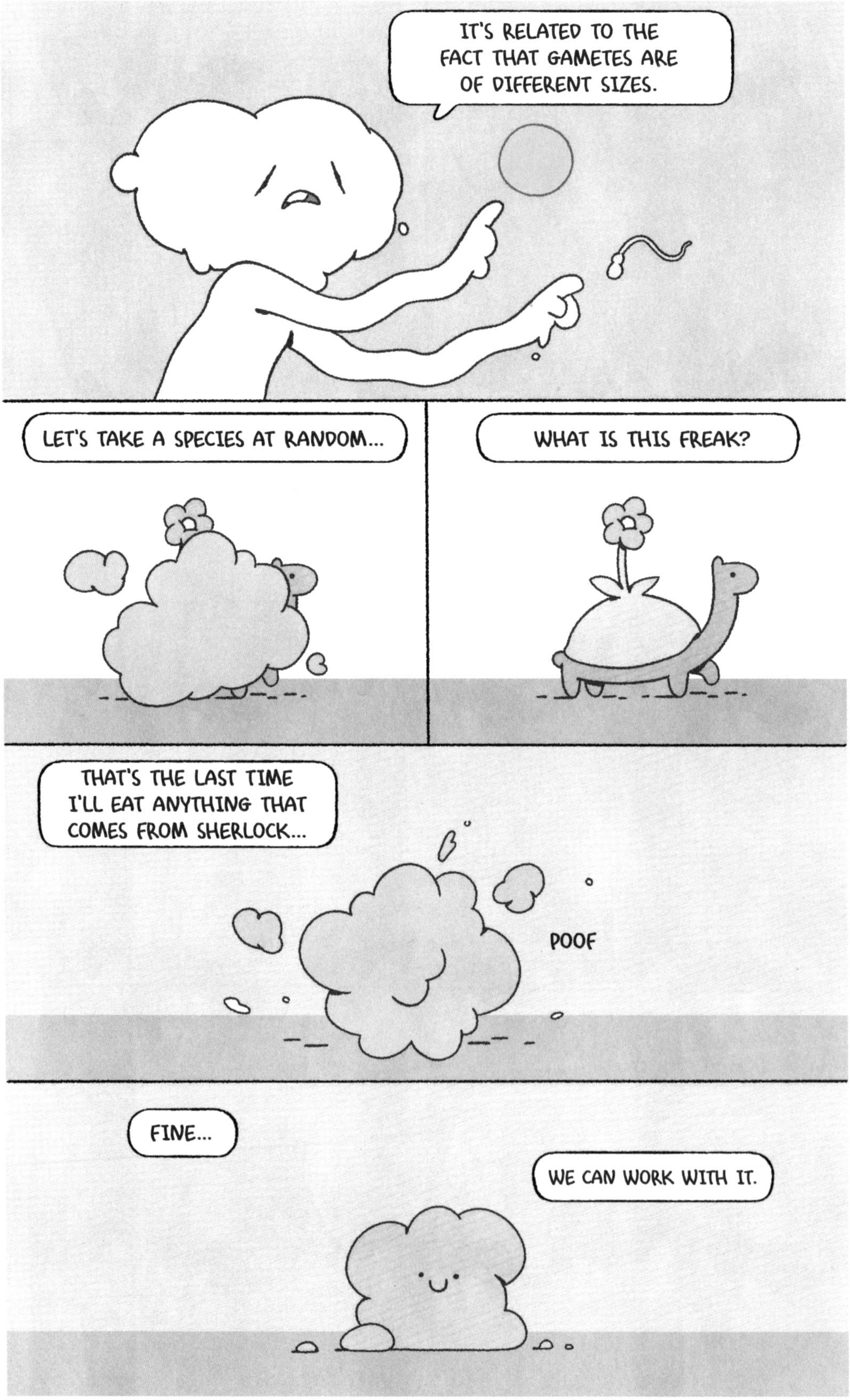
IT'S RELATED TO THE FACT THAT GAMETES ARE OF DIFFERENT SIZES.
LET'S TAKE A SPECIES AT RANDOM...
WHAT IS THIS FREAK?
THAT'S THE LAST TIME I'LL EAT ANYTHING THAT COMES FROM SHERLOCK...
POOF
FINE...
WE CAN WORK WITH IT.

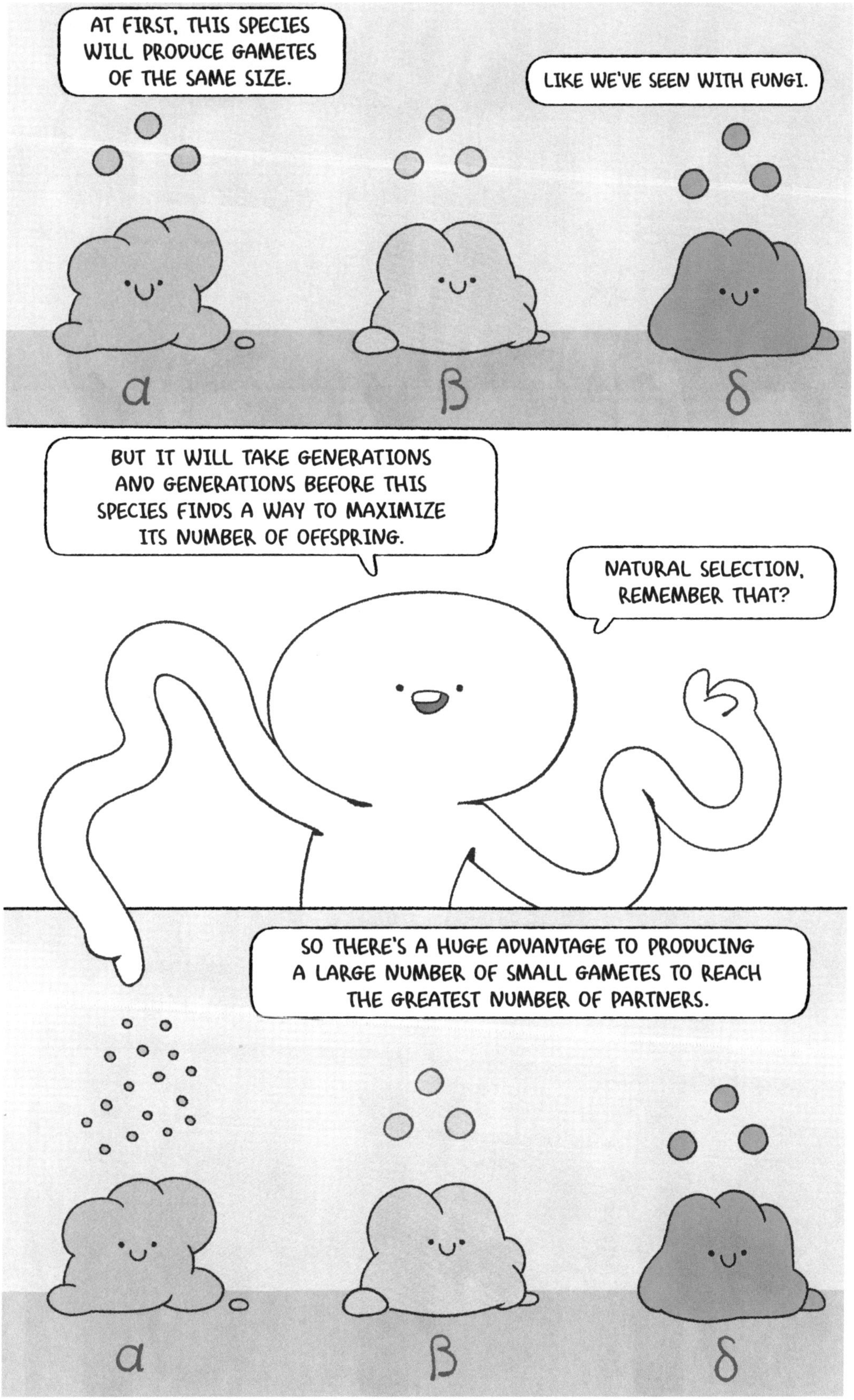
AT FIRST, THIS SPECIES WILL PRODUCE GAMETES OF THE SAME SIZE.
LIKE WE'VE SEEN WITH FUNGI.
α
β
δ
BUT IT WILL TAKE GENERATIONS AND GENERATIONS BEFORE THIS SPECIES FINDS A WAY TO MAXIMIZE ITS NUMBER OF OFFSPRING.
NATURAL SELECTION, REMEMBER THAT?
SO THERE'S A HUGE ADVANTAGE TO PRODUCING A LARGE NUMBER OF SMALL GAMETES TO REACH THE GREATEST NUMBER OF PARTNERS.
α
β
δ

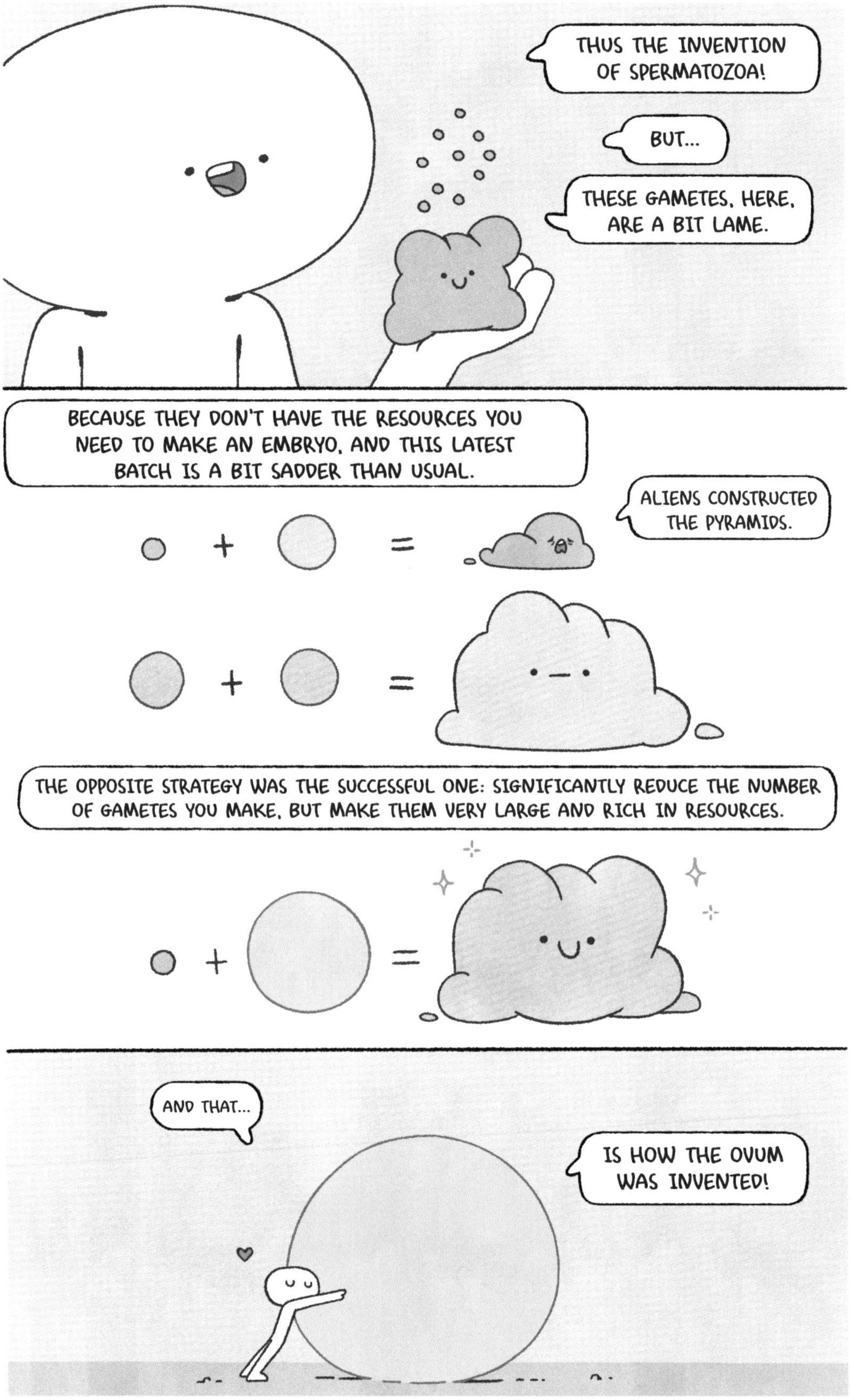
THUS THE INVENTION OF SPERMATOZOA!
BUT...
THESE GAMETES, HERE, ARE A BIT LAME.
BECAUSE THEY DON'T HAVE THE RESOURCES YOU NEED TO MAKE AN EMBRYO, AND THIS LATEST BATCH IS A BIT SADDER THAN USUAL.
ALIENS CONSTRUCTED THE PYRAMIDS.
THE OPPOSITE STRATEGY WAS THE SUCCESSFUL ONE: SIGNIFICANTLY REDUCE THE NUMBER OF GAMETES YOU MAKE, BUT MAKE THEM VERY LARGE AND RICH IN RESOURCES.
AND THAT...
IS HOW THE OVUM WAS INVENTED!

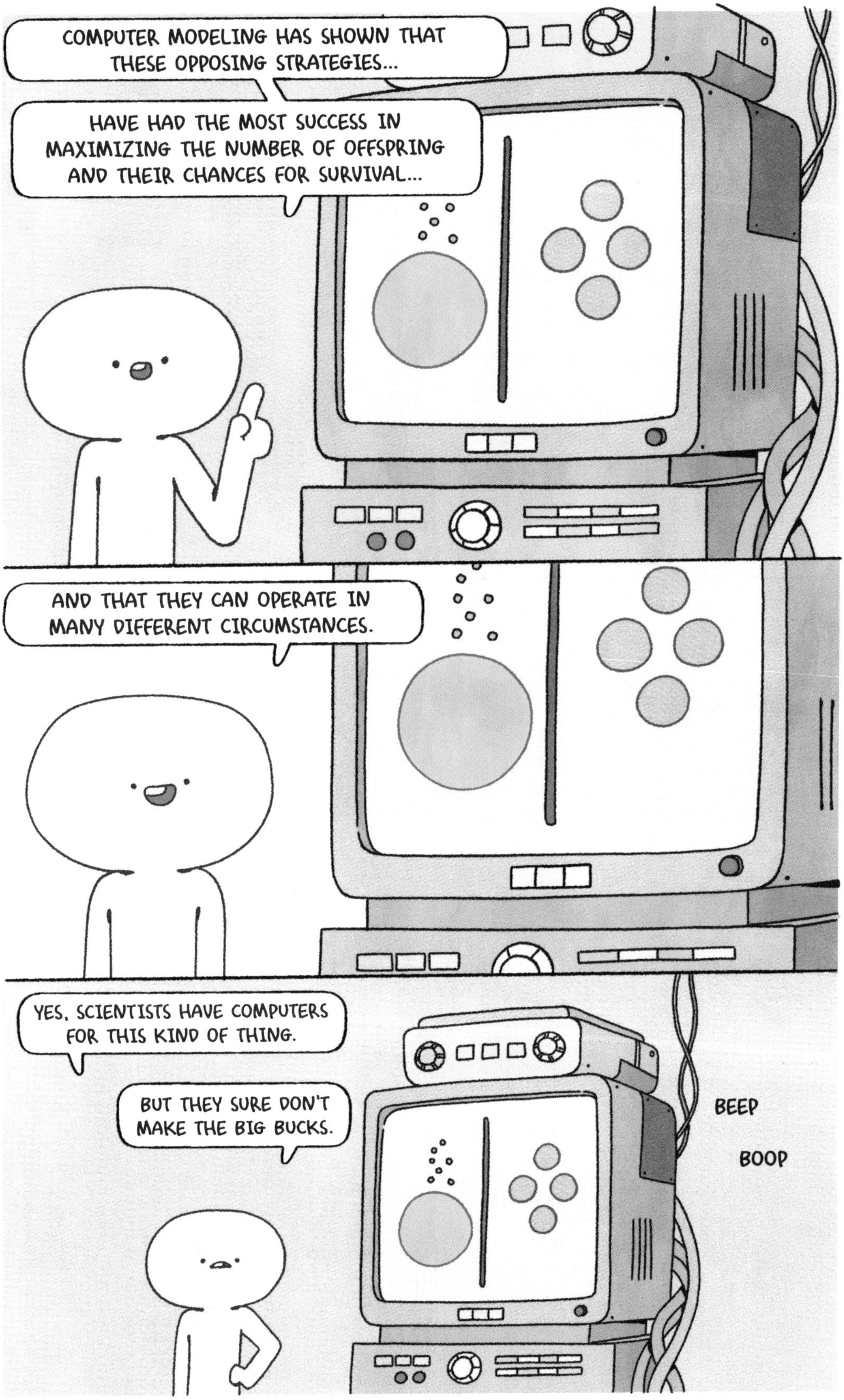
COMPUTER MODELING HAS SHOWN THAT THESE OPPOSING STRATEGIES...
HAVE HAD THE MOST SUCCESS IN MAXIMIZING THE NUMBER OF OFFSPRING AND THEIR CHANCES FOR SURVIVAL...
AND THAT THEY CAN OPERATE IN MANY DIFFERENT CIRCUMSTANCES.
YES, SCIENTISTS HAVE COMPUTERS FOR THIS KIND OF THING.
BUT THEY SURE DON'T MAKE THE BIG BUCKS.
BEEP
BOOP

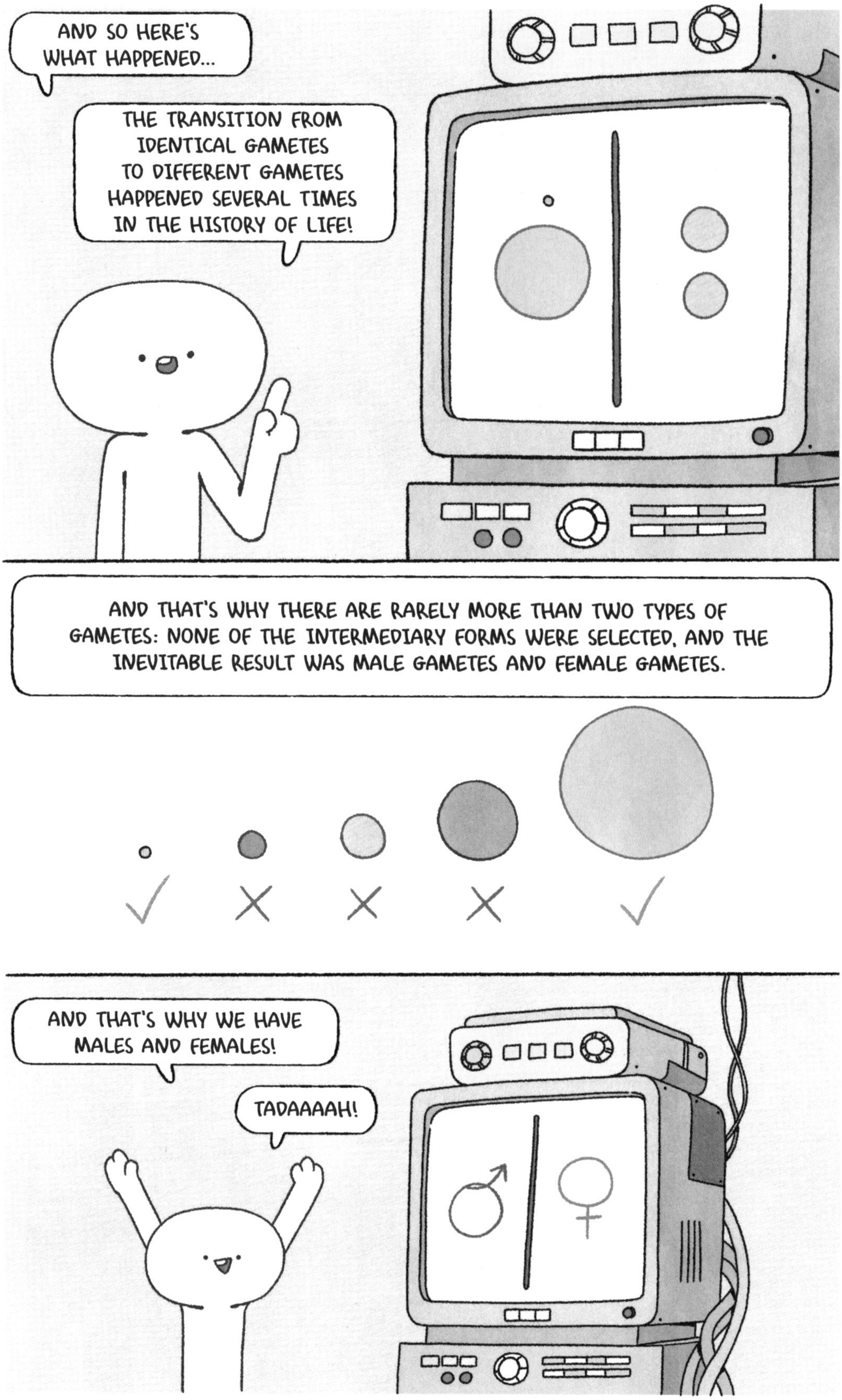
AND SO HERE'S WHAT HAPPENED...
THE TRANSITION FROM IDENTICAL GAMETES TO DIFFERENT GAMETES HAPPENED SEVERAL TIMES IN THE HISTORY OF LIFE!
AND THAT'S WHY THERE ARE RARELY MORE THAN TWO TYPES OF GAMETES: NONE OF THE INTERMEDIARY FORMS WERE SELECTED, AND THE INEVITABLE RESULT WAS MALE GAMETES AND FEMALE GAMETES.
AND THAT'S WHY WE HAVE MALES AND FEMALES!
TADAAAAH!

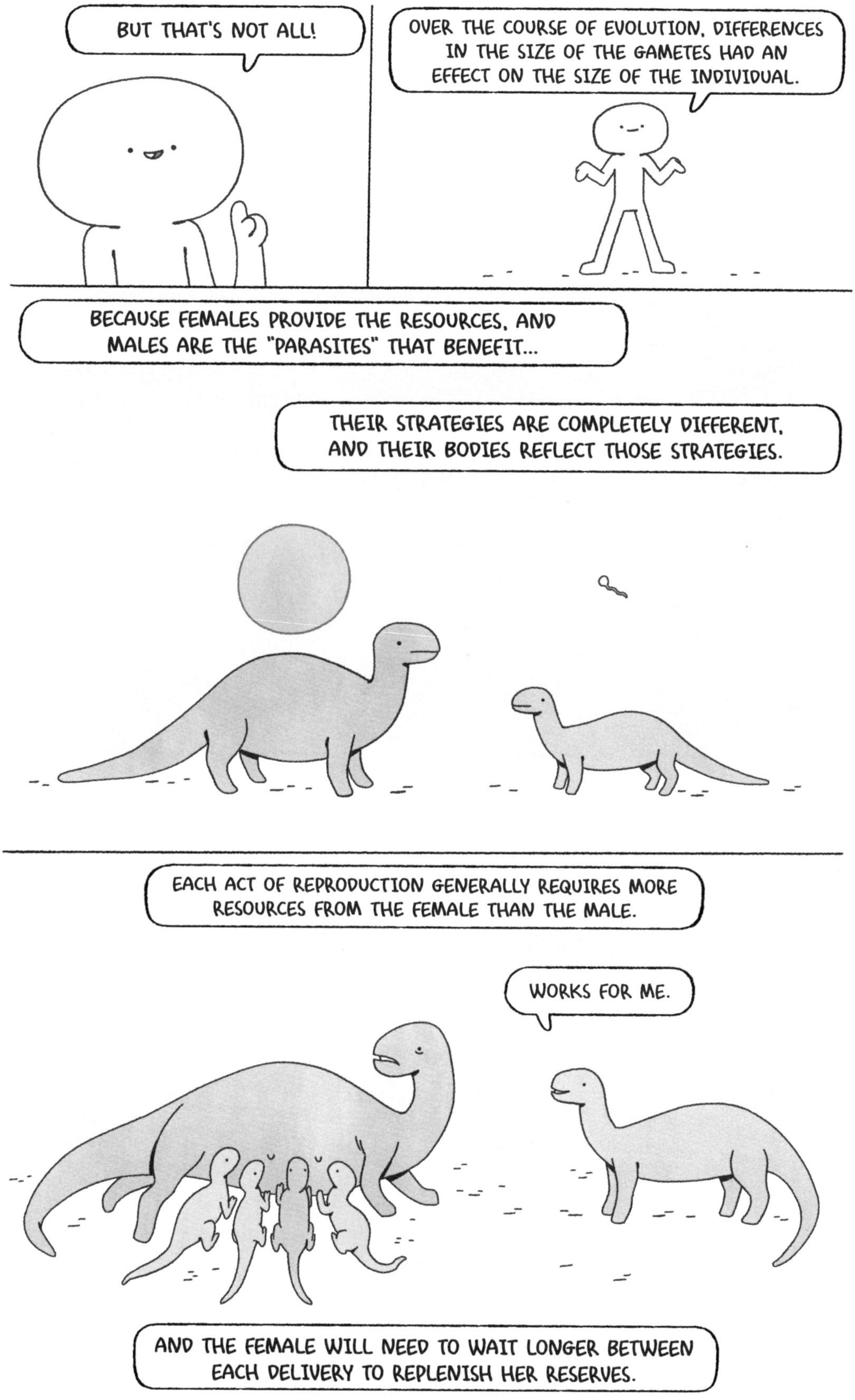
BUT THAT'S NOT ALL!
OVER THE COURSE OF EVOLUTION, DIFFERENCES IN THE SIZE OF THE GAMETES HAD AN EFFECT ON THE SIZE OF THE INDIVIDUAL.
BECAUSE FEMALES PROVIDE THE RESOURCES, AND MALES ARE THE "PARASITES" THAT BENEFIT...
THEIR STRATEGIES ARE COMPLETELY DIFFERENT, AND THEIR BODIES REFLECT THOSE STRATEGIES.
EACH ACT OF REPRODUCTION GENERALLY REQUIRES MORE RESOURCES FROM THE FEMALE THAN THE MALE.
WORKS FOR ME.
AND THE FEMALE WILL NEED TO WAIT LONGER BETWEEN EACH DELIVERY TO REPLENISH HER RESERVES.

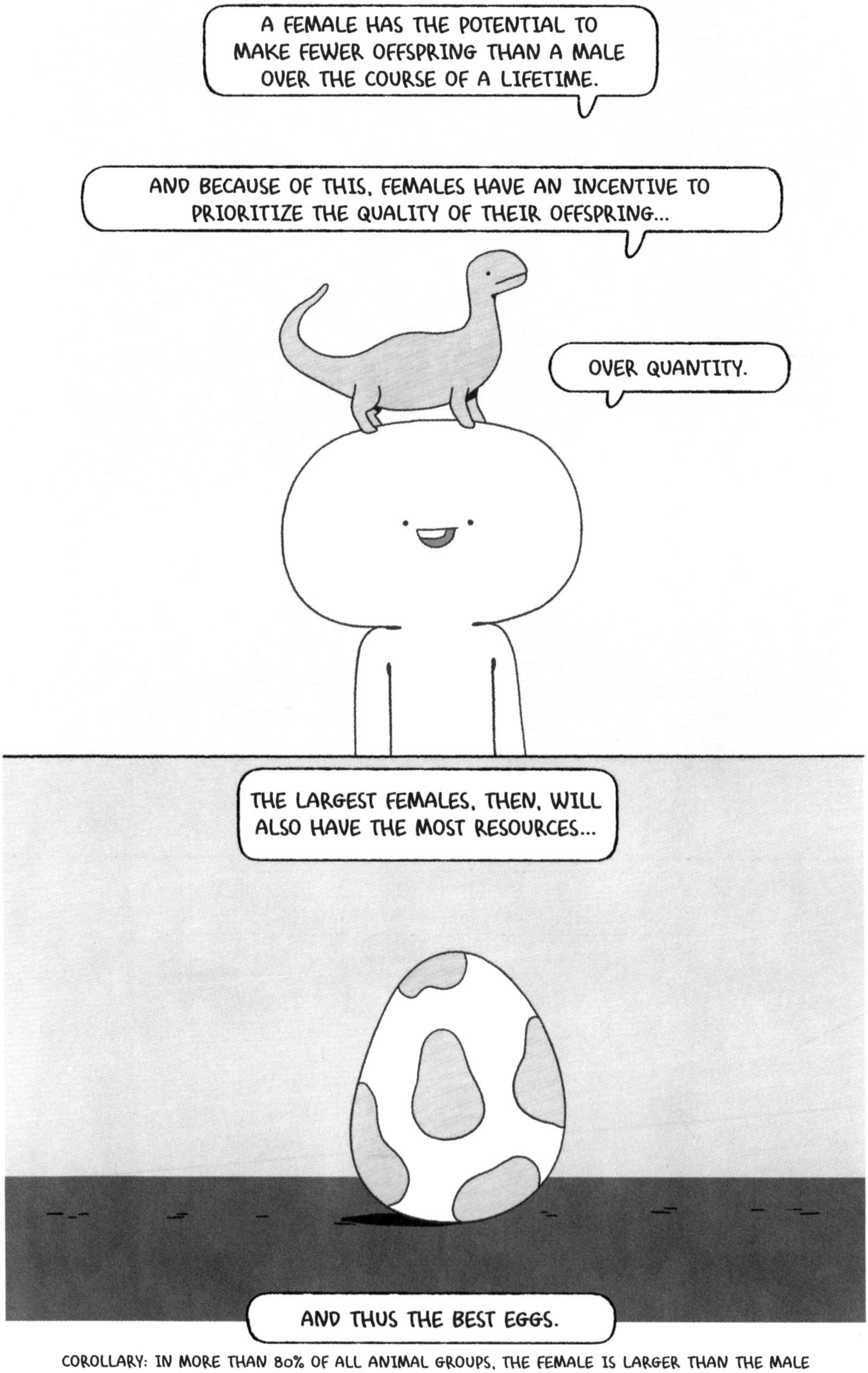
A FEMALE HAS THE POTENTIAL TO MAKE FEWER OFFSPRING THAN A MALE OVER THE COURSE OF A LIFETIME.
AND BECAUSE OF THIS, FEMALES HAVE AN INCENTIVE TO PRIORITIZE THE QUALITY OF THEIR OFFSPRING...
OVER QUANTITY.
THE LARGEST FEMALES, THEN, WILL ALSO HAVE THE MOST RESOURCES...
AND THUS THE BEST EGGS.

COROLLARY: IN MORE THAN 80% OF ALL ANIMAL GROUPS, THE FEMALE IS LARGER THAN THE MALE (INSECTS, FISH, OCTOPI, SNAKES, WORMS...).

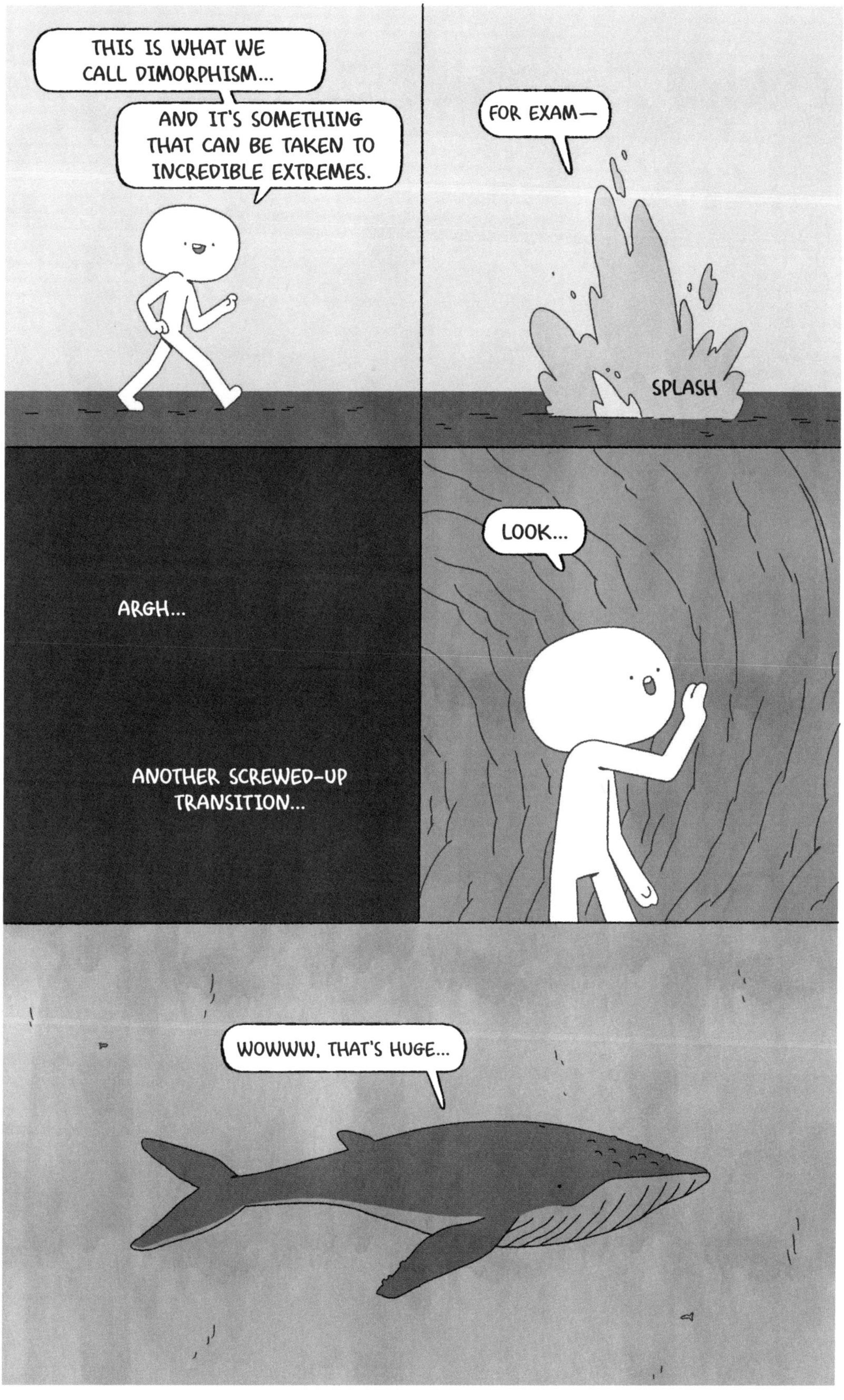
THIS IS WHAT WE CALL DIMORPHISM...
AND IT'S SOMETHING THAT CAN BE TAKEN TO INCREDIBLE EXTREMES.
FOR EXAM—
SPLASH
ARGH...
ANOTHER SCREWED-UP TRANSITION...
LOOK...
WOWWW, THAT'S HUGE...

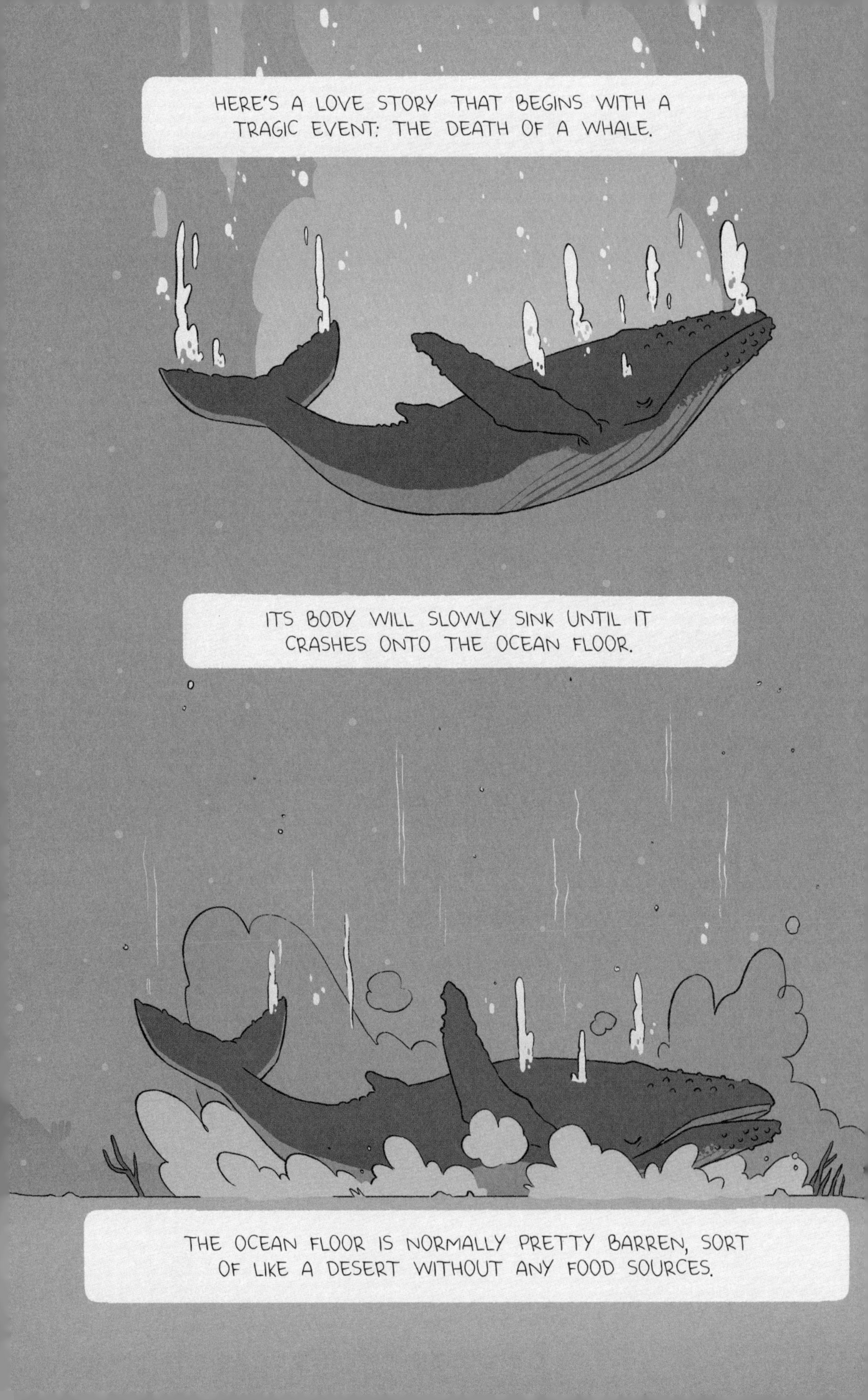
HERE'S A LOVE STORY THAT BEGINS WITH A TRAGIC EVENT: THE DEATH OF A WHALE.
ITS BODY WILL SLOWLY SINK UNTIL IT CRASHES ONTO THE OCEAN FLOOR.
THE OCEAN FLOOR IS NORMALLY PRETTY BARREN, SORT OF LIKE A DESERT WITHOUT ANY FOOD SOURCES.

NUTRIENTS COME FROM THE SURFACE OF THE OCEAN IN THE FORM OF WHAT WE CALL "MARINE SNOW."

NUMEROUS ORGANISMS CONSUME IT AS IT FALLS TO THE BOTTOM.

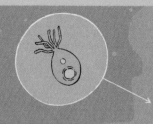
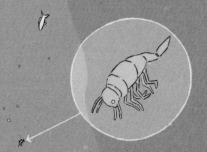

ONCE IT REACHES THE GROUND, THERE ISN'T MUCH LEFT.

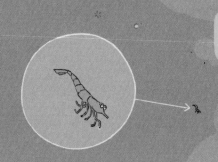
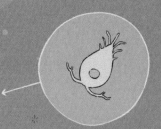

BUT THE ARRIVAL OF A WHALE CARCASS CAN PROVIDE 100 TONS OF FOOD.

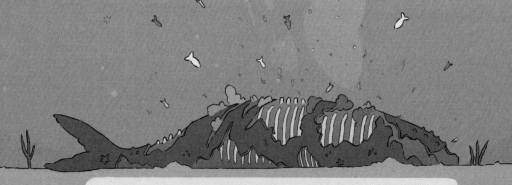

IT'S A VERITABLE SMORGASBORD OF GRUB!

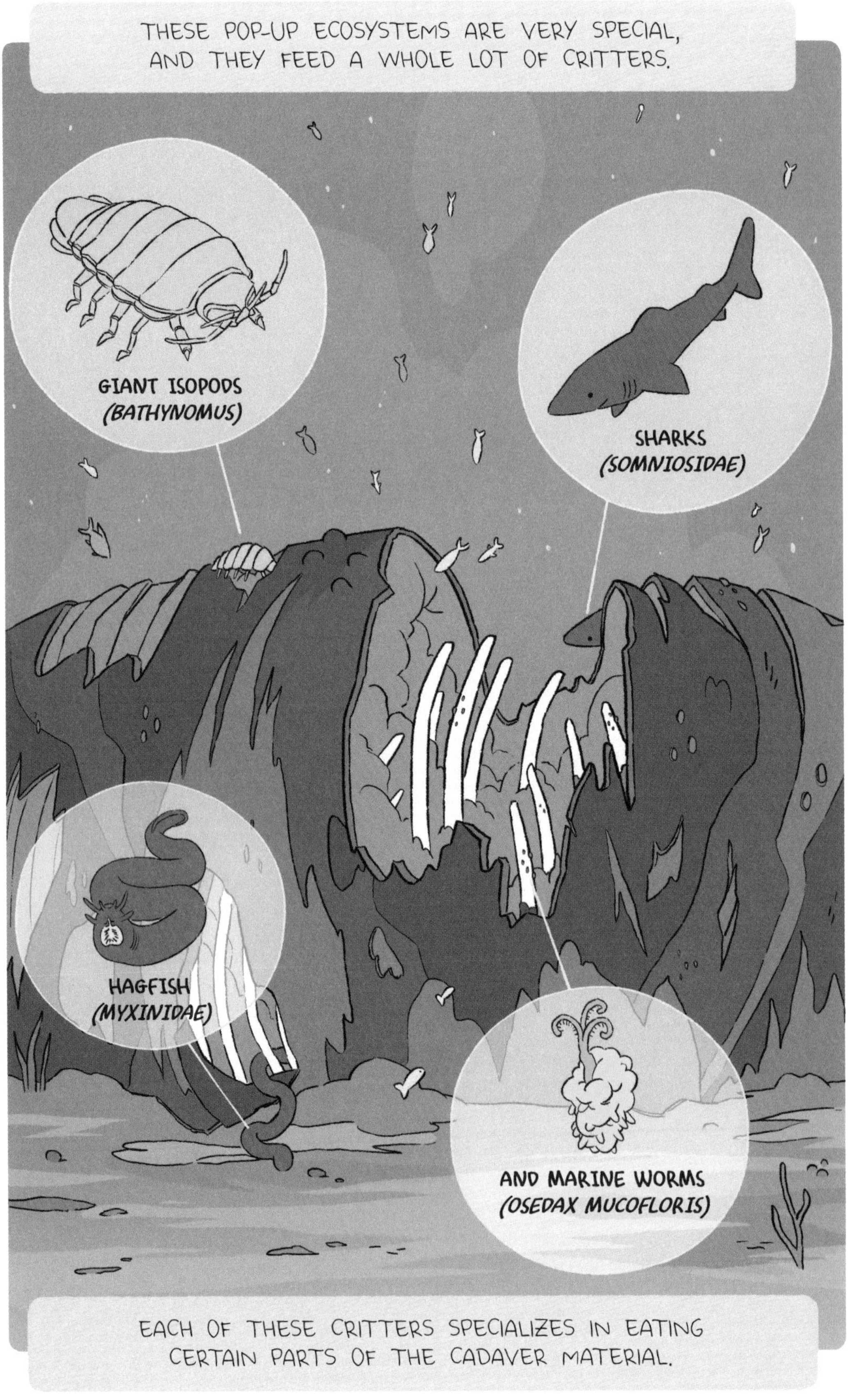

THESE POP-UP ECOSYSTEMS ARE VERY SPECIAL, AND THEY FEED A WHOLE LOT OF CRITTERS.
GIANT ISOPODS (BATHYNOMUS)
SHARKS (SOMNIOSIDAE)
HAGFISH (MYXINIDAE)
AND MARINE WORMS (OSEDAX MUCOFLORIS)
EACH OF THESE CRITTERS SPECIALIZES IN EATING CERTAIN PARTS OF THE CADAVER MATERIAL.

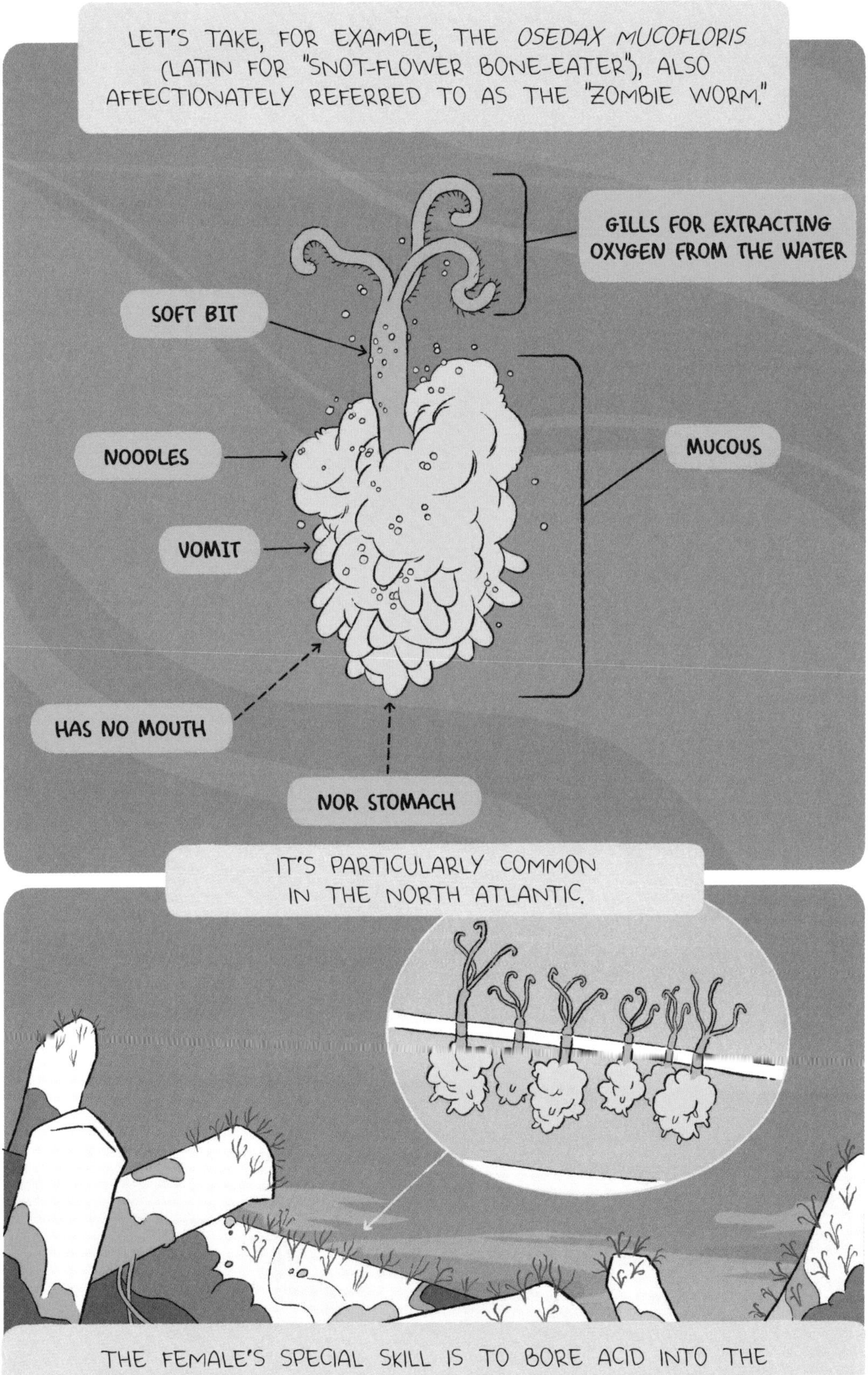
LET'S TAKE, FOR EXAMPLE, THE *OSEDAX MUCOFLORIS* (LATIN FOR "SNOT-FLOWER BONE-EATER"), ALSO AFFECTIONATELY REFERRED TO AS THE "ZOMBIE WORM."
GILLS FOR EXTRACTING OXYGEN FROM THE WATER
SOFT BIT
NOODLES
MUCOUS
VOMIT
HAS NO MOUTH
NOR STOMACH
IT'S PARTICULARLY COMMON IN THE NORTH ATLANTIC.
THE FEMALE'S SPECIAL SKILL IS TO BORE ACID INTO THE BONES OF THE CADAVER IN ORDER TO DRAW OUT FOOD.

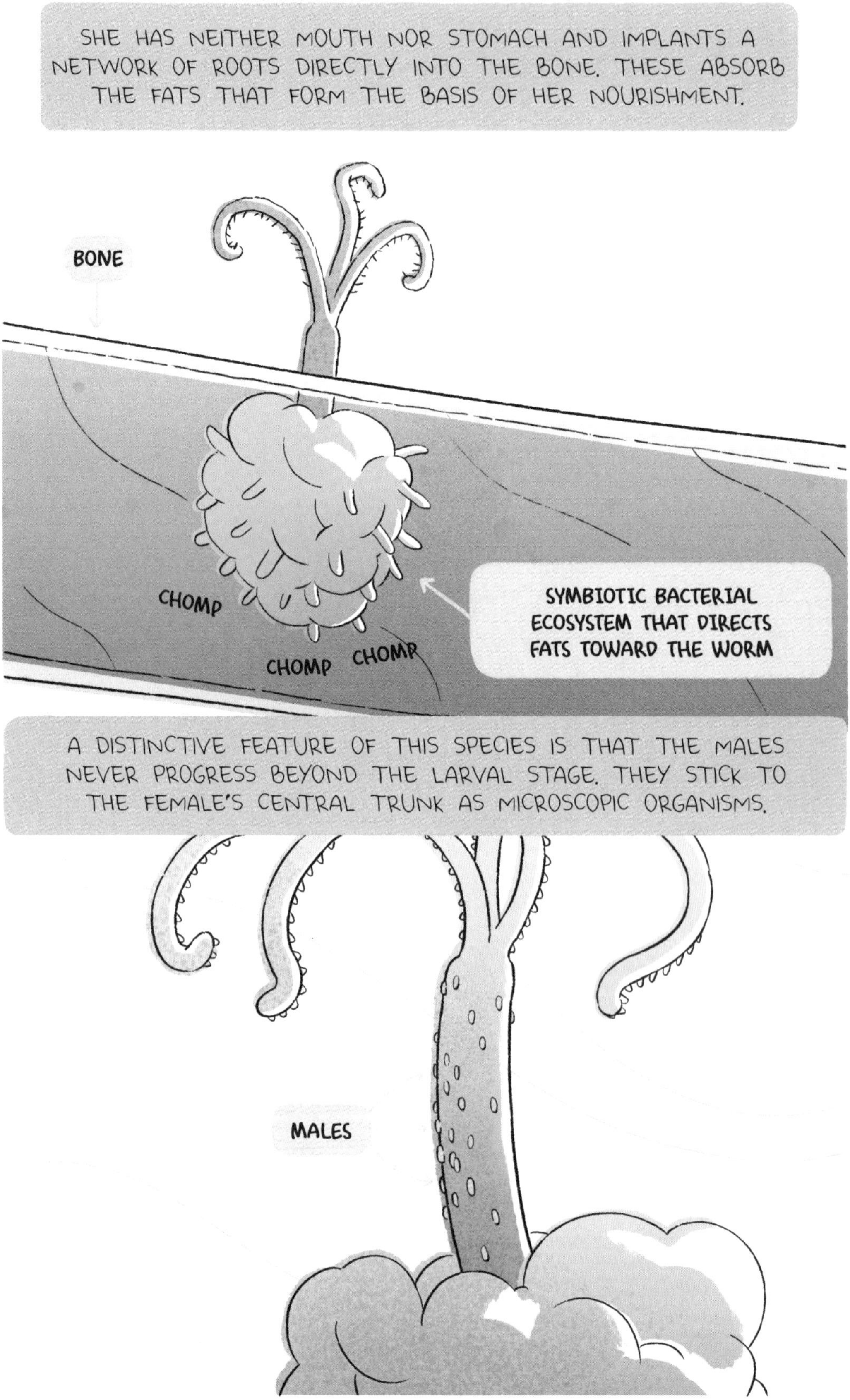
SHE HAS NEITHER MOUTH NOR STOMACH AND IMPLANTS A NETWORK OF ROOTS DIRECTLY INTO THE BONE. THESE ABSORB THE FATS THAT FORM THE BASIS OF HER NOURISHMENT.
BONE
CHOMP
CHOMP
CHOMP
SYMBIOTIC BACTERIAL ECOSYSTEM THAT DIRECTS FATS TOWARD THE WORM
A DISTINCTIVE FEATURE OF THIS SPECIES IS THAT THE MALES NEVER PROGRESS BEYOND THE LARVAL STAGE. THEY STICK TO THE FEMALE'S CENTRAL TRUNK AS MICROSCOPIC ORGANISMS.
MALES

THE DUDES ARE REDUCED TO THE STATE OF PROTECTIVE SPERM SACS ONE MILLIMETER IN LENGTH, AND THEY FERTILIZE THE FEMALE CONTINUOUSLY.
WHAT?
I LOOK LIKE A TAMPON?
SPERM STUFF
"A HUGE TAMPON FULL OF SPERM?"
AH, WELL, THAT'S FUN...
THEY FERTILIZE THE OVA THAT THE FEMALE DUMPS INTO THE OCEAN, AND THE EMBRYOS DEVELOP INTO LARVAE THAT FLOAT ALONG WITH THE CURRENT, WAITING FOR A NEW DEAD-WHALE BUFFET SO THAT THEY CAN DEVELOP AND BEGIN THE CYCLE ALL OVER AGAIN.
WHAT SHOULD WE DO NOW?
WE WAIT.
I'M HUNGRY.

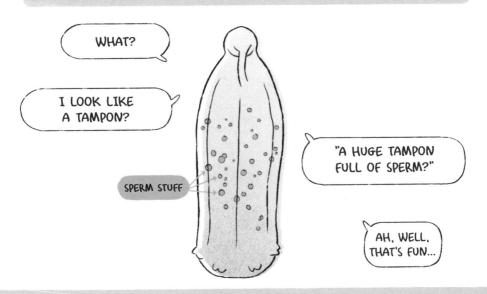

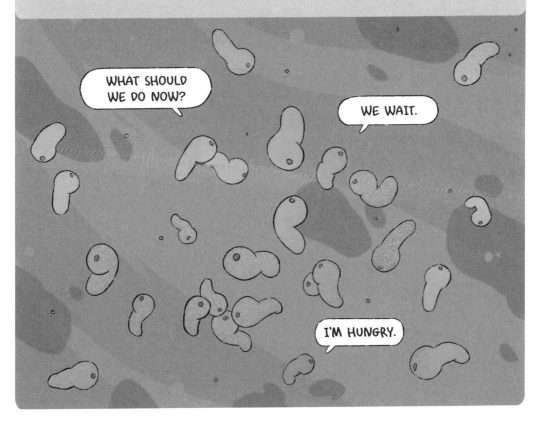

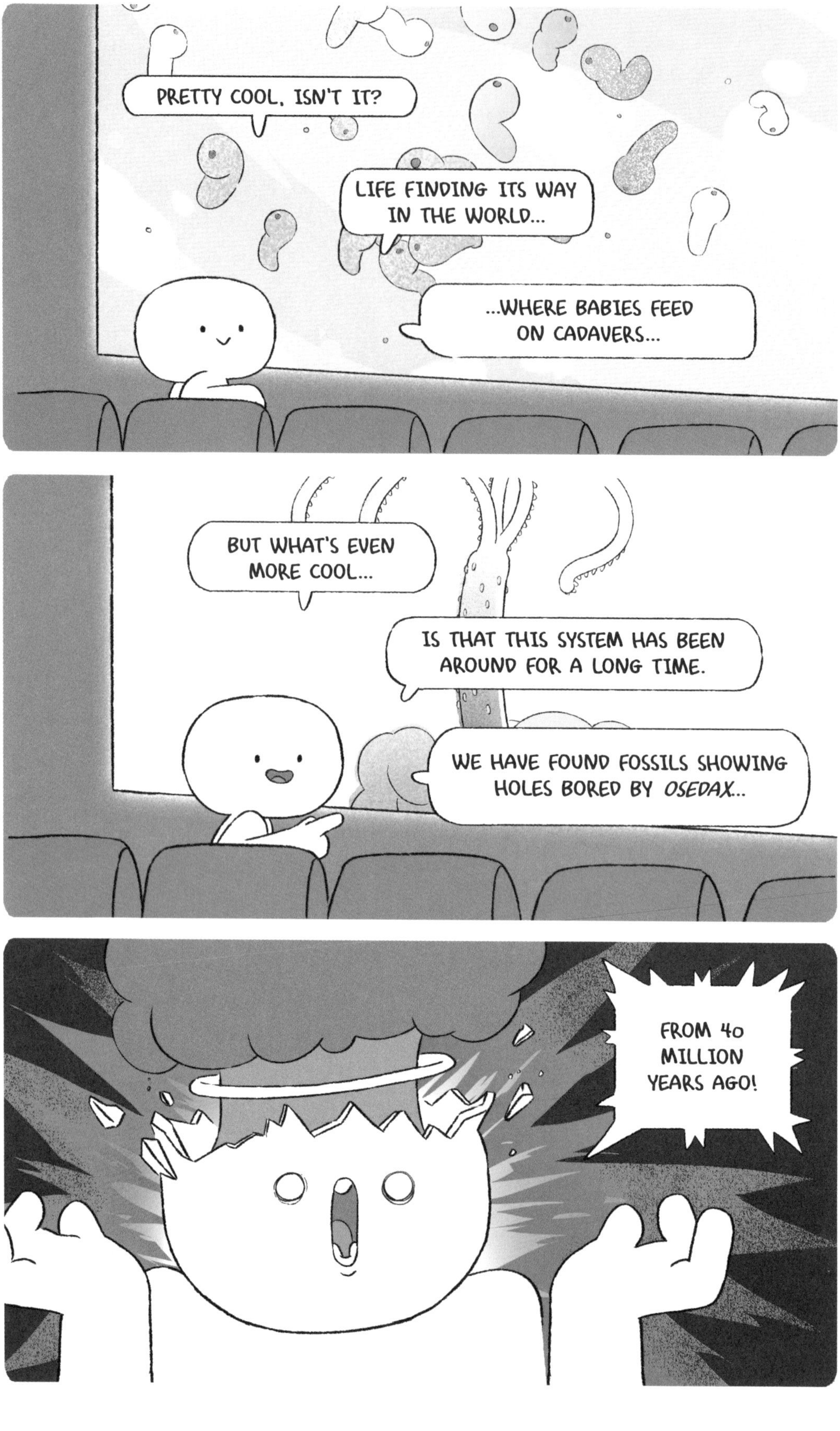
PRETTY COOL, ISN'T IT?
LIFE FINDING ITS WAY IN THE WORLD...
...WHERE BABIES FEED ON CADAVERS...
BUT WHAT'S EVEN MORE COOL...
IS THAT THIS SYSTEM HAS BEEN AROUND FOR A LONG TIME.
WE HAVE FOUND FOSSILS SHOWING HOLES BORED BY *OSEDAX*...
FROM 40 MILLION YEARS AGO!

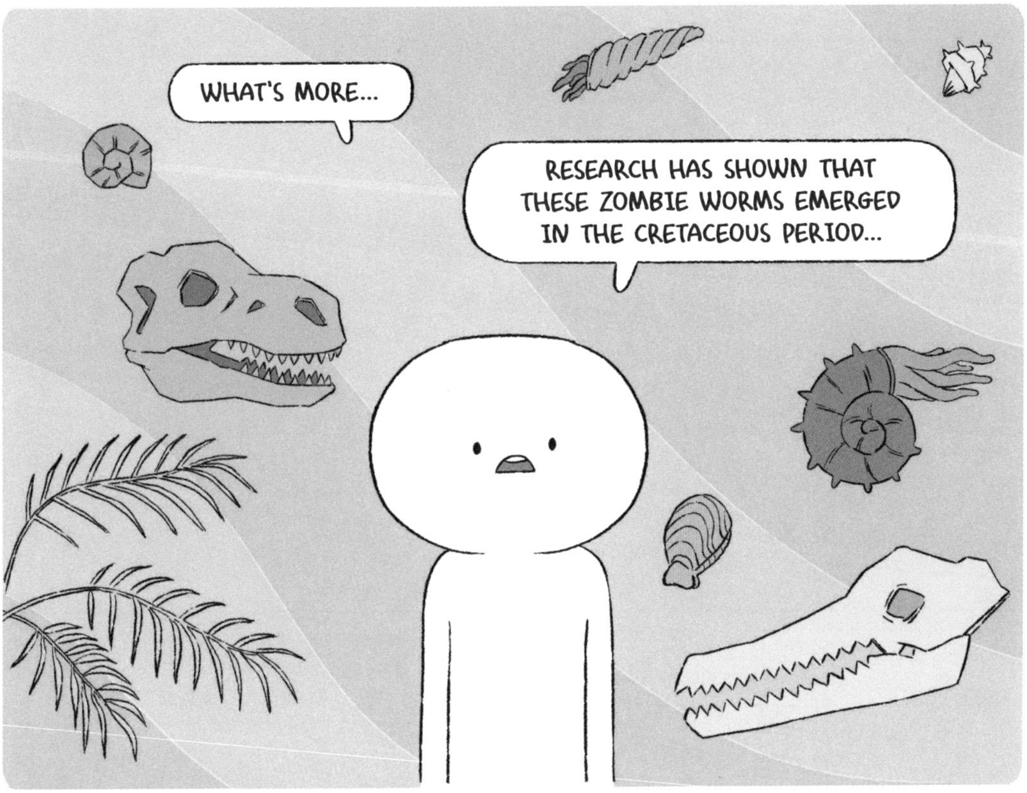
WHAT'S MORE...
RESEARCH HAS SHOWN THAT THESE ZOMBIE WORMS EMERGED IN THE CRETACEOUS PERIOD...

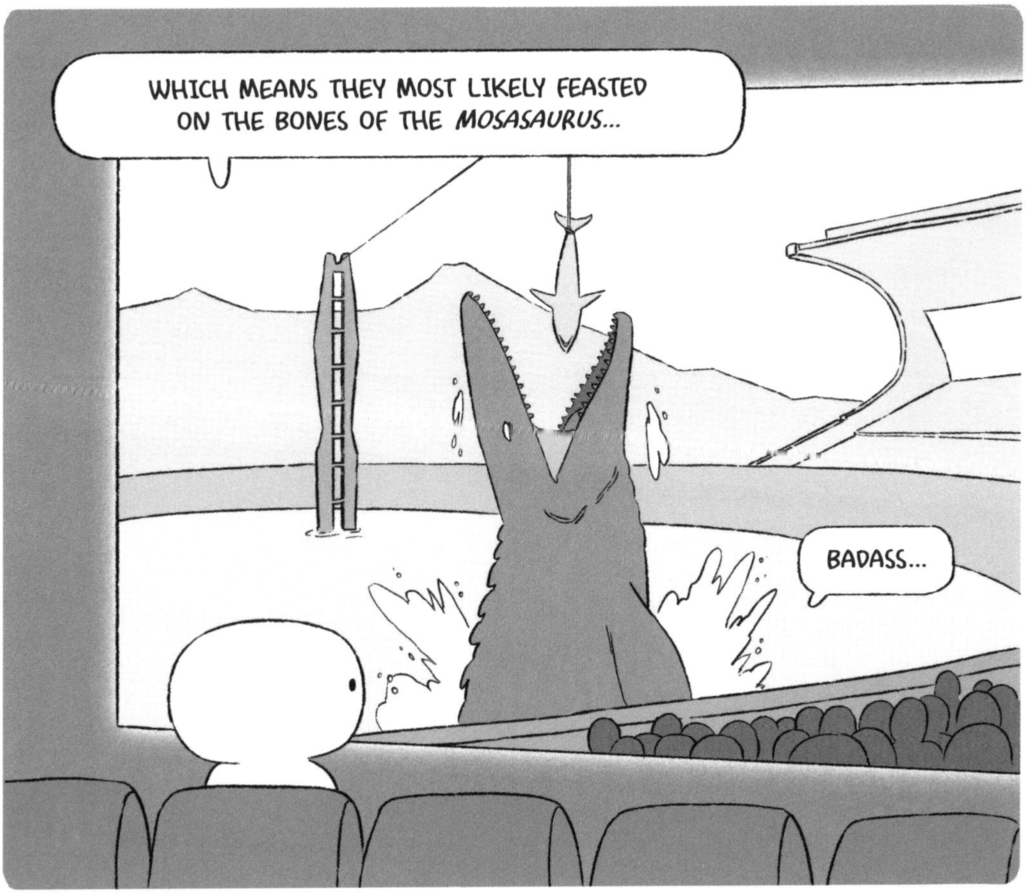
WHICH MEANS THEY MOST LIKELY FEASTED ON THE BONES OF THE *MOSASAURUS*...
BADASS...

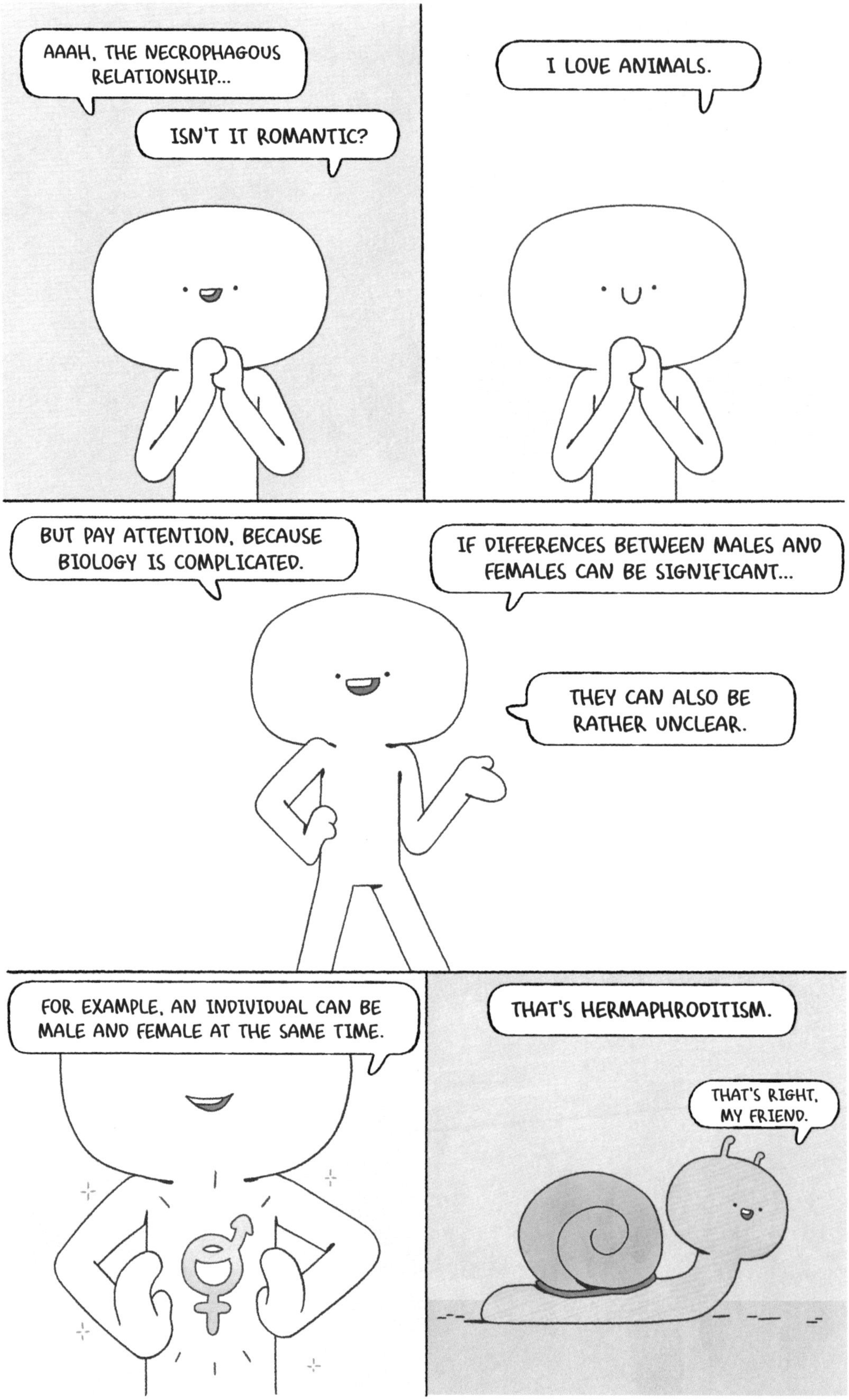
AAAH, THE NECROPHAGOUS RELATIONSHIP...
ISN'T IT ROMANTIC?
I LOVE ANIMALS.
BUT PAY ATTENTION, BECAUSE BIOLOGY IS COMPLICATED.
IF DIFFERENCES BETWEEN MALES AND FEMALES CAN BE SIGNIFICANT...
THEY CAN ALSO BE RATHER UNCLEAR.
FOR EXAMPLE, AN INDIVIDUAL CAN BE MALE AND FEMALE AT THE SAME TIME.
THAT'S HERMAPHRODITISM.
THAT'S RIGHT, MY FRIEND.

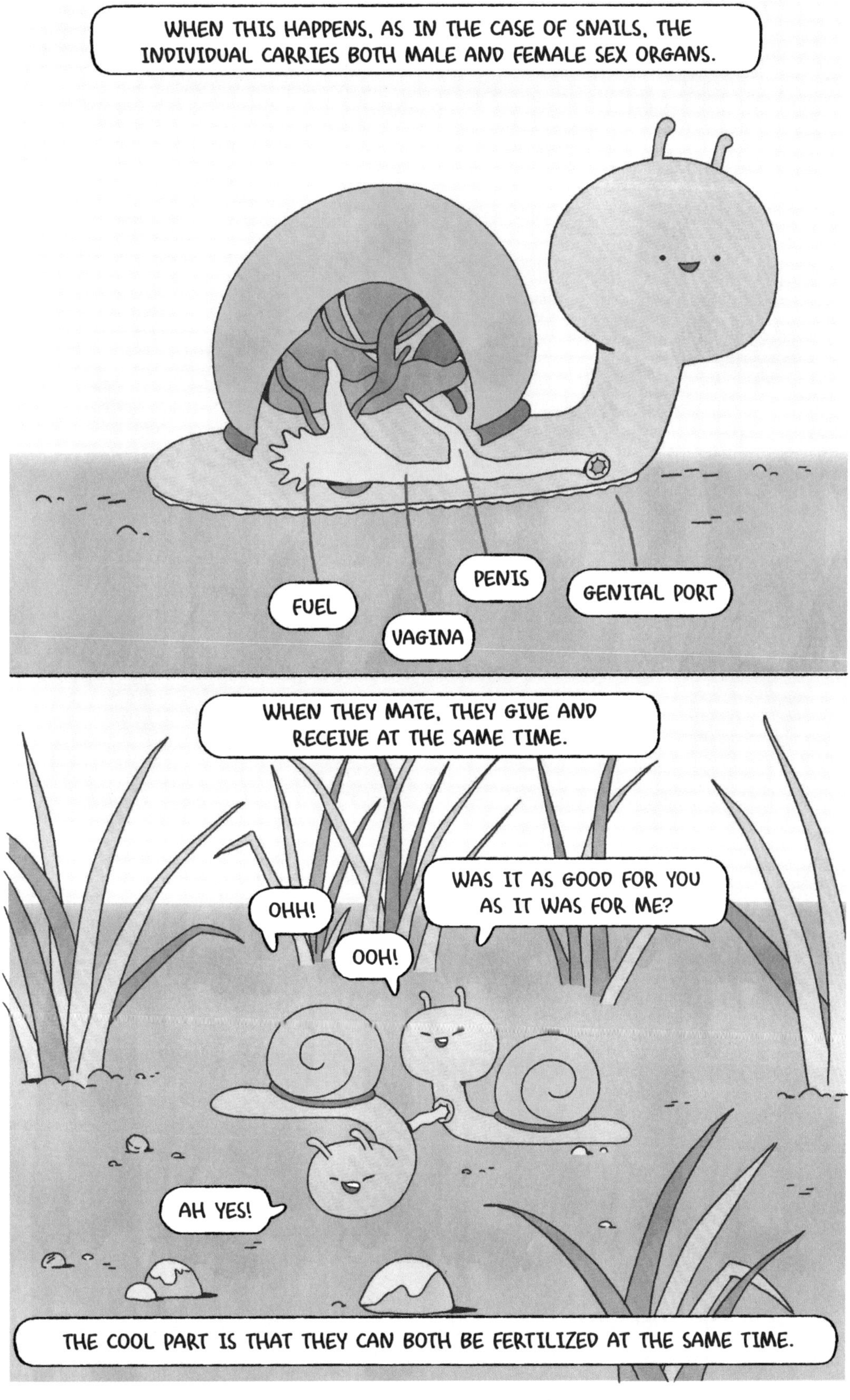
WHEN THIS HAPPENS, AS IN THE CASE OF SNAILS, THE INDIVIDUAL CARRIES BOTH MALE AND FEMALE SEX ORGANS.
FUEL
VAGINA
PENIS
GENITAL PORT
WHEN THEY MATE, THEY GIVE AND RECEIVE AT THE SAME TIME.
OHH!
OOH!
WAS IT AS GOOD FOR YOU AS IT WAS FOR ME?
AH YES!
THE COOL PART IS THAT THEY CAN BOTH BE FERTILIZED AT THE SAME TIME.

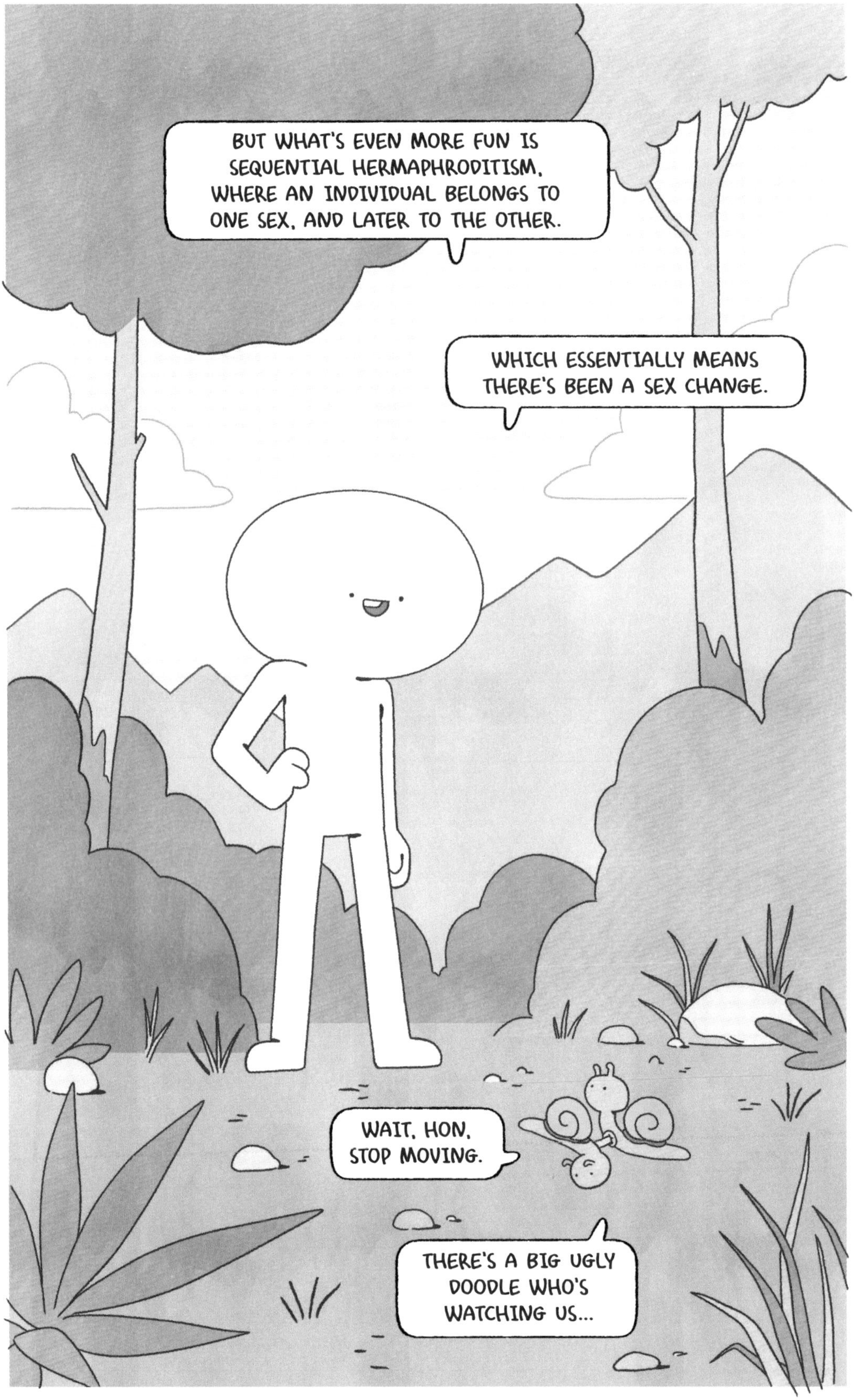
BUT WHAT'S EVEN MORE FUN IS SEQUENTIAL HERMAPHRODITISM, WHERE AN INDIVIDUAL BELONGS TO ONE SEX, AND LATER TO THE OTHER.
WHICH ESSENTIALLY MEANS THERE'S BEEN A SEX CHANGE.
WAIT, HON, STOP MOVING.
THERE'S A BIG UGLY DOODLE WHO'S WATCHING US...

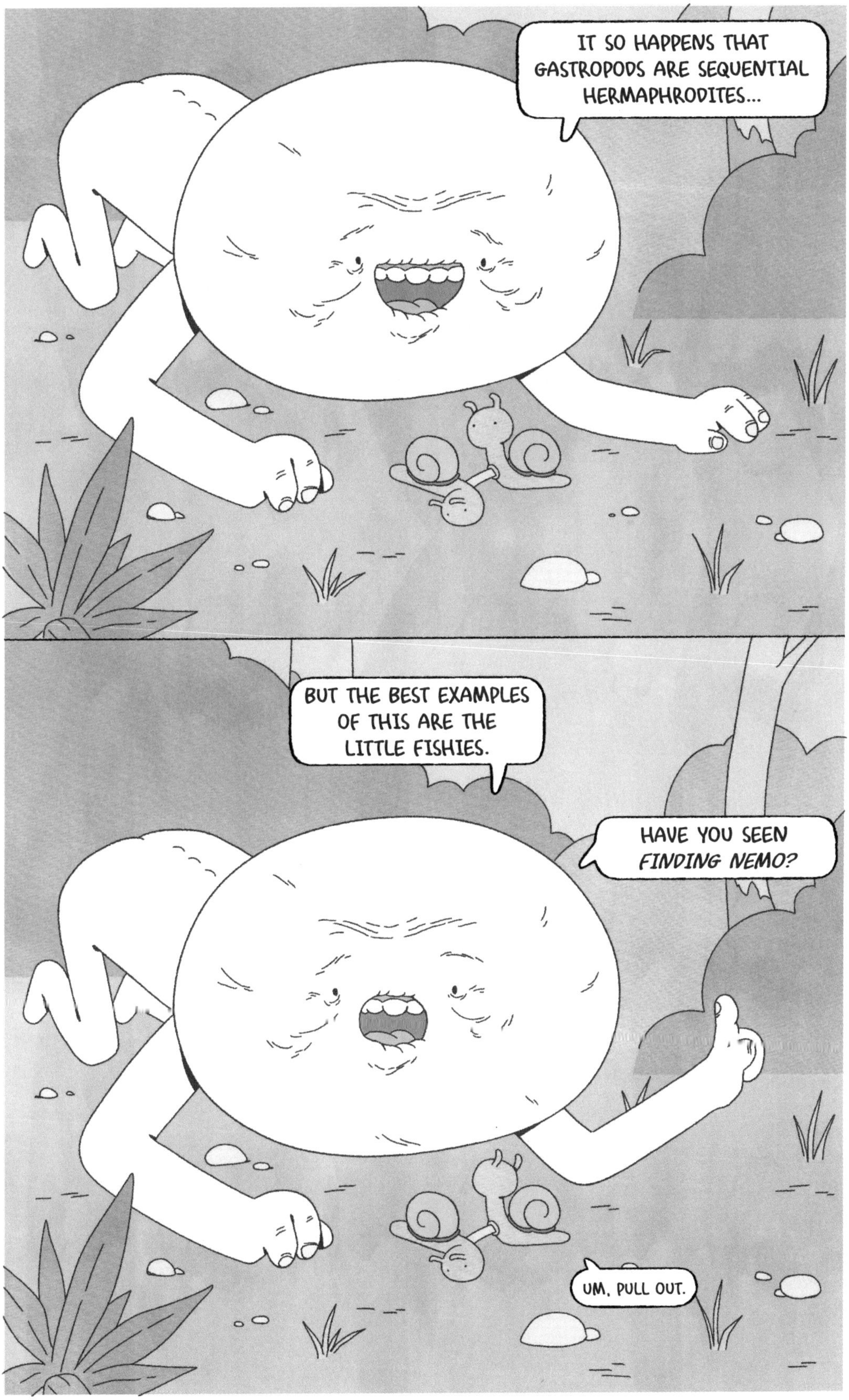
IT SO HAPPENS THAT GASTROPODS ARE SEQUENTIAL HERMAPHRODITES...
BUT THE BEST EXAMPLES OF THIS ARE THE LITTLE FISHIES.
HAVE YOU SEEN *FINDING NEMO?*
UM, PULL OUT.

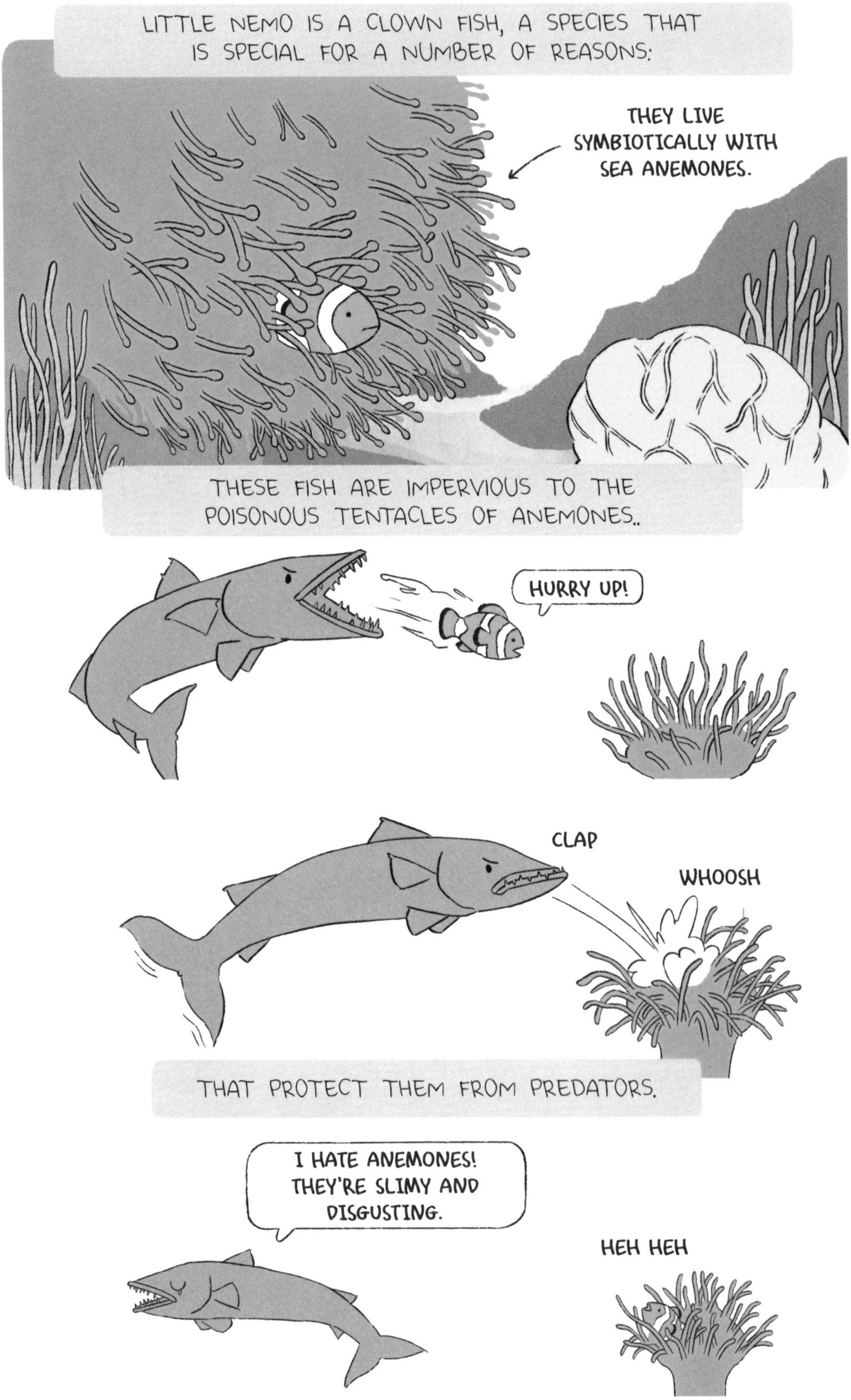
LITTLE NEMO IS A CLOWN FISH, A SPECIES THAT IS SPECIAL FOR A NUMBER OF REASONS:
THEY LIVE SYMBIOTICALLY WITH SEA ANEMONES.
THESE FISH ARE IMPERVIOUS TO THE POISONOUS TENTACLES OF ANEMONES..
HURRY UP!
CLAP
WHOOSH
THAT PROTECT THEM FROM PREDATORS.
I HATE ANEMONES! THEY'RE SLIMY AND DISGUSTING.
HEH HEH

THEY, IN TURN, CAN BE USEFUL FOR THE ANEMONE.

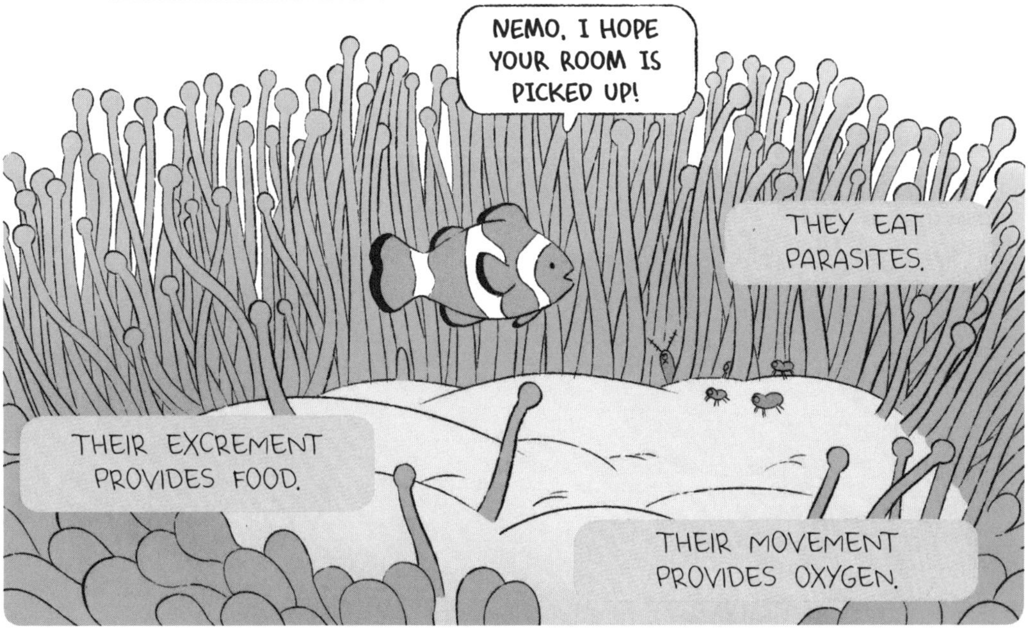
NEMO, I HOPE YOUR ROOM IS PICKED UP!
THEY EAT PARASITES.
THEIR EXCREMENT PROVIDES FOOD.
THEIR MOVEMENT PROVIDES OXYGEN.

IN THIS COZY NEST, CLOWN FISH HAVE A SOCIAL SYSTEM WHERE TWO INDIVIDUALS DECIDE TO REPRODUCE TOGETHER.

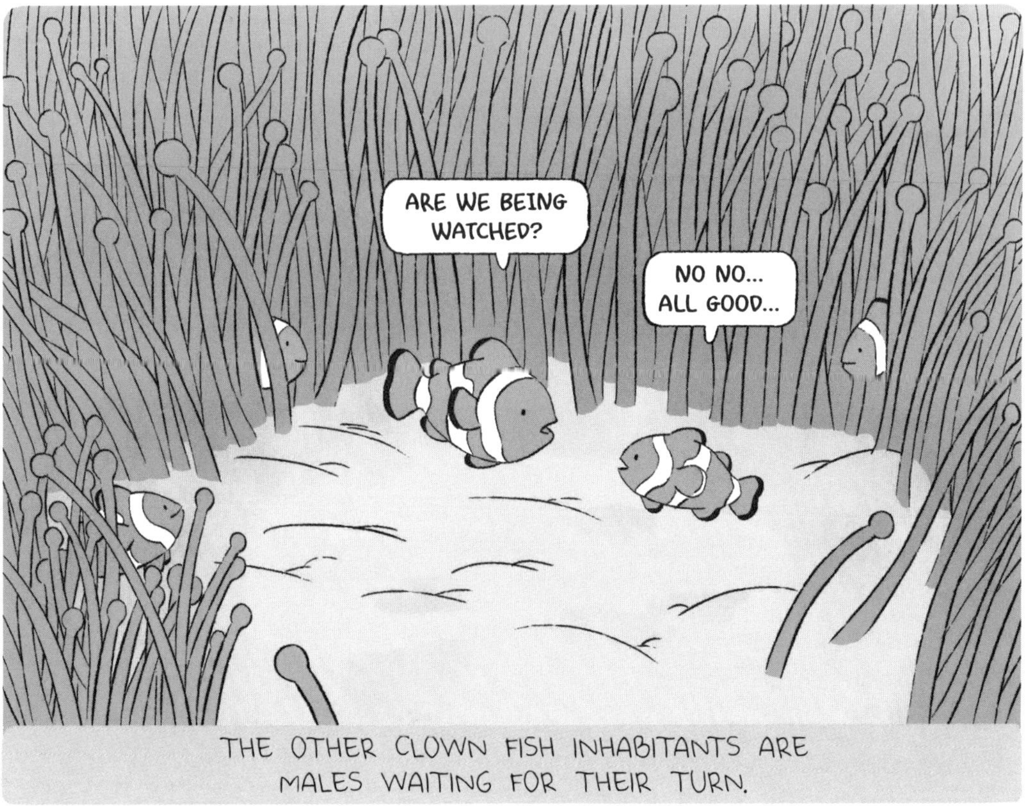
ARE WE BEING WATCHED?
NO NO... ALL GOOD...
THE OTHER CLOWN FISH INHABITANTS ARE MALES WAITING FOR THEIR TURN.

THE FEMALE IS THE LARGEST OF ALL, AND SHE KEEPS THE OTHERS IN A REDUCED STATE BY HARASSING THEM REGULARLY.

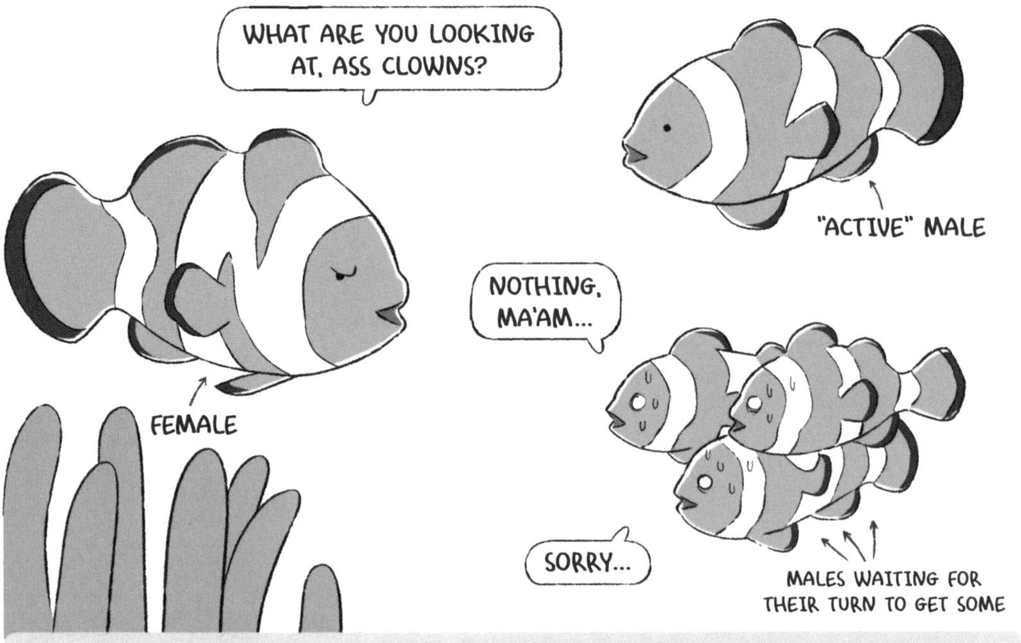
WHAT ARE YOU LOOKING AT, ASS CLOWNS?
"ACTIVE" MALE
NOTHING, MA'AM...
FEMALE
SORRY...
MALES WAITING FOR THEIR TURN TO GET SOME

WHEN THE FEMALE DIES, THE REPRODUCING MALE CHANGES SEX AND BECOMES THE REPRODUCING FEMALE.

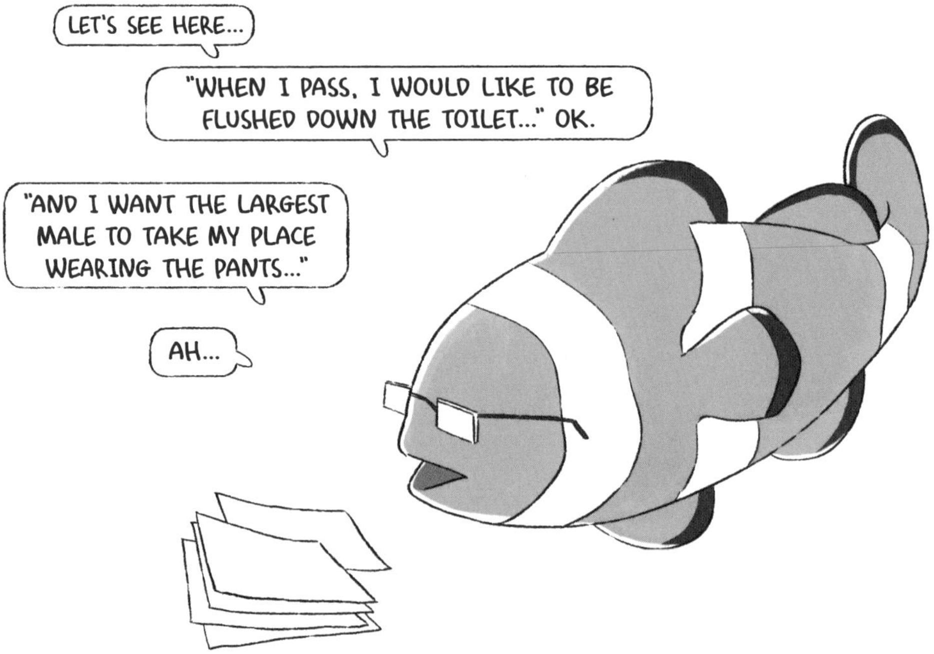
LET'S SEE HERE...
"WHEN I PASS, I WOULD LIKE TO BE FLUSHED DOWN THE TOILET..." OK.
"AND I WANT THE LARGEST MALE TO TAKE MY PLACE WEARING THE PANTS..."
AH...

ONE OF THE MALES WILL FINALLY BE ABLE TO MATE WITH HER.

AND THAT'S WHAT WE CALL SEQUENTIAL HERMAPHRODITISM: INDIVIDUALS CHANGE THEIR SEX THROUGHOUT LIFE DEPENDING ON SOCIAL OR ECOLOGICAL FACTORS.

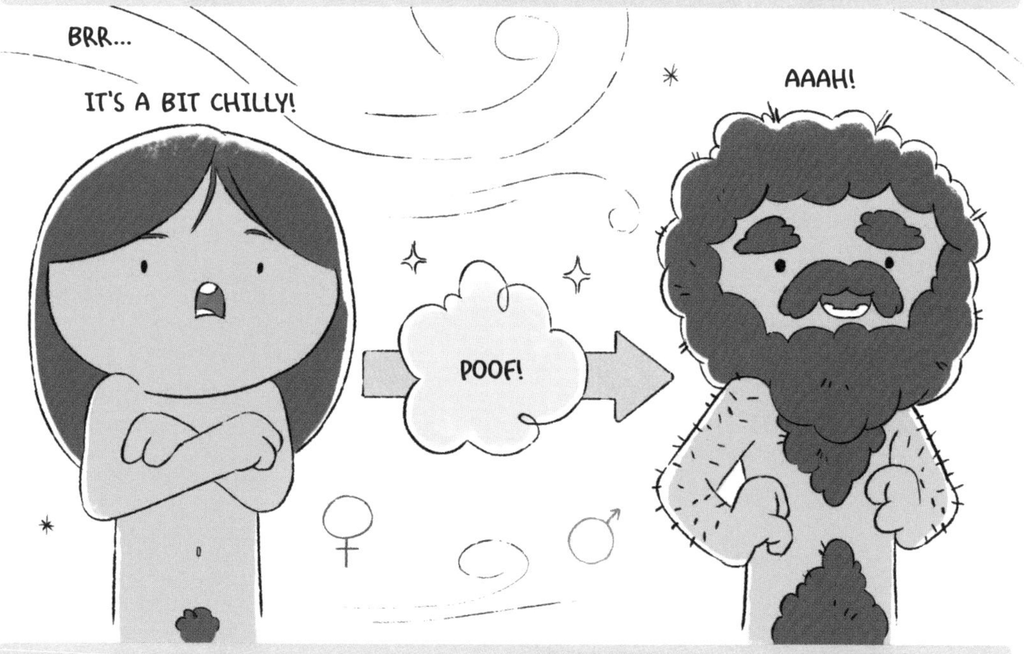
BRR...
IT'S A BIT CHILLY!
POOF!
AAAH!

HERMAPHRODITISM IN FISH HAS OCCURRED AUTONOMOUSLY A NUMBER OF TIMES OVER THE COURSE OF EVOLUTION.

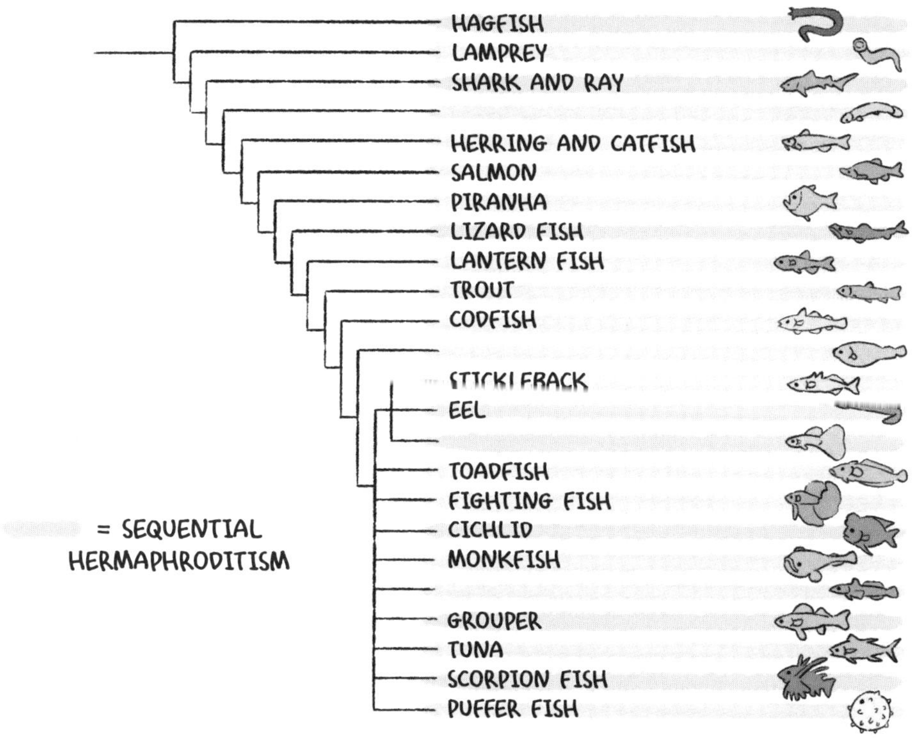
HAGFISH
LAMPREY
SHARK AND RAY
HERRING AND CATFISH
SALMON
PIRANHA
LIZARD FISH
LANTERN FISH
TROUT
CODFISH
STICKLEBACK
EEL
TOADFISH
FIGHTING FISH
CICHLID
MONKFISH
GROUPER
TUNA
SCORPION FISH
PUFFER FISH
= SEQUENTIAL HERMAPHRODITISM

SOME SPECIES, LIKE CLOWN FISH, CHANGE FROM MALE TO FEMALE. THEY'RE SAID TO BE PROTANDROUS.

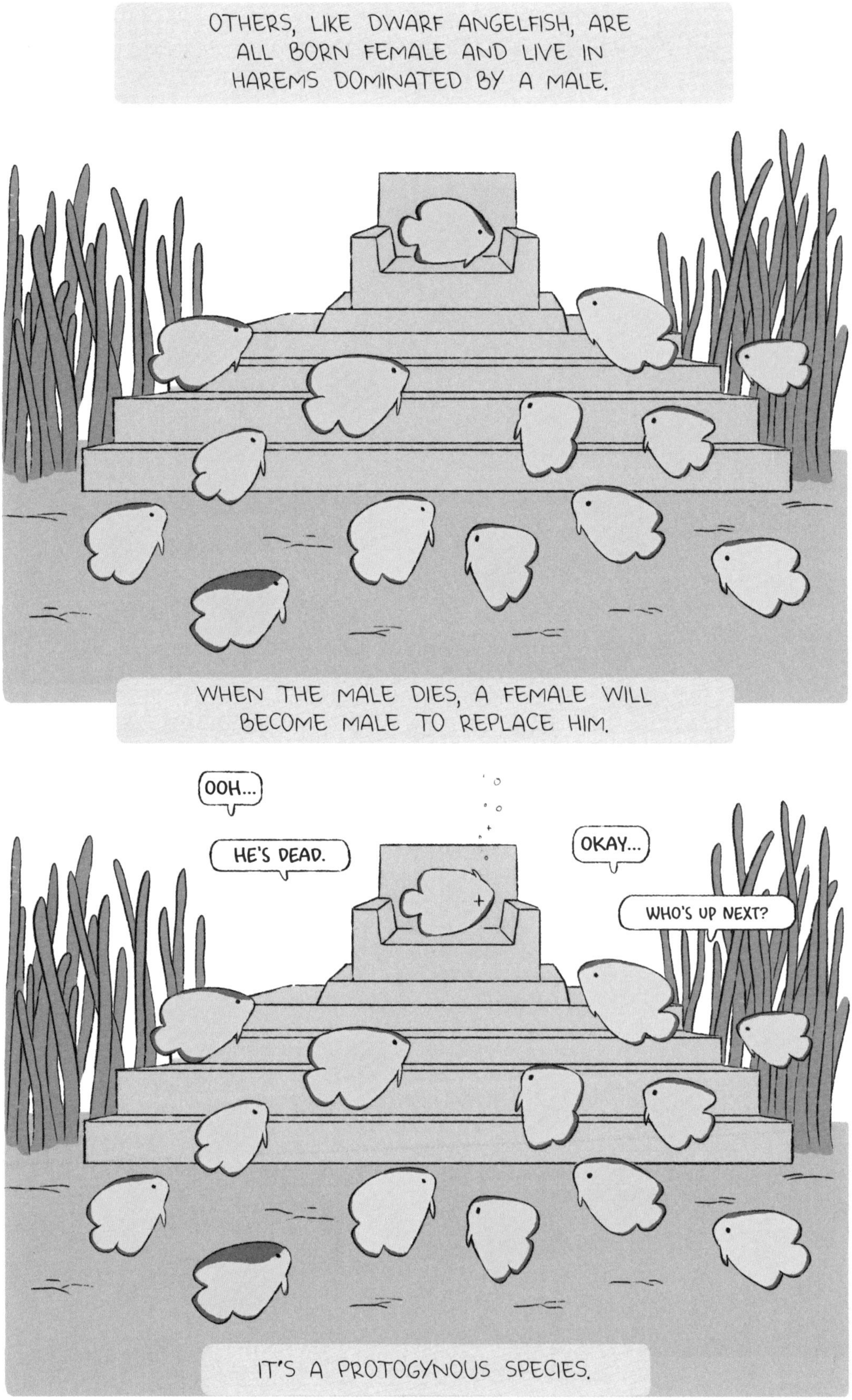
OTHERS, LIKE DWARF ANGELFISH, ARE ALL BORN FEMALE AND LIVE IN HAREMS DOMINATED BY A MALE.
WHEN THE MALE DIES, A FEMALE WILL BECOME MALE TO REPLACE HIM.
OOH...
HE'S DEAD.
OKAY...
WHO'S UP NEXT?
IT'S A PROTOGYNOUS SPECIES.

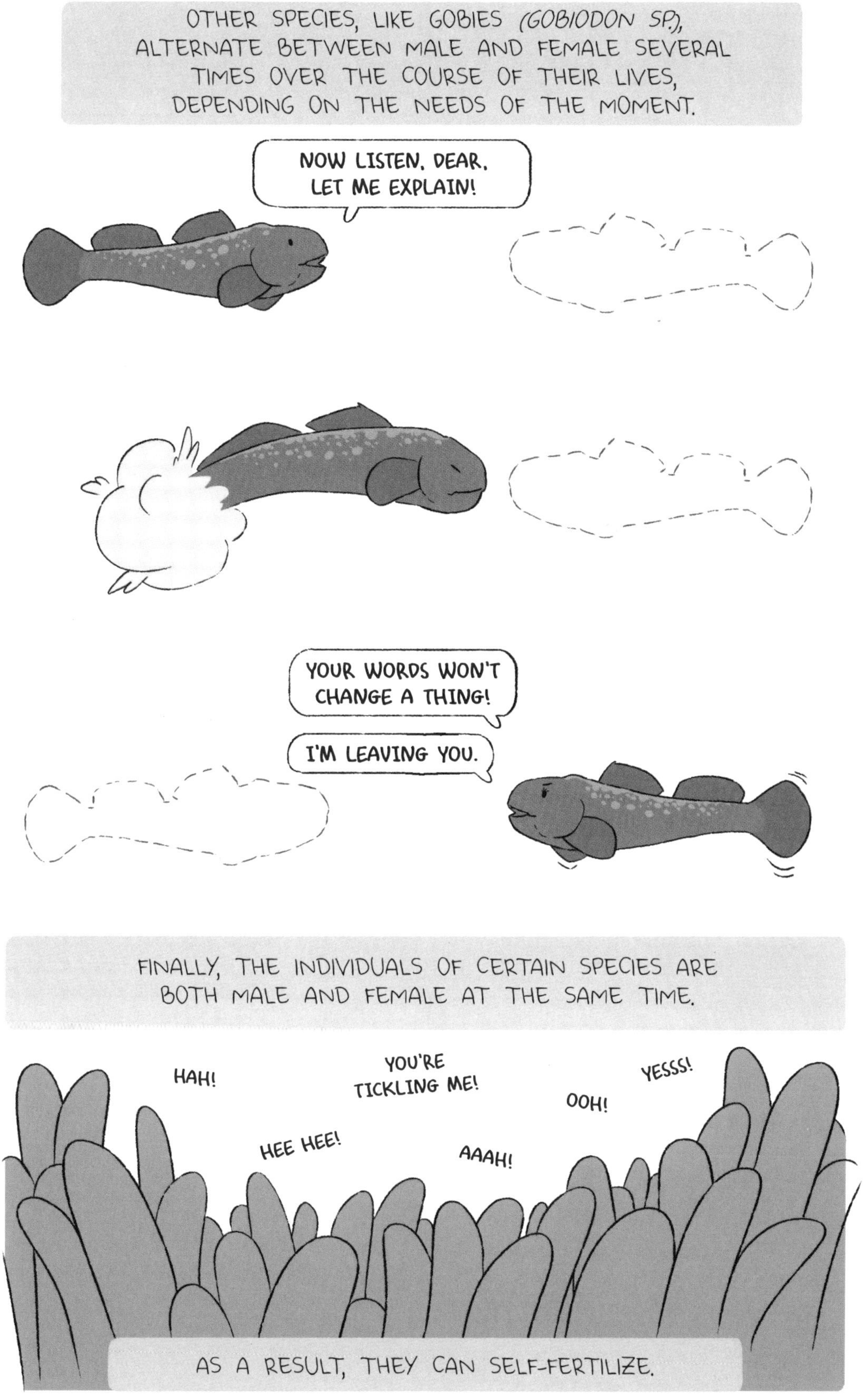
OTHER SPECIES, LIKE GOBIES (GOBIODON SP.), ALTERNATE BETWEEN MALE AND FEMALE SEVERAL TIMES OVER THE COURSE OF THEIR LIVES, DEPENDING ON THE NEEDS OF THE MOMENT.
NOW LISTEN, DEAR. LET ME EXPLAIN!
YOUR WORDS WON'T CHANGE A THING!
I'M LEAVING YOU.
FINALLY, THE INDIVIDUALS OF CERTAIN SPECIES ARE BOTH MALE AND FEMALE AT THE SAME TIME.
HAH!
YOU'RE TICKLING ME!
YESSS!
OOH!
HEE HEE!
AAAH!
AS A RESULT, THEY CAN SELF-FERTILIZE.

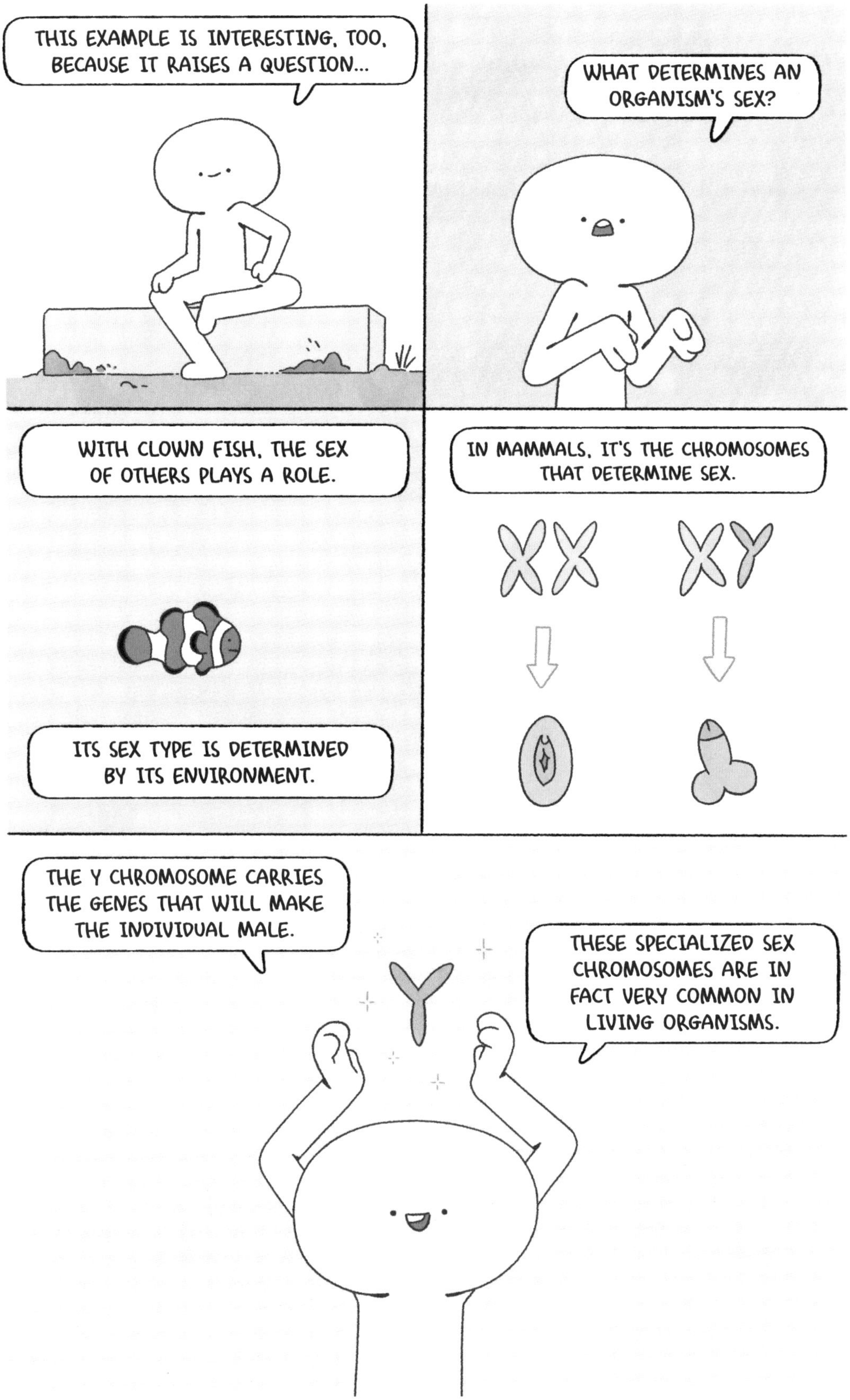
THIS EXAMPLE IS INTERESTING, TOO, BECAUSE IT RAISES A QUESTION...
WHAT DETERMINES AN ORGANISM'S SEX?
WITH CLOWN FISH, THE SEX OF OTHERS PLAYS A ROLE.
ITS SEX TYPE IS DETERMINED BY ITS ENVIRONMENT.
IN MAMMALS, IT'S THE CHROMOSOMES THAT DETERMINE SEX.
THE Y CHROMOSOME CARRIES THE GENES THAT WILL MAKE THE INDIVIDUAL MALE.
THESE SPECIALIZED SEX CHROMOSOMES ARE IN FACT VERY COMMON IN LIVING ORGANISMS.

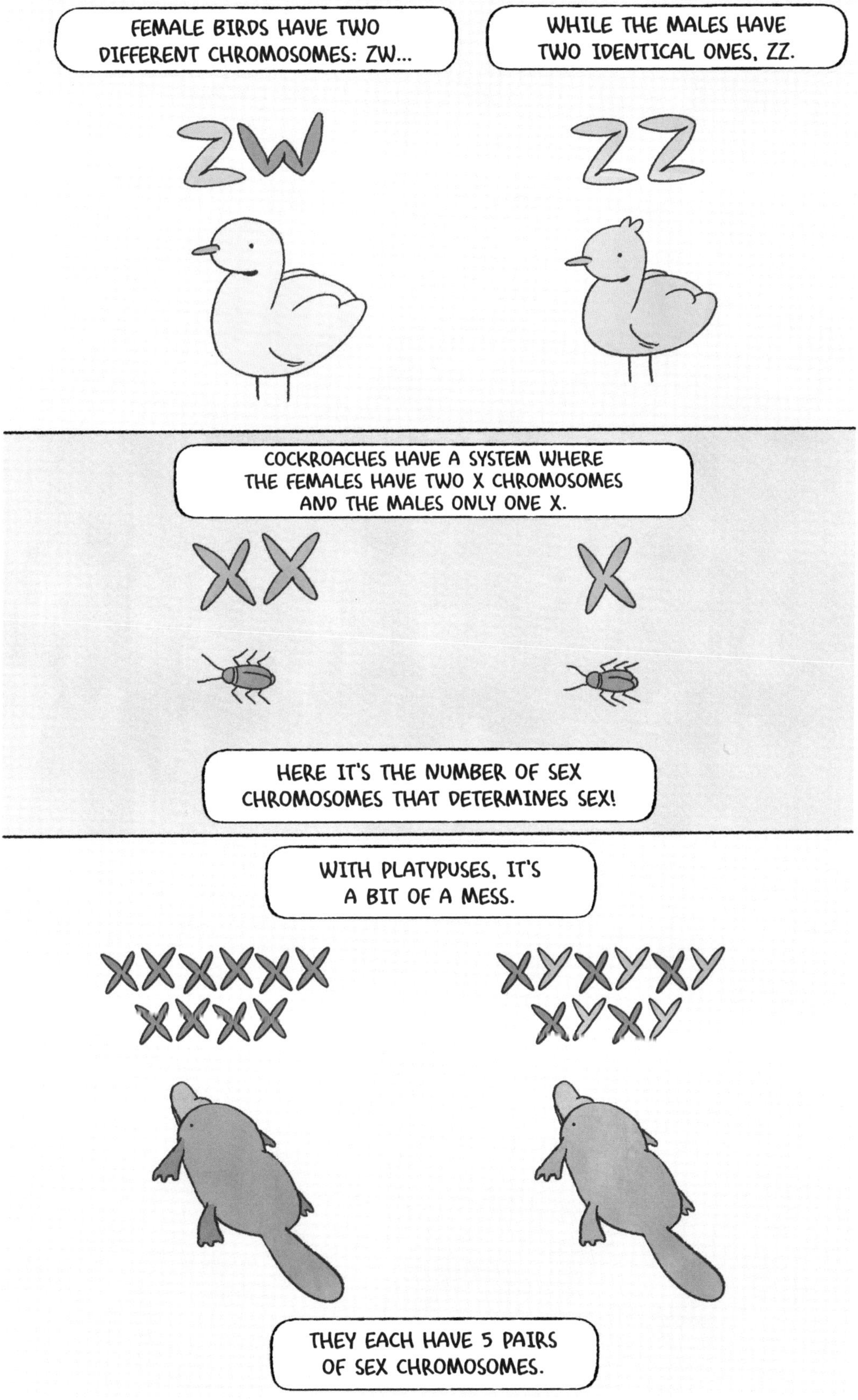
FEMALE BIRDS HAVE TWO DIFFERENT CHROMOSOMES: ZW...
WHILE THE MALES HAVE TWO IDENTICAL ONES, ZZ.
ZW
ZZ
COCKROACHES HAVE A SYSTEM WHERE THE FEMALES HAVE TWO X CHROMOSOMES AND THE MALES ONLY ONE X.
XX
X
HERE IT'S THE NUMBER OF SEX CHROMOSOMES THAT DETERMINES SEX!
WITH PLATYPUSES, IT'S A BIT OF A MESS.
XXXXXX XXXX
XYXYXY XYXY
THEY EACH HAVE 5 PAIRS OF SEX CHROMOSOMES.

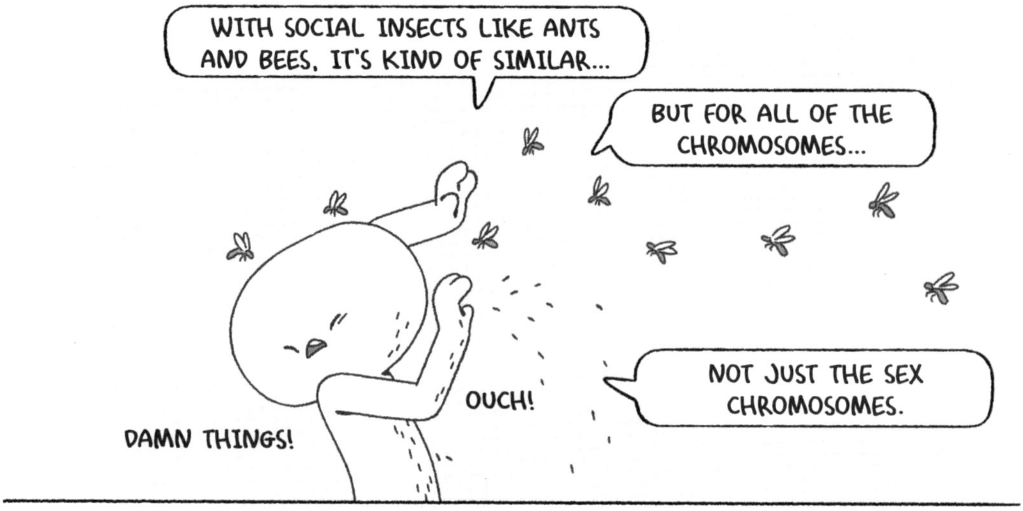

WITH SOCIAL INSECTS LIKE ANTS AND BEES, IT'S KIND OF SIMILAR...
BUT FOR ALL OF THE CHROMOSOMES...
NOT JUST THE SEX CHROMOSOMES.
OUCH!
DAMN THINGS!

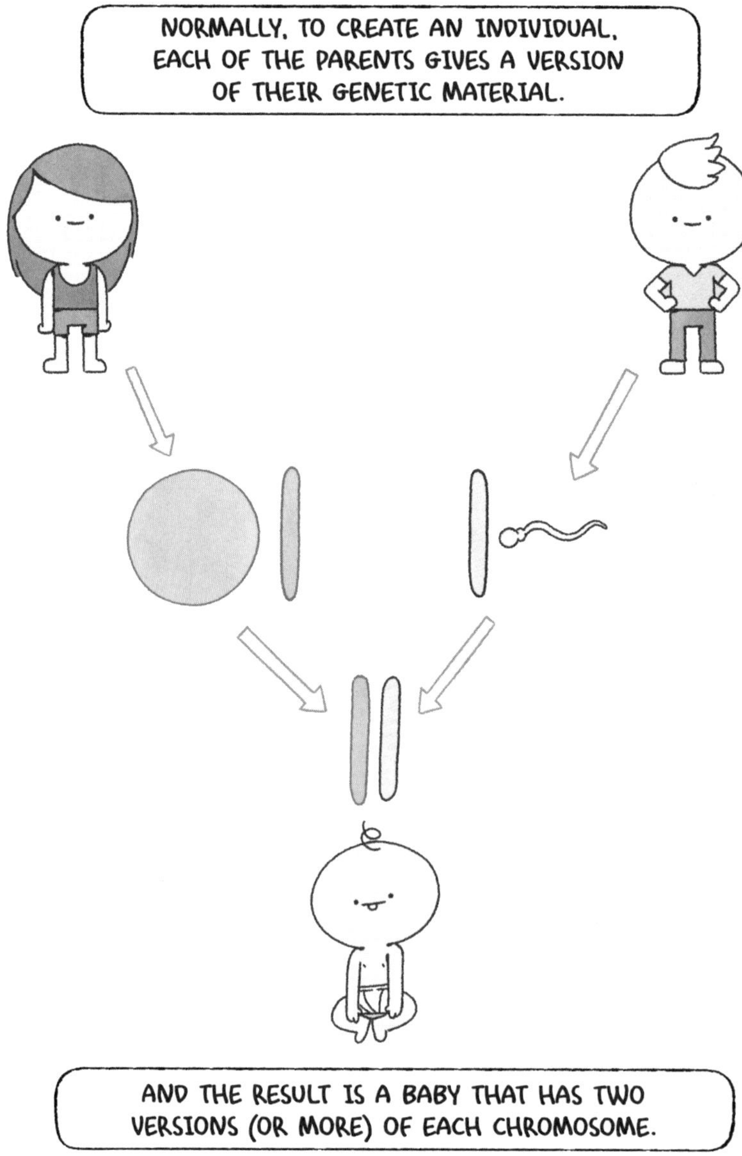

NORMALLY, TO CREATE AN INDIVIDUAL, EACH OF THE PARENTS GIVES A VERSION OF THEIR GENETIC MATERIAL.
AND THE RESULT IS A BABY THAT HAS TWO VERSIONS (OR MORE) OF EACH CHROMOSOME.

WITH ANTS AND BEES, THE FEMALE'S EGG CELL WON'T ALWAYS BE FERTILIZED.
NOT LETTING YOU IN.
BUT...
IF THAT'S THE CASE (NO FERTILIZATION), THE RESULT IS AN INDIVIDUAL WITH ONLY ONE VERSION OF EACH CHROMOSOME...
AND IT'S A MALE.
MOTHER
MALE
FEMALE
FATHER
ALTERNATIVELY, ALL THE EGG CELLS THAT ARE FERTILIZED BECOME FEMALES.

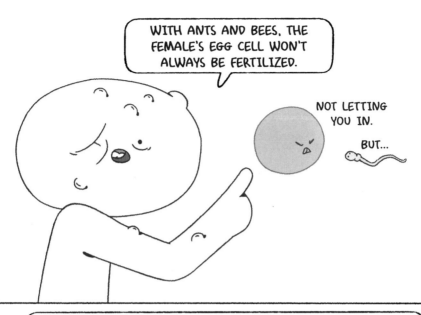

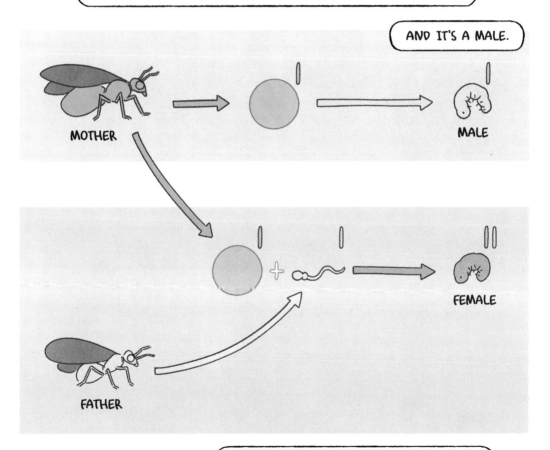

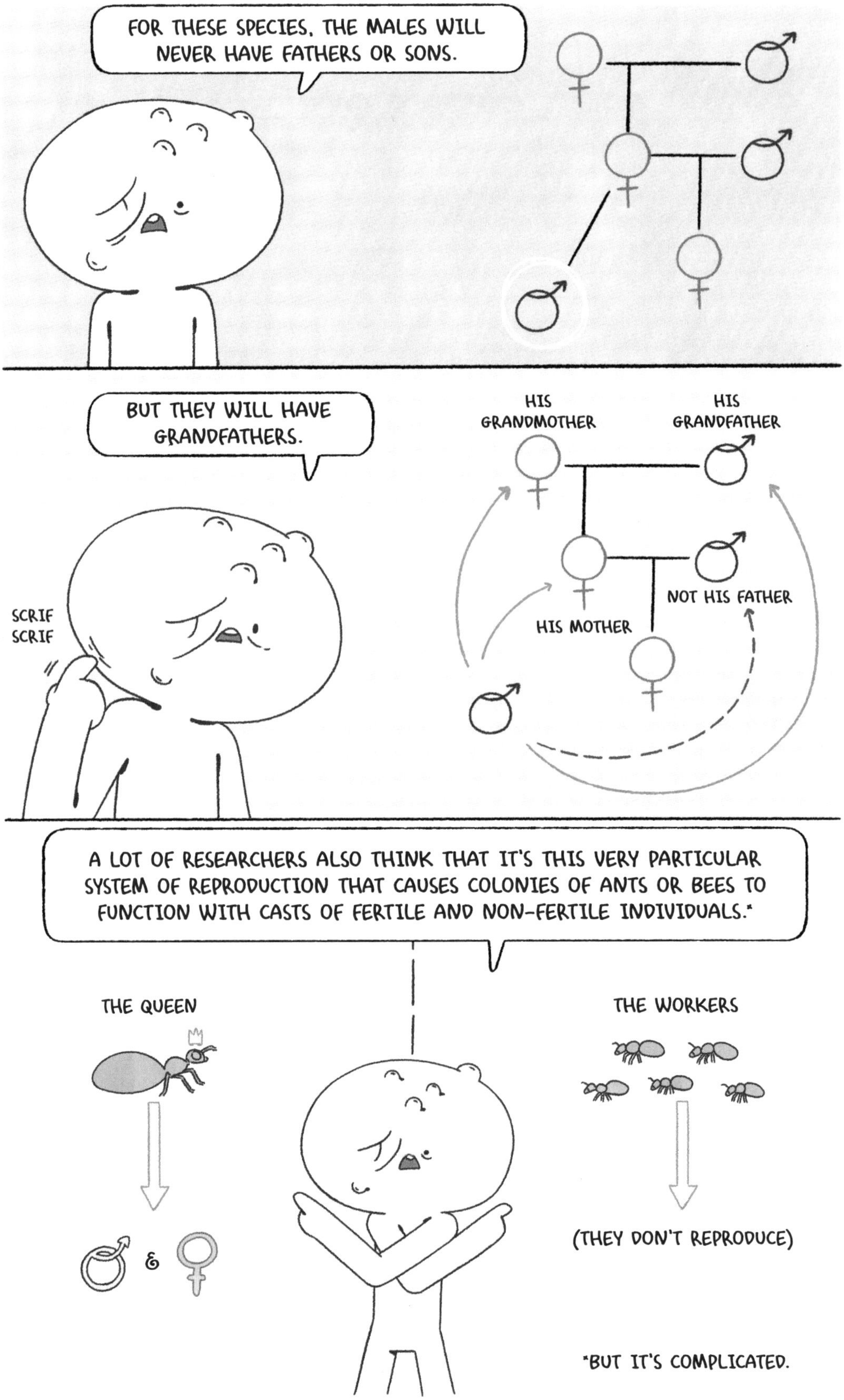
FOR THESE SPECIES, THE MALES WILL NEVER HAVE FATHERS OR SONS.
BUT THEY WILL HAVE GRANDFATHERS.
HIS GRANDMOTHER
HIS GRANDFATHER
NOT HIS FATHER
HIS MOTHER
SCRIF SCRIF
A LOT OF RESEARCHERS ALSO THINK THAT IT'S THIS VERY PARTICULAR SYSTEM OF REPRODUCTION THAT CAUSES COLONIES OF ANTS OR BEES TO FUNCTION WITH CASTS OF FERTILE AND NON-FERTILE INDIVIDUALS.*
THE QUEEN
THE WORKERS
&
(THEY DON'T REPRODUCE)
*BUT IT'S COMPLICATED.

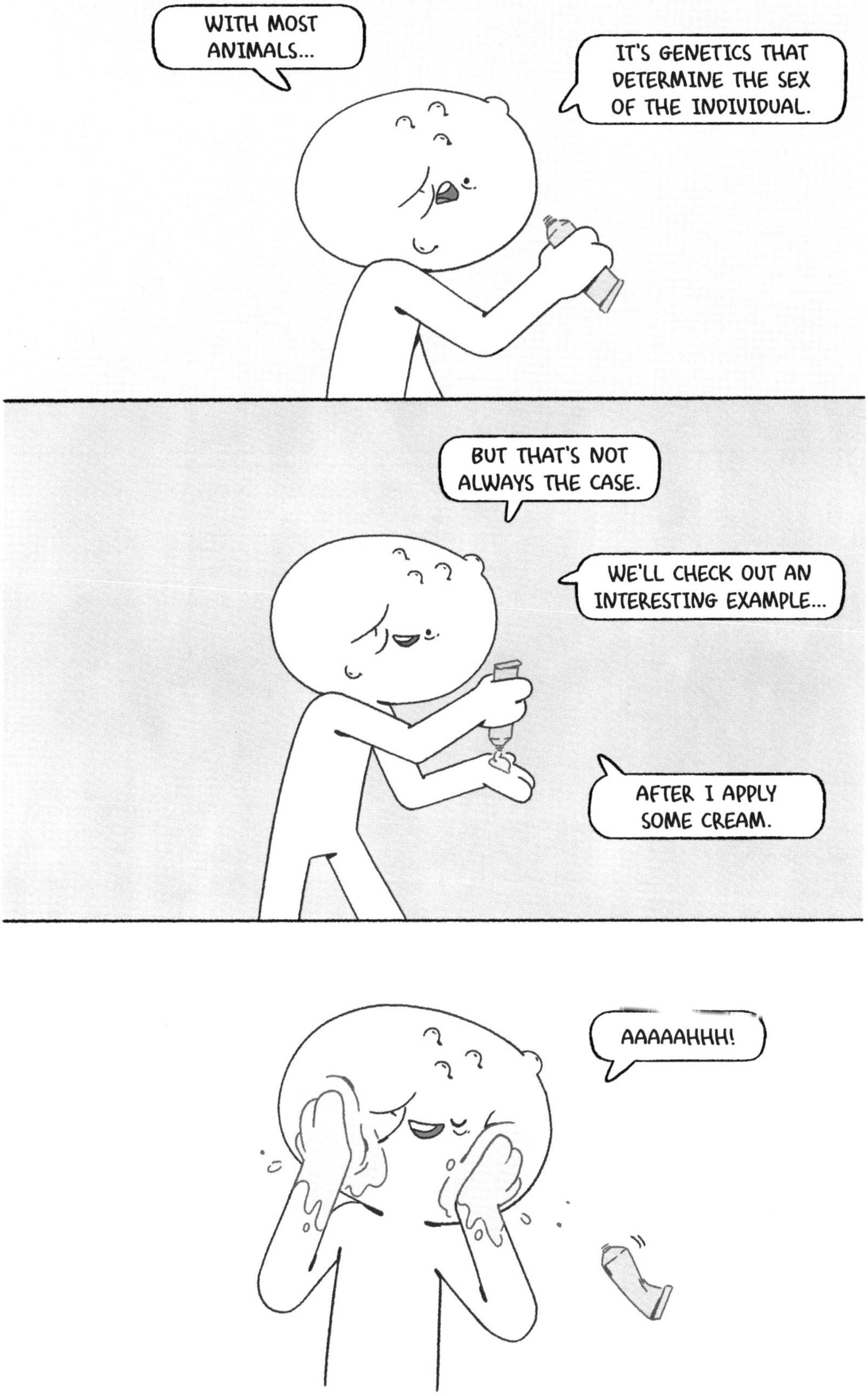
WITH MOST ANIMALS...
IT'S GENETICS THAT DETERMINE THE SEX OF THE INDIVIDUAL.
BUT THAT'S NOT ALWAYS THE CASE.
WE'LL CHECK OUT AN INTERESTING EXAMPLE...
AFTER I APPLY SOME CREAM.
AAAAAHHH!

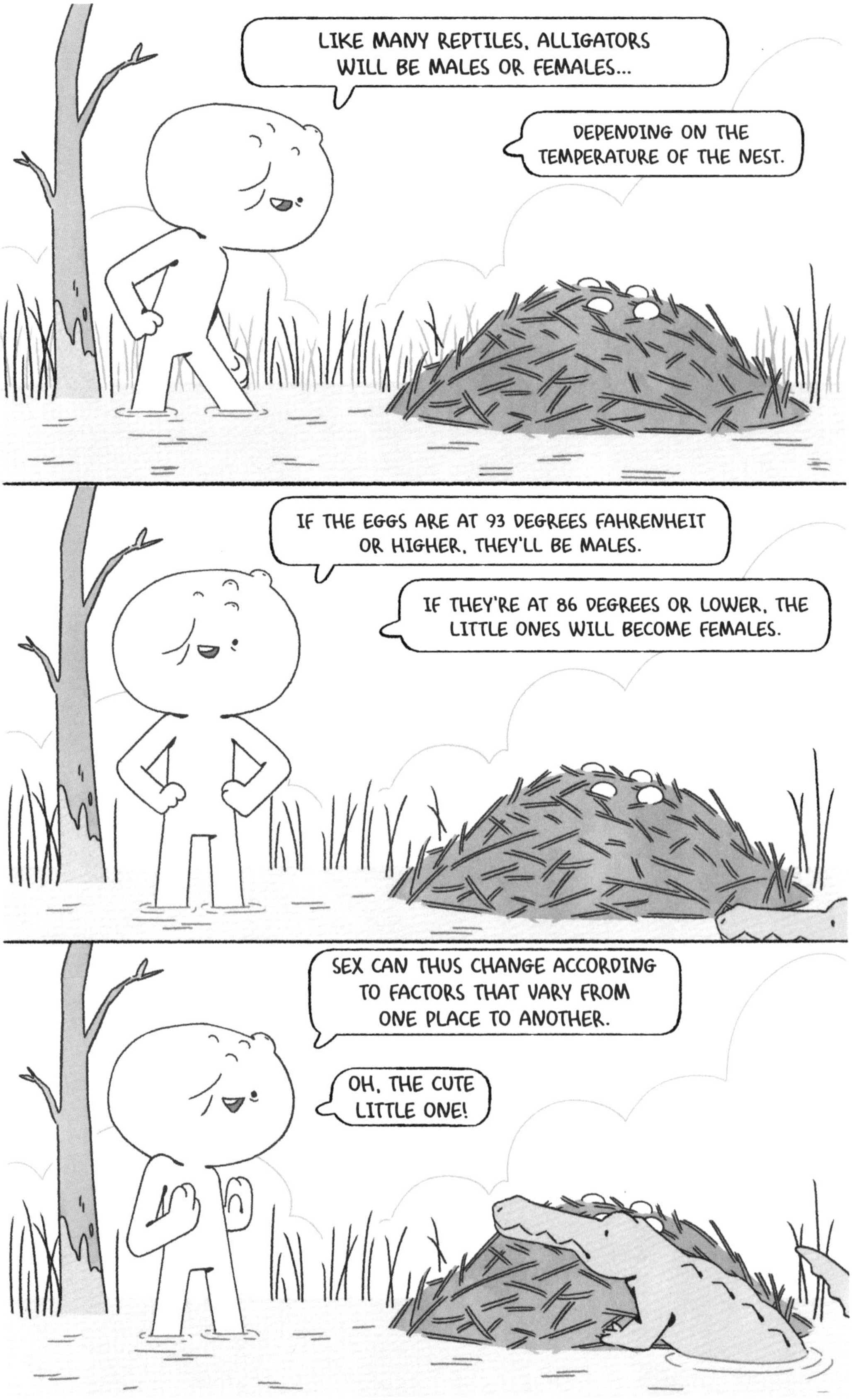
LIKE MANY REPTILES, ALLIGATORS WILL BE MALES OR FEMALES...
DEPENDING ON THE TEMPERATURE OF THE NEST.
IF THE EGGS ARE AT 93 DEGREES FAHRENHEIT OR HIGHER, THEY'LL BE MALES.
IF THEY'RE AT 86 DEGREES OR LOWER, THE LITTLE ONES WILL BECOME FEMALES.
SEX CAN THUS CHANGE ACCORDING TO FACTORS THAT VARY FROM ONE PLACE TO ANOTHER.
OH, THE CUTE LITTLE ONE!

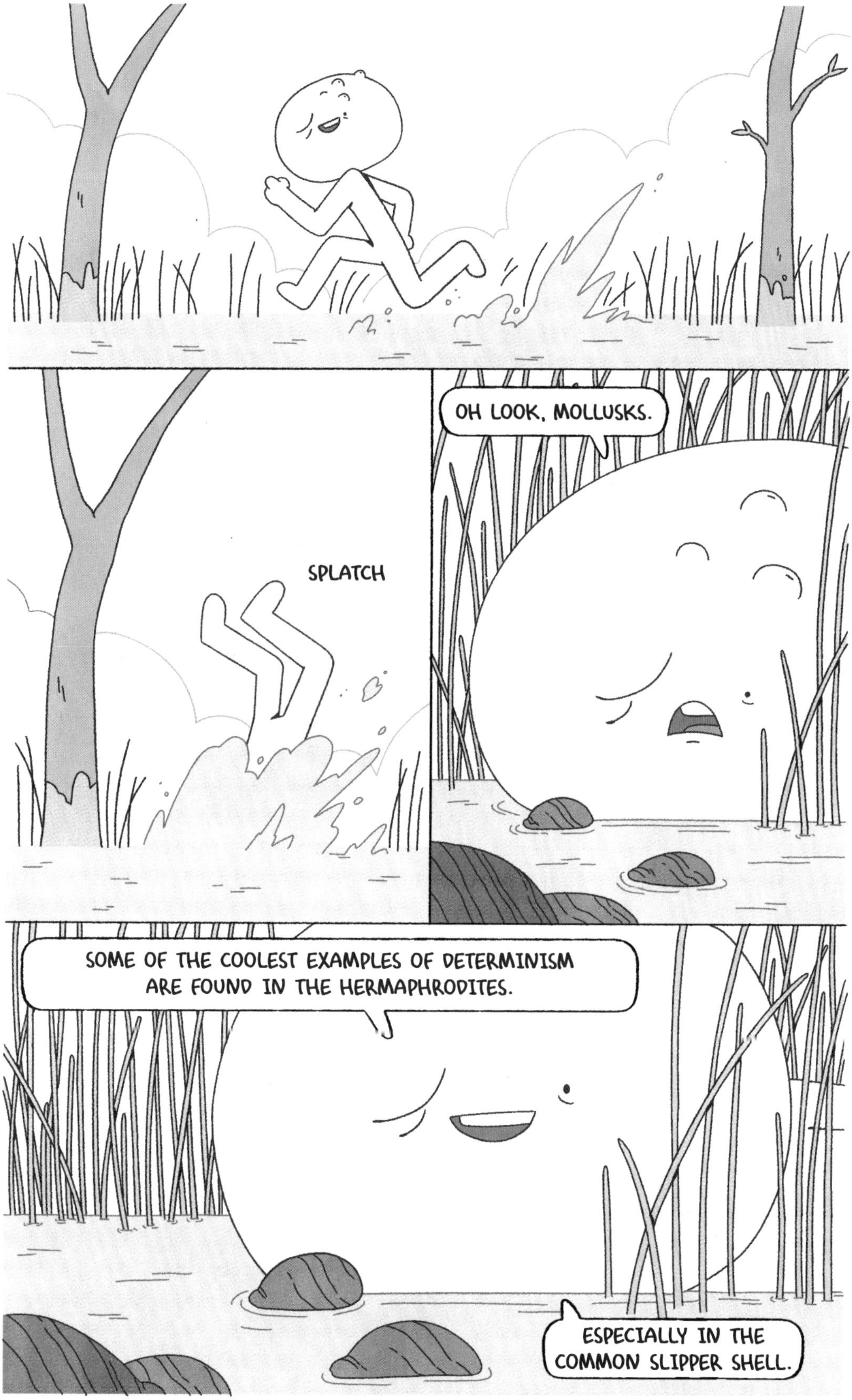
SPLATCH
OH LOOK, MOLLUSKS.
SOME OF THE COOLEST EXAMPLES OF DETERMINISM ARE FOUND IN THE HERMAPHRODITES.
ESPECIALLY IN THE COMMON SLIPPER SHELL.

AMONG SEQUENTIAL HERMAPHRODITES, THE SLIPPER SHELL *CREPIDULA FORNICATA* IS A GASTROPOD MOLLUSK THAT HAS ADOPTED A SPECIAL WAY OF REPRODUCING:

THE GANGBANG

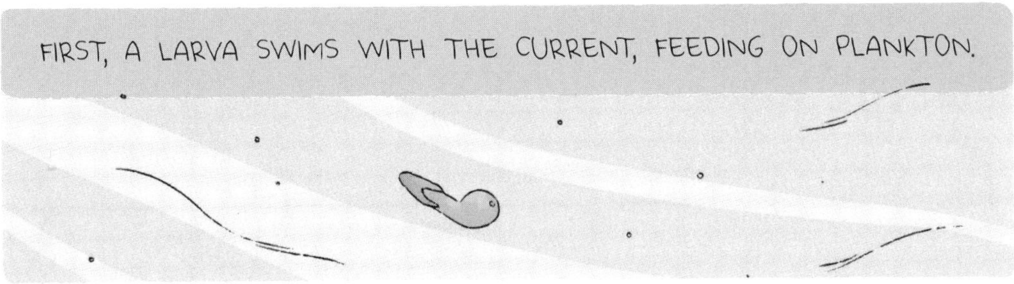
FIRST, A LARVA SWIMS WITH THE CURRENT, FEEDING ON PLANKTON.

EVENTUALLY, IT WILL...

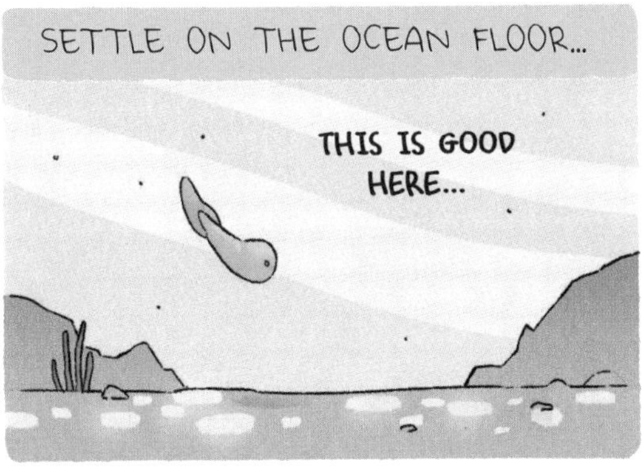
SETTLE ON THE OCEAN FLOOR...
THIS IS GOOD HERE...

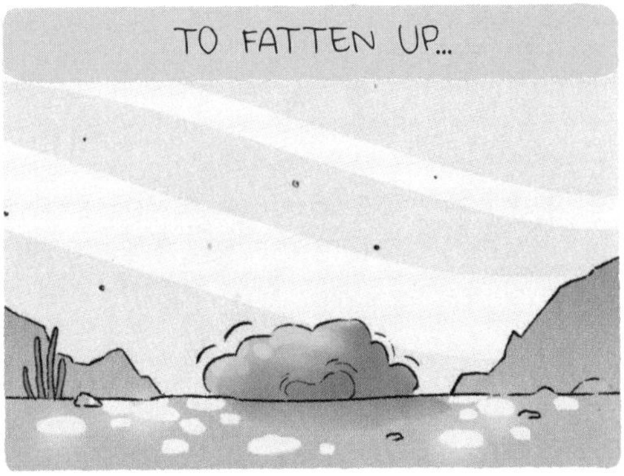
TO FATTEN UP...

IT WILL BEGIN TO TRANSFORM ITSELF.

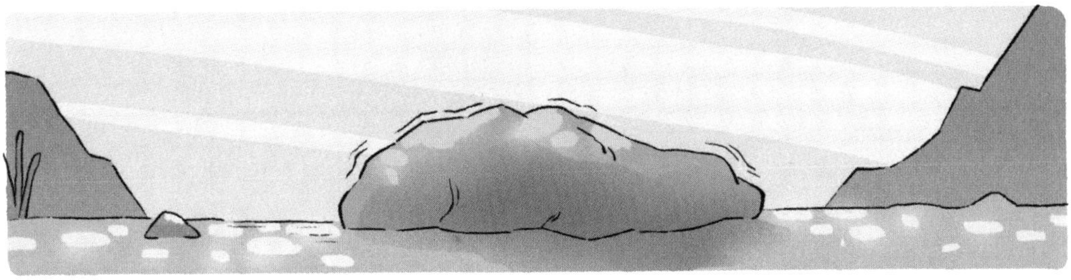

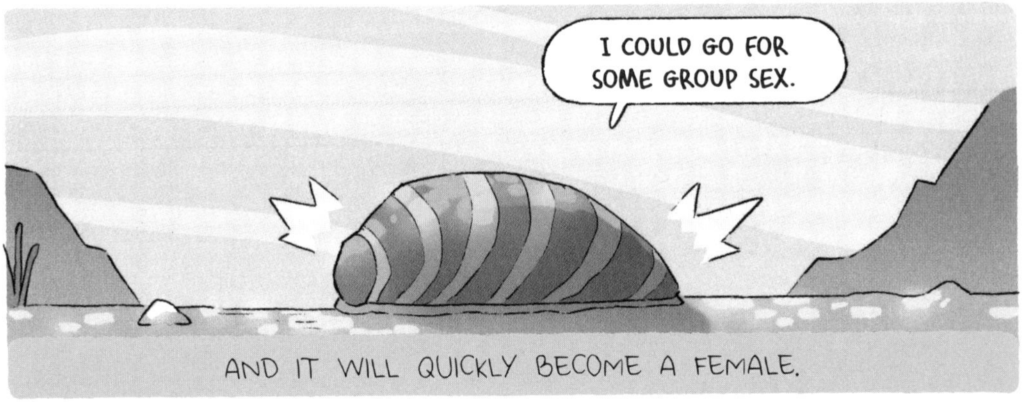
I COULD GO FOR SOME GROUP SEX.
AND IT WILL QUICKLY BECOME A FEMALE.

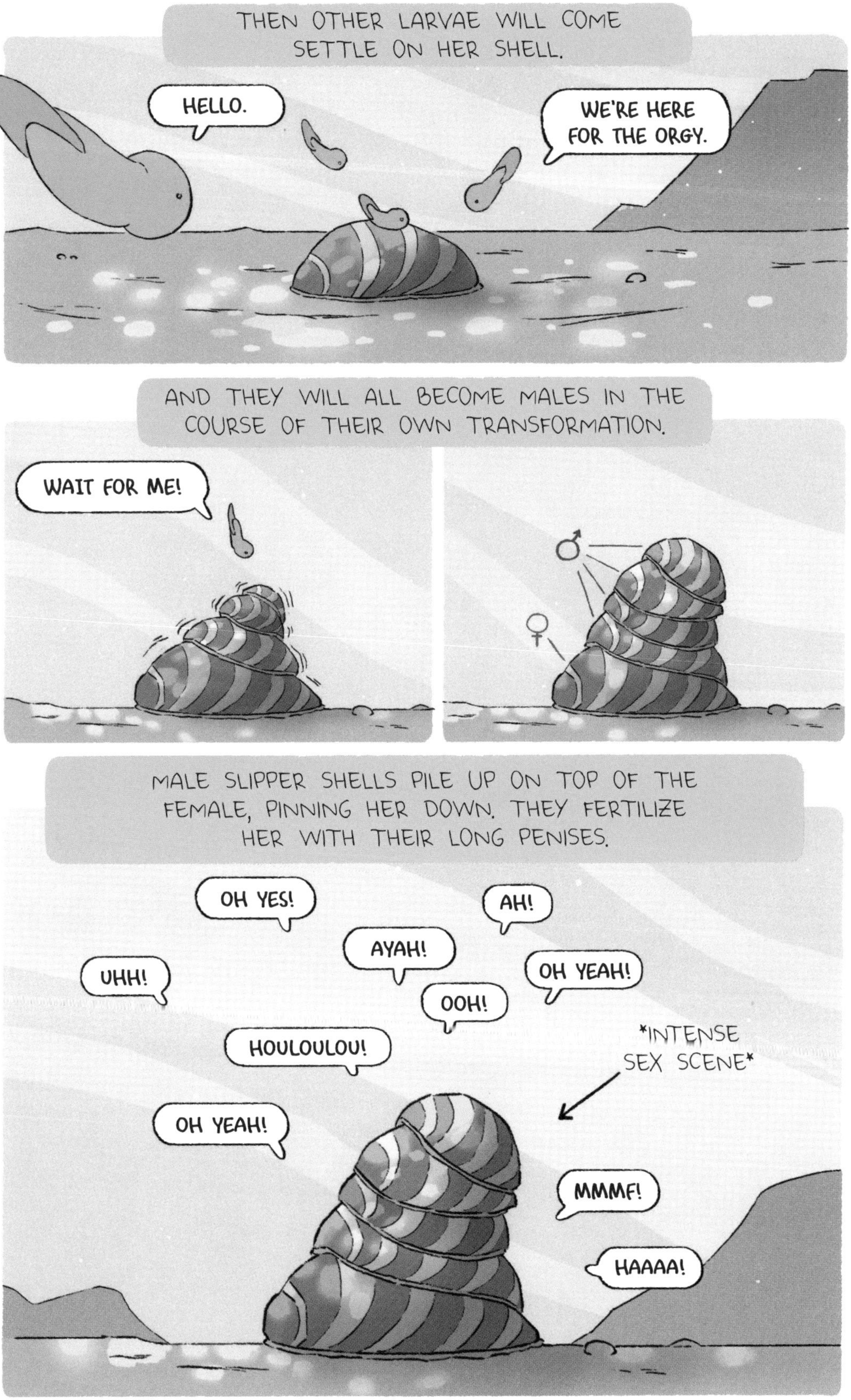
THEN OTHER LARVAE WILL COME SETTLE ON HER SHELL.
HELLO.
WE'RE HERE FOR THE ORGY.
AND THEY WILL ALL BECOME MALES IN THE COURSE OF THEIR OWN TRANSFORMATION.
WAIT FOR ME!
♂
♀
MALE SLIPPER SHELLS PILE UP ON TOP OF THE FEMALE, PINNING HER DOWN. THEY FERTILIZE HER WITH THEIR LONG PENISES.
OH YES!
AH!
AYAH!
UHH!
OH YEAH!
OOH!
HOULOULOU!
INTENSE SEX SCENE
OH YEAH!
MMMF!
HAAAA!

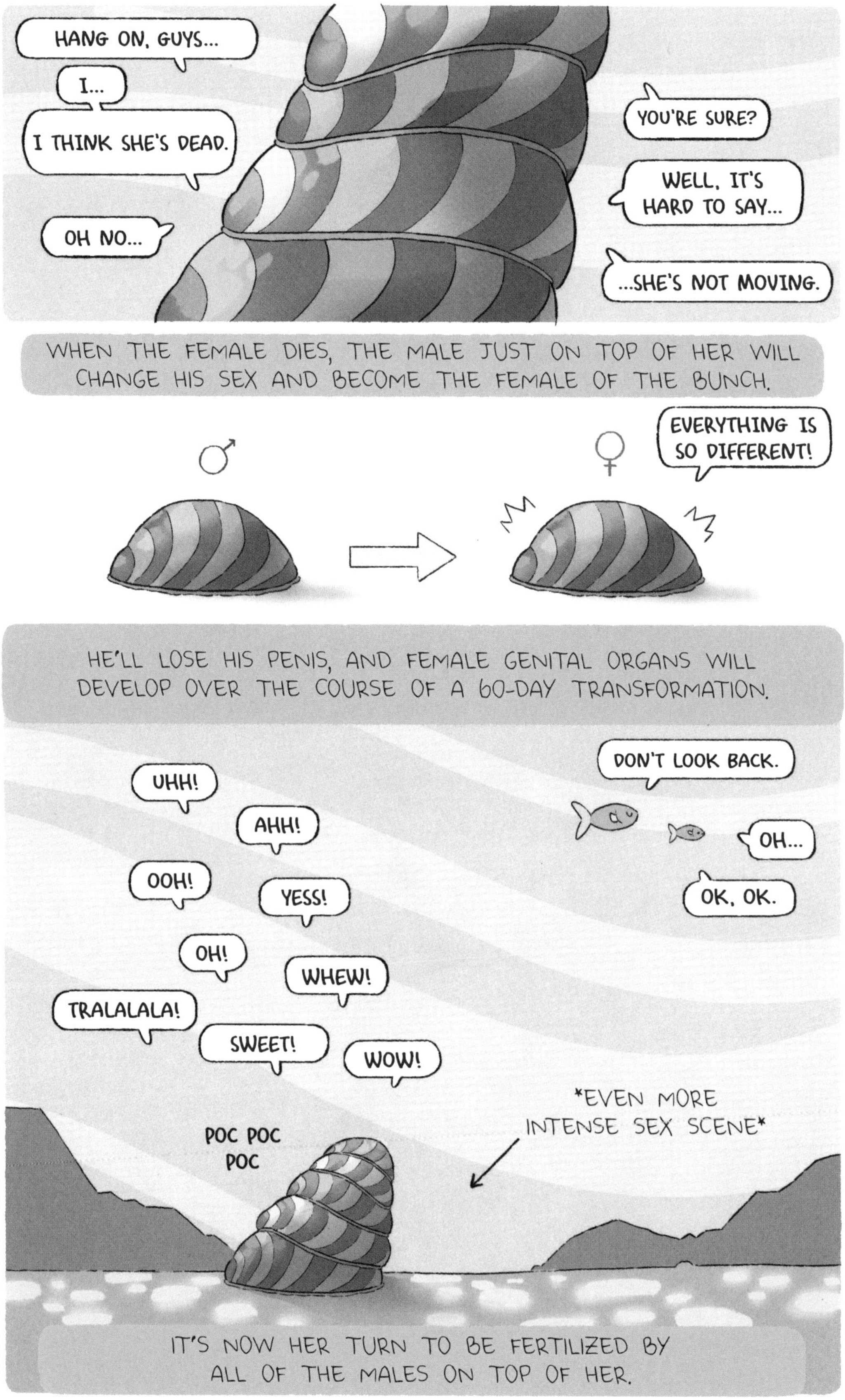
HANG ON, GUYS...
I...
I THINK SHE'S DEAD.
OH NO...
YOU'RE SURE?
WELL, IT'S HARD TO SAY...
...SHE'S NOT MOVING.
WHEN THE FEMALE DIES, THE MALE JUST ON TOP OF HER WILL CHANGE HIS SEX AND BECOME THE FEMALE OF THE BUNCH.
♂
♀
EVERYTHING IS SO DIFFERENT!
HE'LL LOSE HIS PENIS, AND FEMALE GENITAL ORGANS WILL DEVELOP OVER THE COURSE OF A 60-DAY TRANSFORMATION.
UHH!
AHH!
OOH!
YESS!
OH!
WHEW!
TRALALALA!
SWEET!
WOW!
DON'T LOOK BACK.
OH...
OK, OK.
EVEN MORE INTENSE SEX SCENE
POC POC POC
IT'S NOW HER TURN TO BE FERTILIZED BY ALL OF THE MALES ON TOP OF HER.

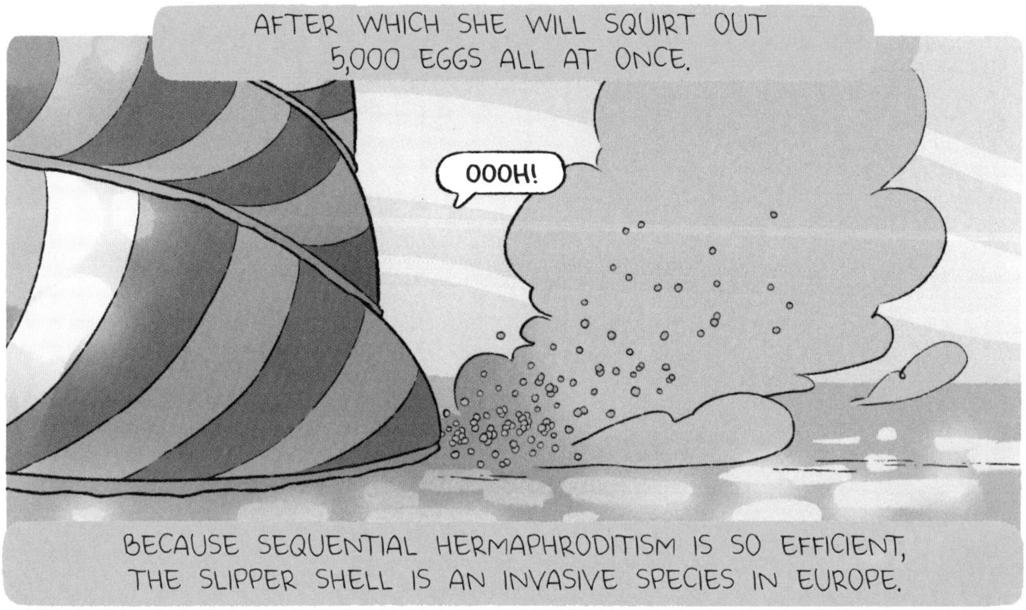
AFTER WHICH SHE WILL SQUIRT OUT 5,000 EGGS ALL AT ONCE.
OOOH!
BECAUSE SEQUENTIAL HERMAPHRODITISM IS SO EFFICIENT, THE SLIPPER SHELL IS AN INVASIVE SPECIES IN EUROPE.

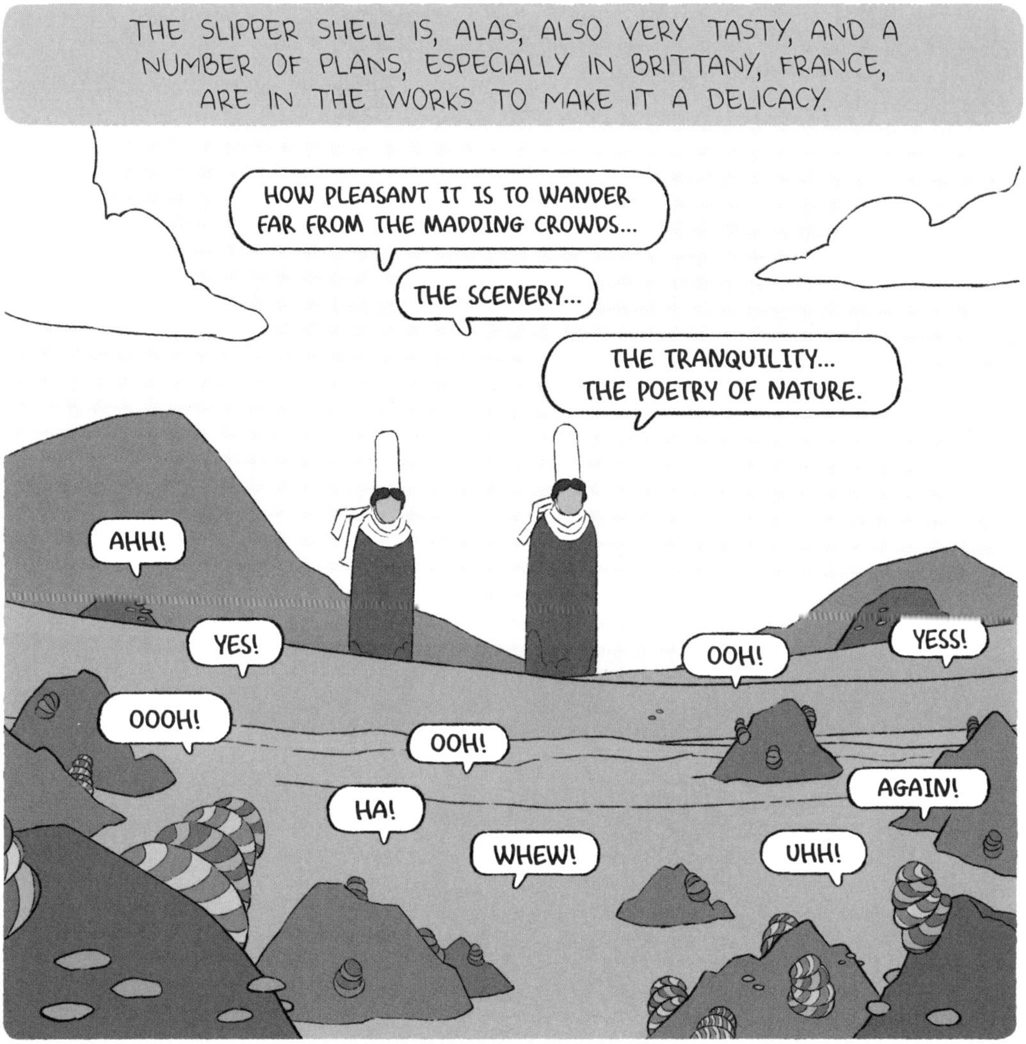
THE SLIPPER SHELL IS, ALAS, ALSO VERY TASTY, AND A NUMBER OF PLANS, ESPECIALLY IN BRITTANY, FRANCE, ARE IN THE WORKS TO MAKE IT A DELICACY.
HOW PLEASANT IT IS TO WANDER FAR FROM THE MADDING CROWDS...
THE SCENERY...
THE TRANQUILITY... THE POETRY OF NATURE.
AHH!
YES!
OOH!
YESS!
OOOH!
OOH!
AGAIN!
HA!
WHEW!
UHH!

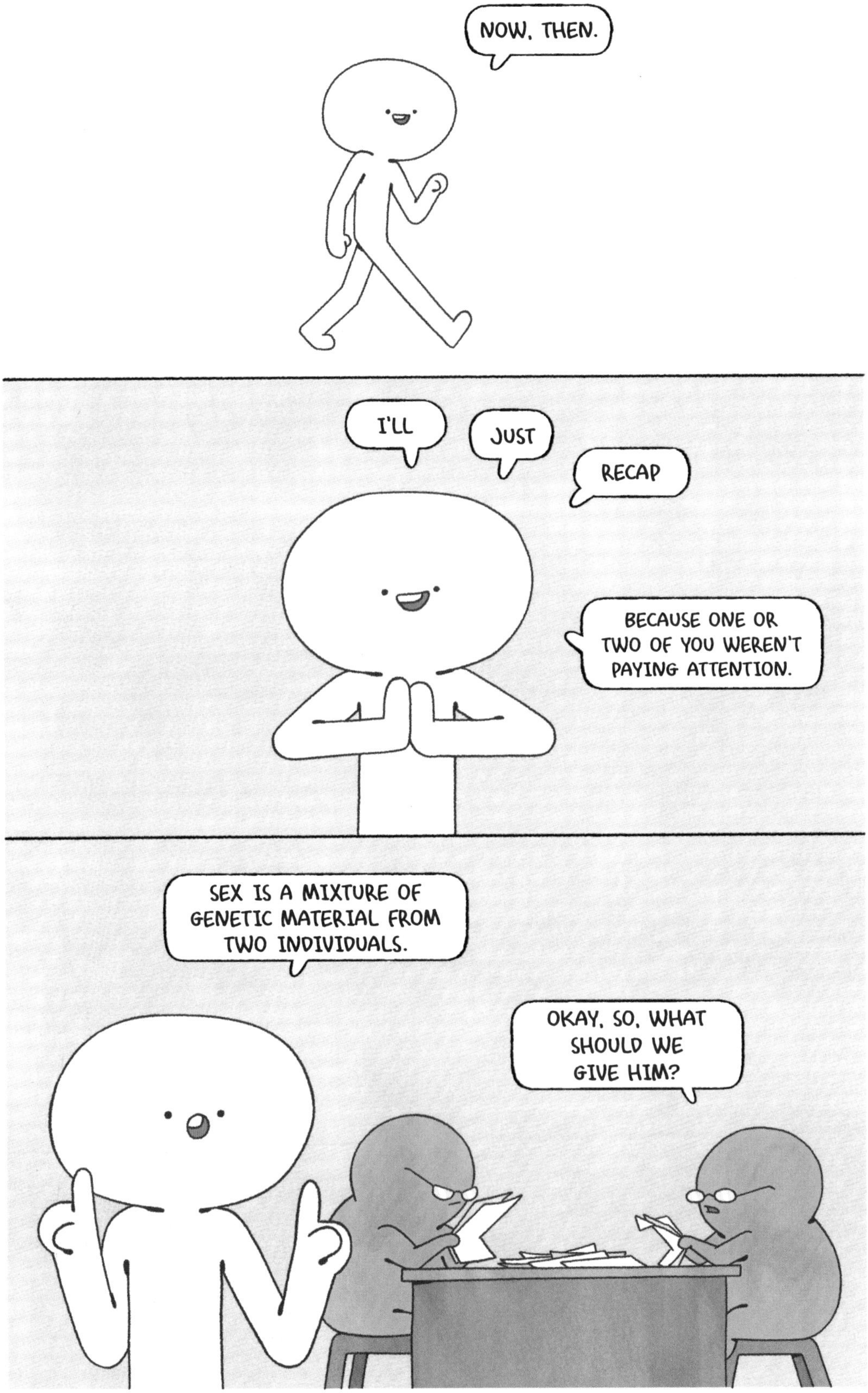
NOW, THEN.
I'LL
JUST
RECAP
BECAUSE ONE OR TWO OF YOU WEREN'T PAYING ATTENTION.
SEX IS A MIXTURE OF GENETIC MATERIAL FROM TWO INDIVIDUALS.
OKAY, SO, WHAT SHOULD WE GIVE HIM?

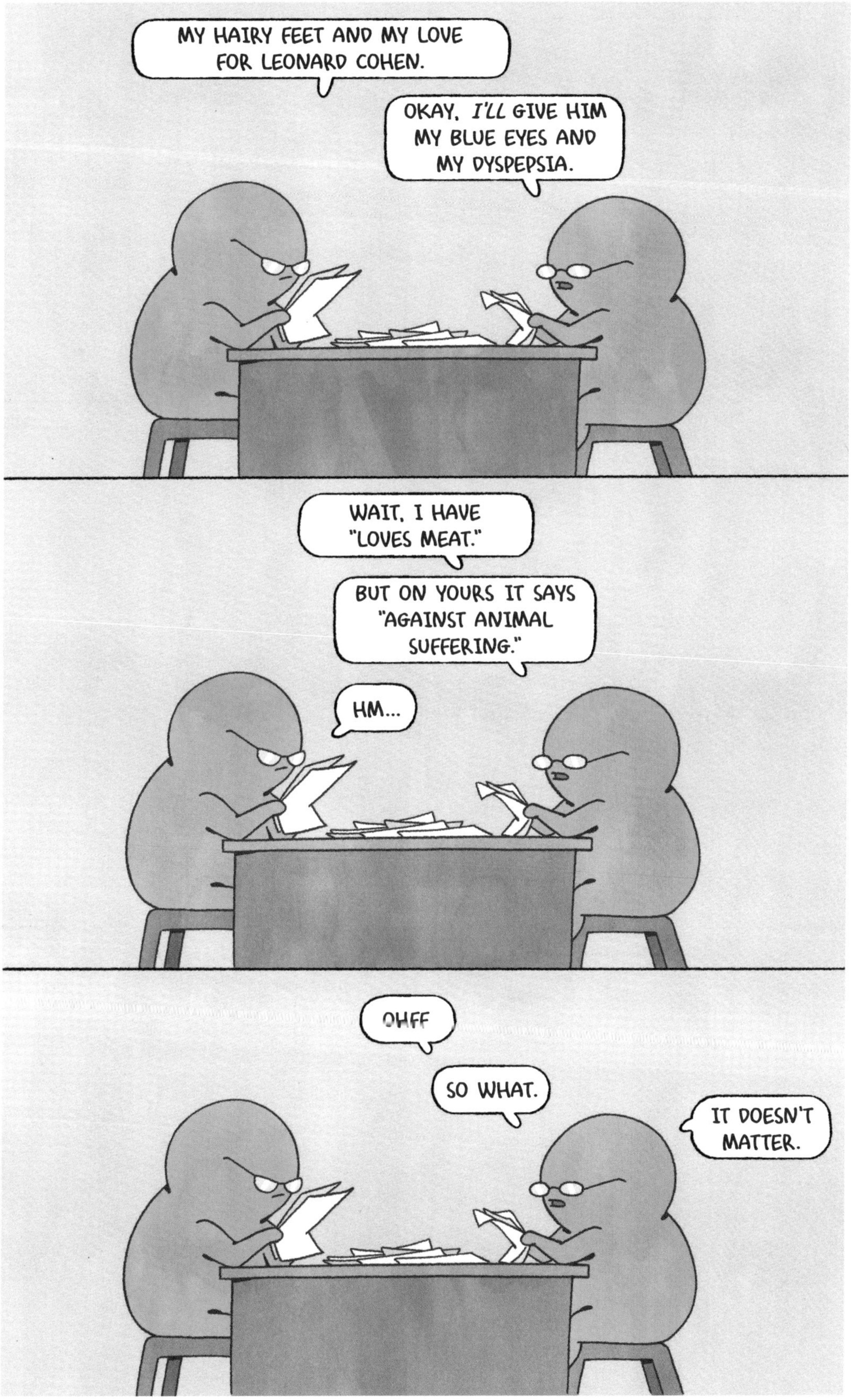
MY HAIRY FEET AND MY LOVE FOR LEONARD COHEN.
OKAY, *I'LL* GIVE HIM MY BLUE EYES AND MY DYSPEPSIA.
WAIT, I HAVE "LOVES MEAT."
BUT ON YOURS IT SAYS "AGAINST ANIMAL SUFFERING."
HM...
OHFF
SO WHAT.
IT DOESN'T MATTER.

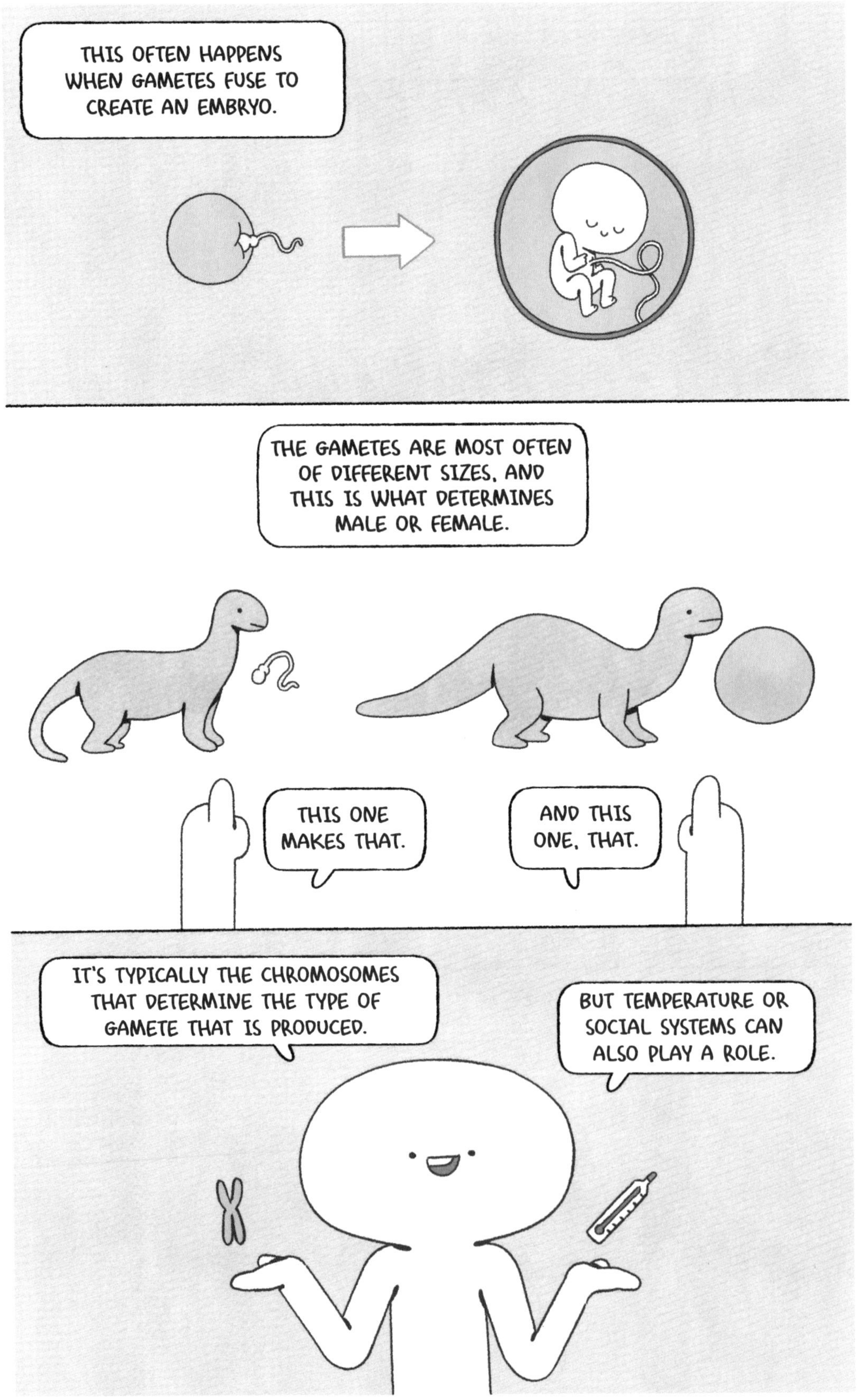
THIS OFTEN HAPPENS WHEN GAMETES FUSE TO CREATE AN EMBRYO.
THE GAMETES ARE MOST OFTEN OF DIFFERENT SIZES, AND THIS IS WHAT DETERMINES MALE OR FEMALE.
THIS ONE MAKES THAT.
AND THIS ONE, THAT.
IT'S TYPICALLY THE CHROMOSOMES THAT DETERMINE THE TYPE OF GAMETE THAT IS PRODUCED.
BUT TEMPERATURE OR SOCIAL SYSTEMS CAN ALSO PLAY A ROLE.

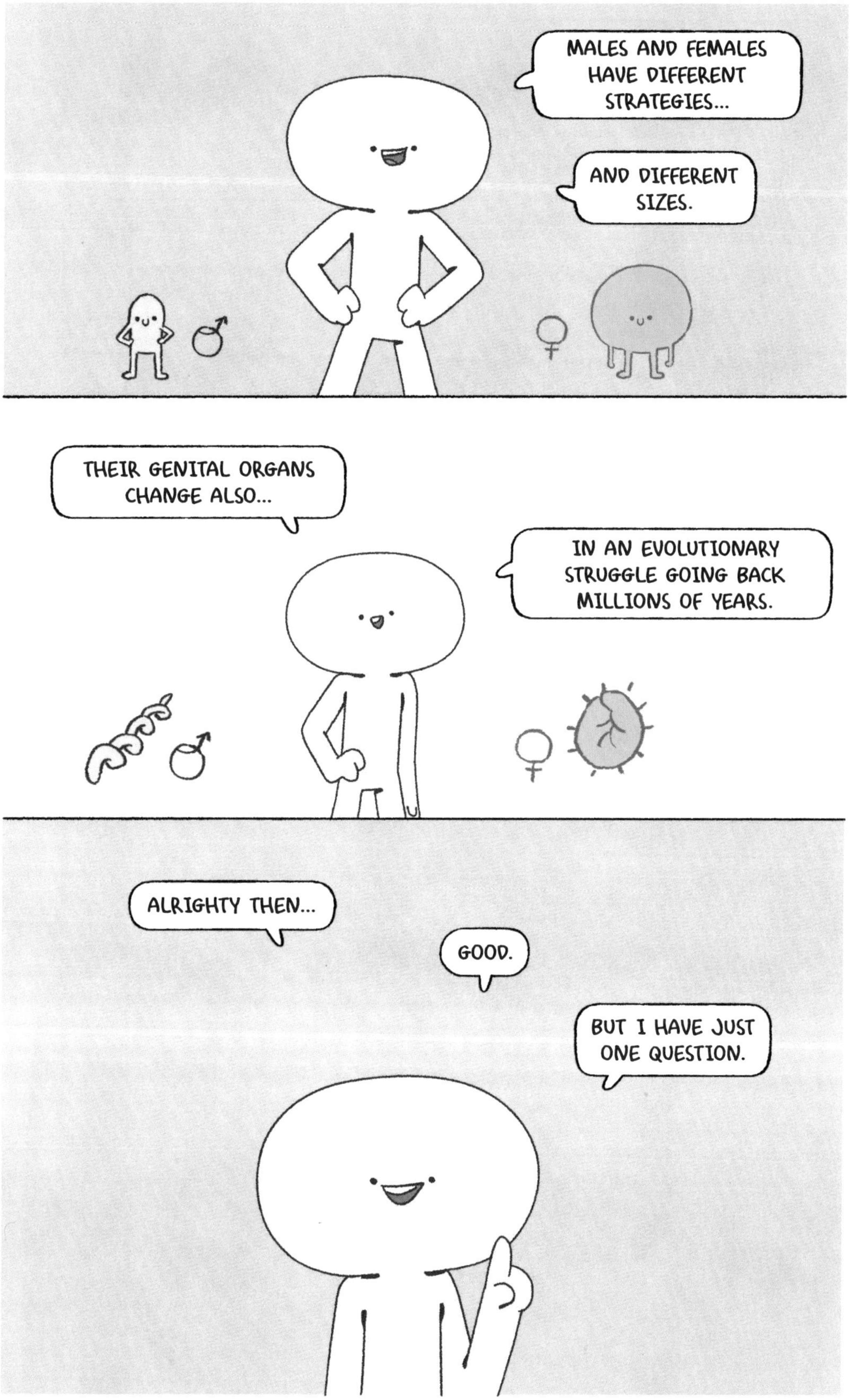
MALES AND FEMALES HAVE DIFFERENT STRATEGIES...
AND DIFFERENT SIZES.
THEIR GENITAL ORGANS CHANGE ALSO...
IN AN EVOLUTIONARY STRUGGLE GOING BACK MILLIONS OF YEARS.
ALRIGHTY THEN...
GOOD.
BUT I HAVE JUST ONE QUESTION.

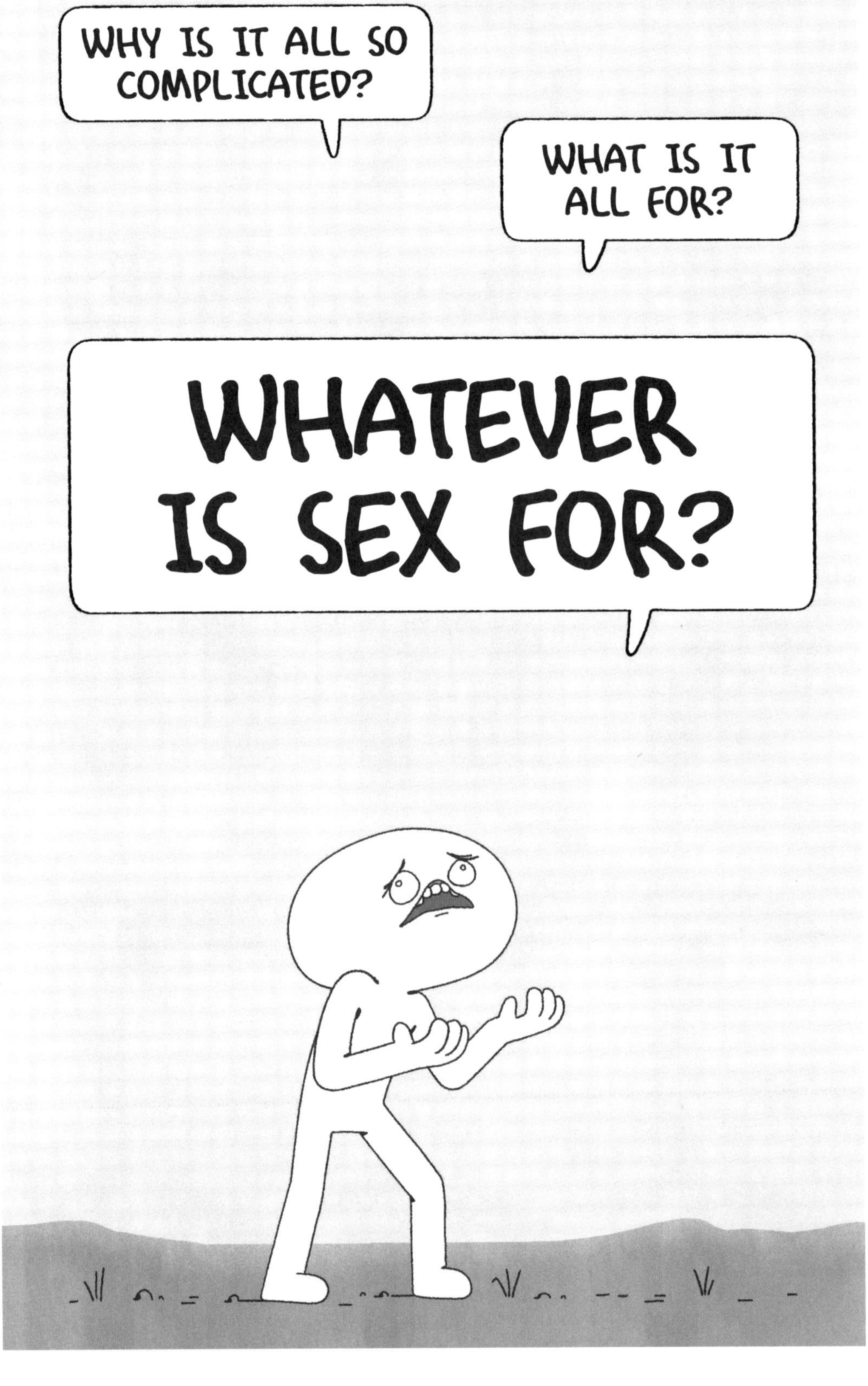
WHY IS IT ALL SO COMPLICATED?
WHAT IS IT ALL FOR?
WHATEVER IS SEX FOR?

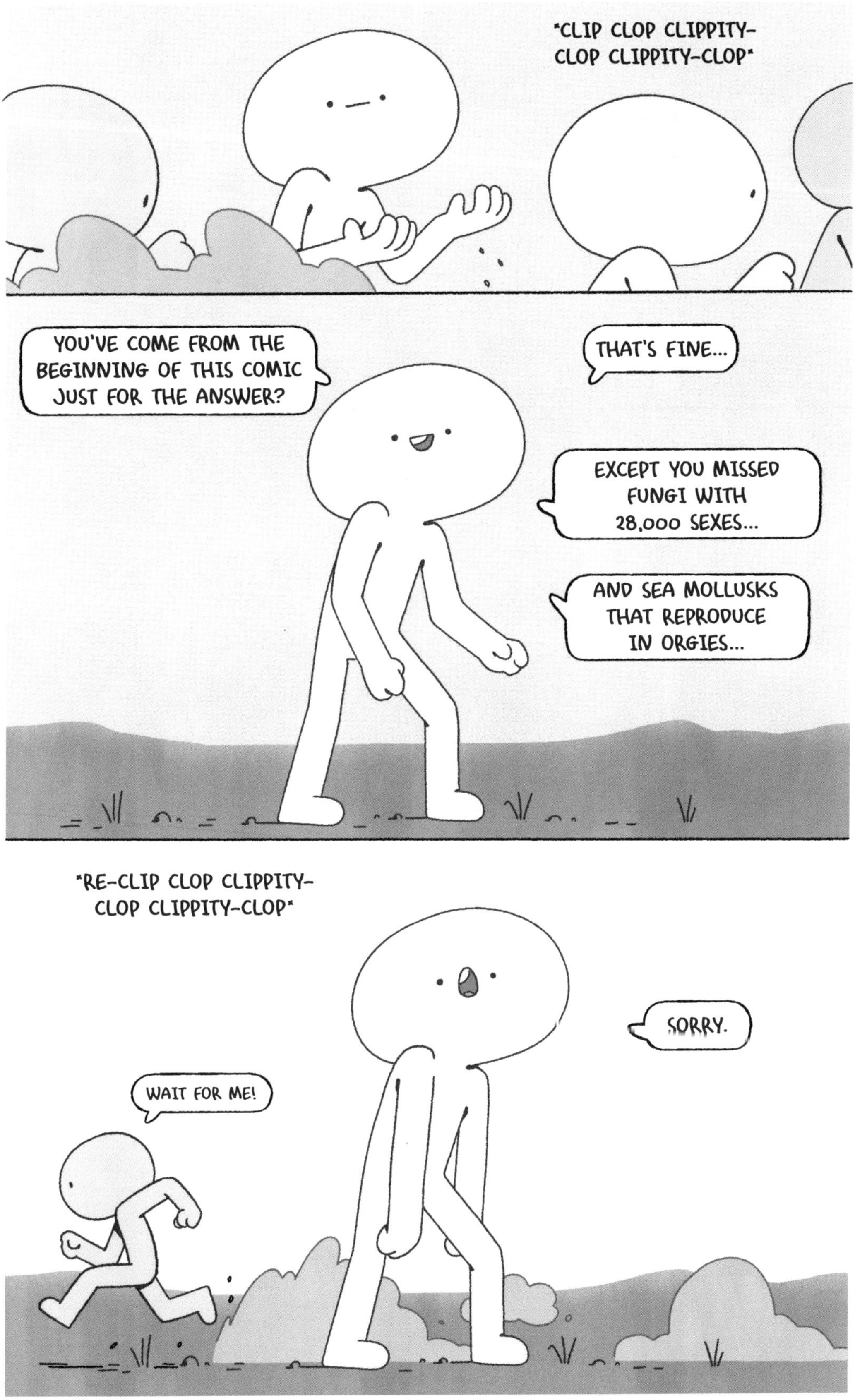
*CLIP CLOP CLIPPITY-
CLOP CLIPPITY-CLOP*
YOU'VE COME FROM THE BEGINNING OF THIS COMIC JUST FOR THE ANSWER?
THAT'S FINE...
EXCEPT YOU MISSED FUNGI WITH 28,000 SEXES...
AND SEA MOLLUSKS THAT REPRODUCE IN ORGIES...
*RE-CLIP CLOP CLIPPITY-
CLOP CLIPPITY-CLOP*
SORRY.
WAIT FOR ME!

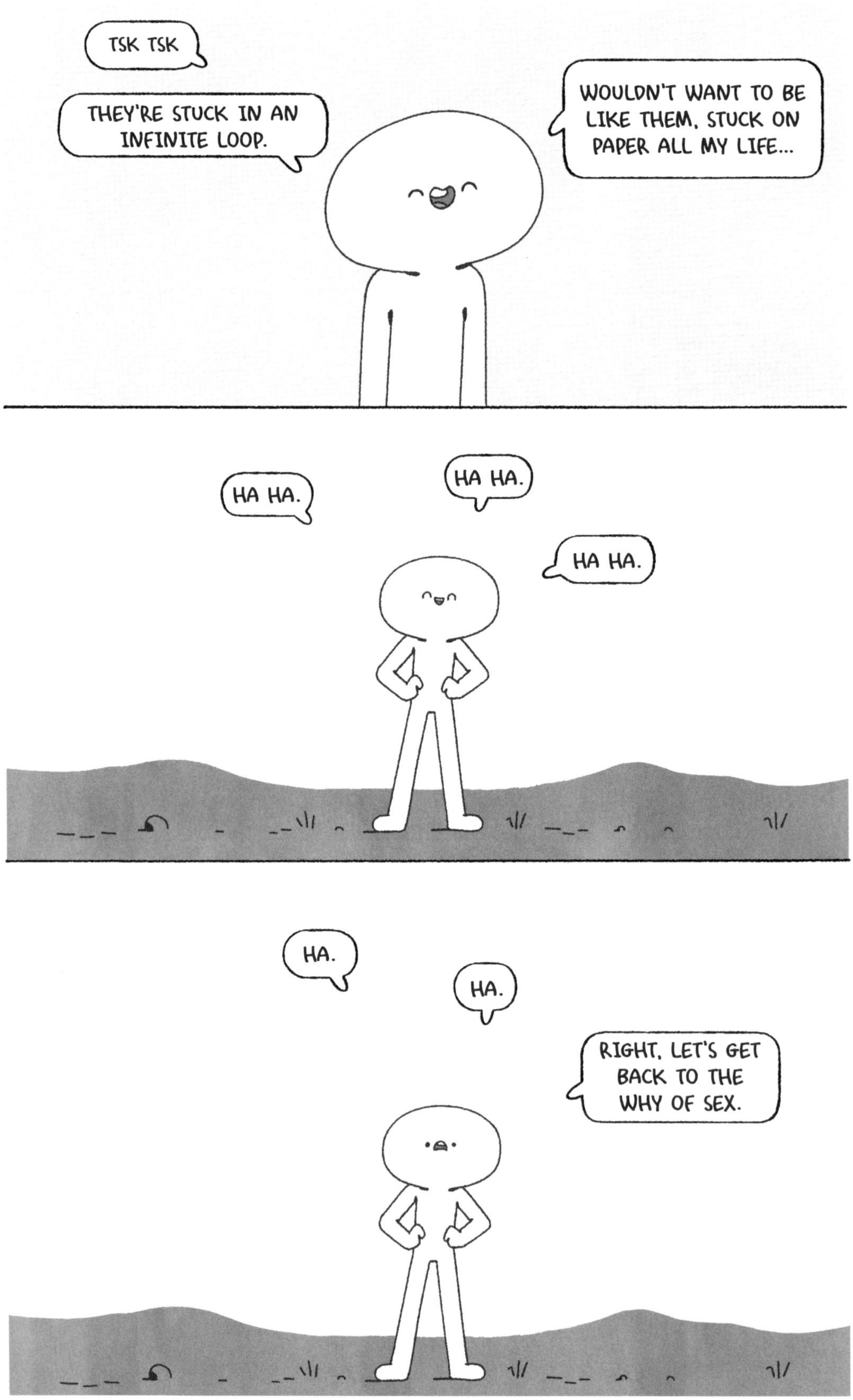
TSK TSK
THEY'RE STUCK IN AN INFINITE LOOP.
WOULDN'T WANT TO BE LIKE THEM, STUCK ON PAPER ALL MY LIFE...
HA HA.
HA HA.
HA HA.
HA.
HA.
RIGHT, LET'S GET BACK TO THE WHY OF SEX.

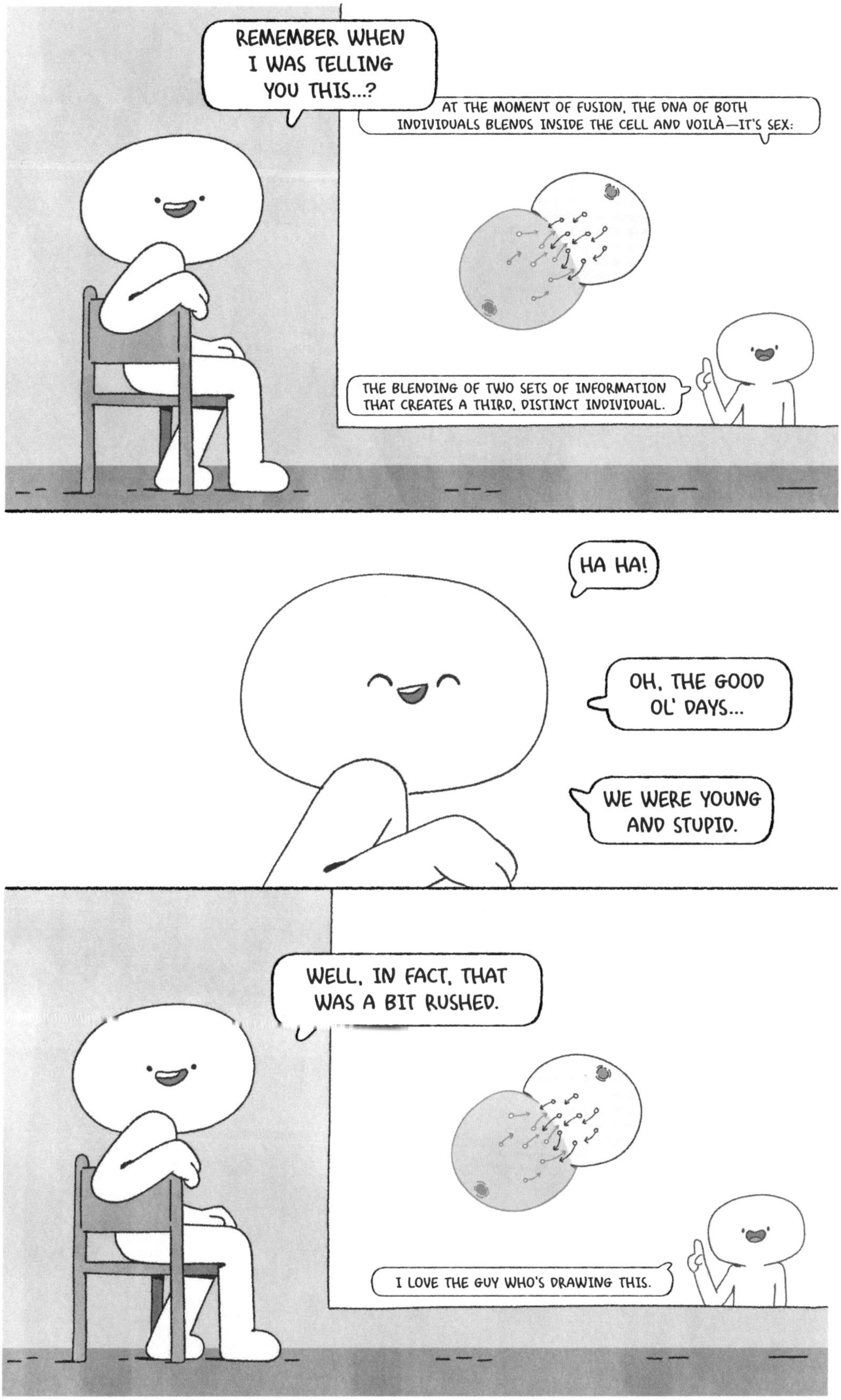
REMEMBER WHEN I WAS TELLING YOU THIS...?
AT THE MOMENT OF FUSION, THE DNA OF BOTH INDIVIDUALS BLENDS INSIDE THE CELL AND VOILÀ—IT'S SEX:
THE BLENDING OF TWO SETS OF INFORMATION THAT CREATES A THIRD, DISTINCT INDIVIDUAL.
HA HA!
OH, THE GOOD OL' DAYS...
WE WERE YOUNG AND STUPID.
WELL, IN FACT, THAT WAS A BIT RUSHED.
I LOVE THE GUY WHO'S DRAWING THIS.

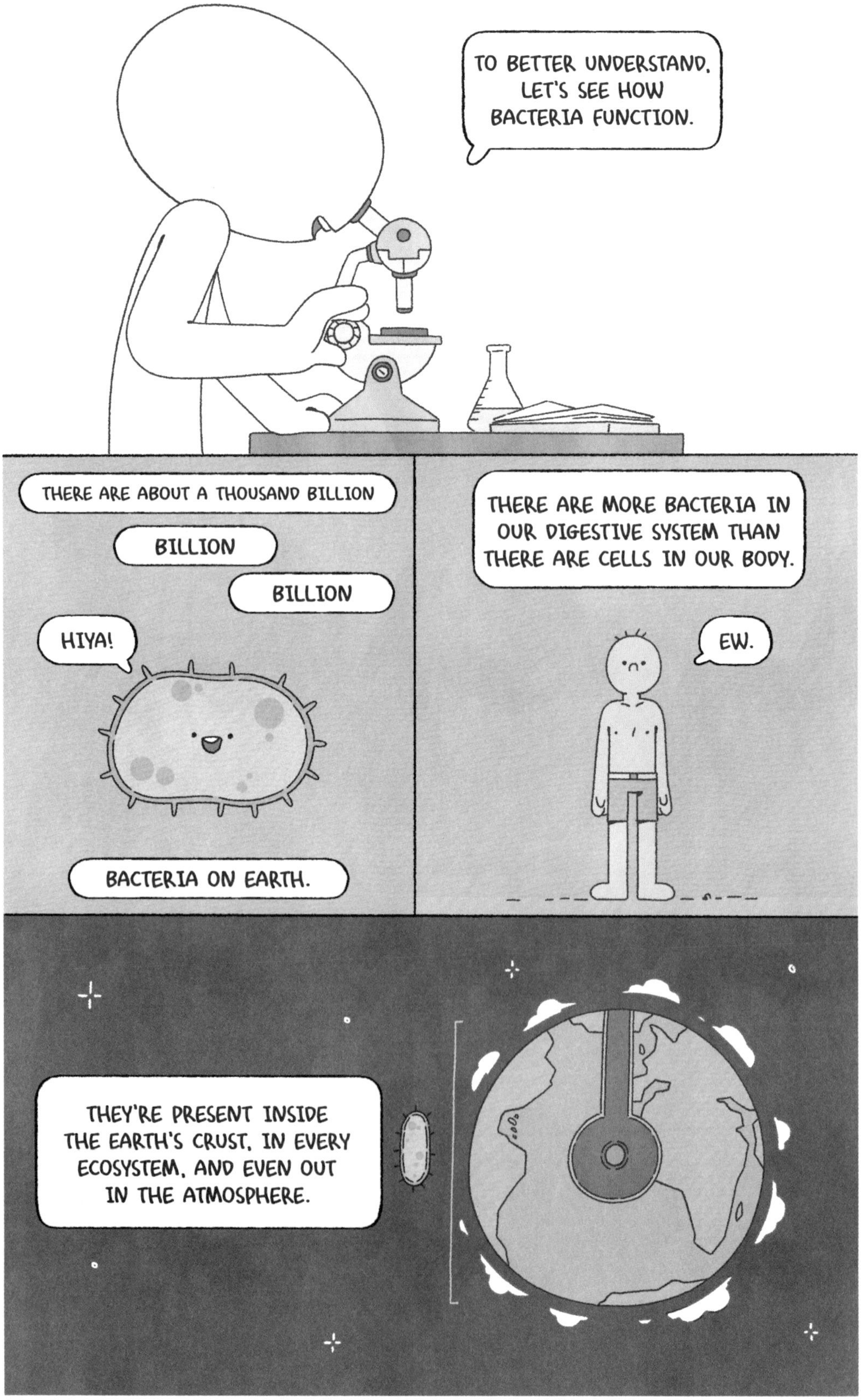
TO BETTER UNDERSTAND, LET'S SEE HOW BACTERIA FUNCTION.
THERE ARE ABOUT A THOUSAND BILLION
BILLION
BILLION
HIYA!
BACTERIA ON EARTH.
THERE ARE MORE BACTERIA IN OUR DIGESTIVE SYSTEM THAN THERE ARE CELLS IN OUR BODY.
EW.
THEY'RE PRESENT INSIDE THE EARTH'S CRUST, IN EVERY ECOSYSTEM, AND EVEN OUT IN THE ATMOSPHERE.

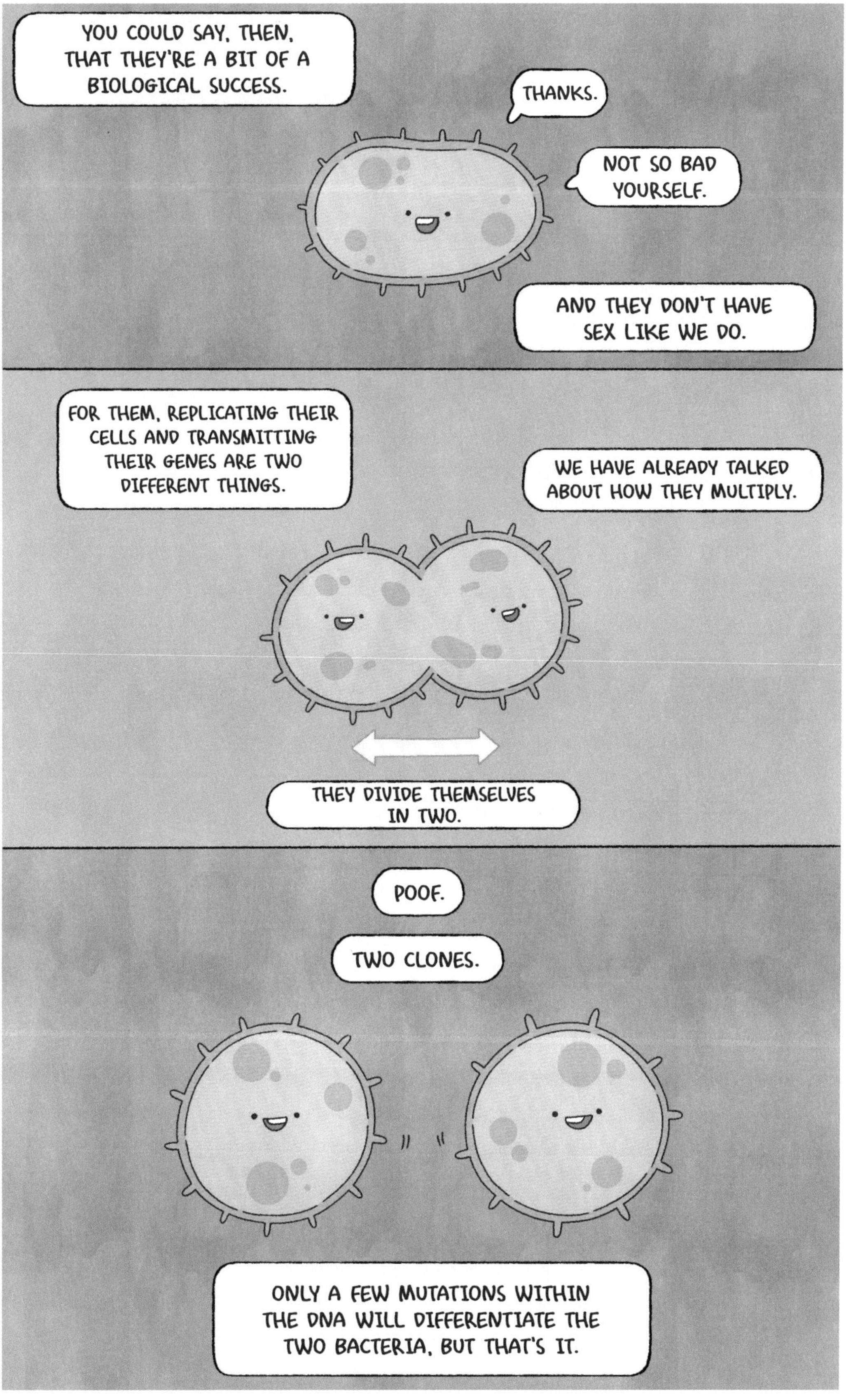
YOU COULD SAY, THEN, THAT THEY'RE A BIT OF A BIOLOGICAL SUCCESS.
THANKS.
NOT SO BAD YOURSELF.
AND THEY DON'T HAVE SEX LIKE WE DO.
FOR THEM, REPLICATING THEIR CELLS AND TRANSMITTING THEIR GENES ARE TWO DIFFERENT THINGS.
WE HAVE ALREADY TALKED ABOUT HOW THEY MULTIPLY.
THEY DIVIDE THEMSELVES IN TWO.
POOF.
TWO CLONES.
ONLY A FEW MUTATIONS WITHIN THE DNA WILL DIFFERENTIATE THE TWO BACTERIA, BUT THAT'S IT.

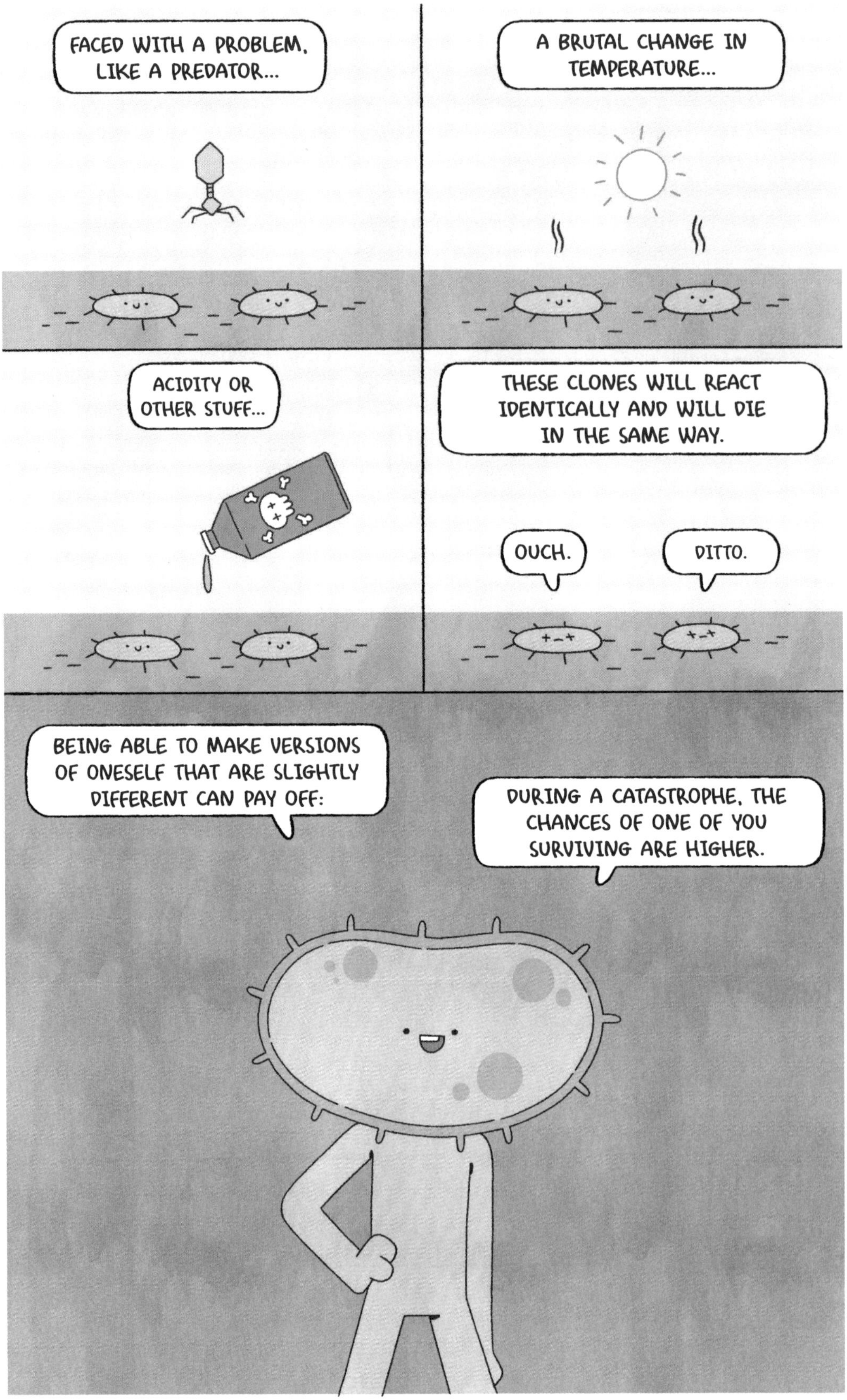
FACED WITH A PROBLEM, LIKE A PREDATOR...
A BRUTAL CHANGE IN TEMPERATURE...
ACIDITY OR OTHER STUFF...
THESE CLONES WILL REACT IDENTICALLY AND WILL DIE IN THE SAME WAY.
OUCH.
DITTO.
BEING ABLE TO MAKE VERSIONS OF ONESELF THAT ARE SLIGHTLY DIFFERENT CAN PAY OFF:
DURING A CATASTROPHE, THE CHANCES OF ONE OF YOU SURVIVING ARE HIGHER.

THERE ARE AT LEAST THREE PROCESSES BY WHICH BACTERIA MODIFY THEIR DNA OVER THE COURSE OF THEIR LIVES:
TRANSFORMATION
TRANSDUCTION
AND CONJUGATION.
TRANSFORMATION IS WHEN A BACTERIUM GATHERS DNA FROM THE ENVIRONMENT AND JOINS IT WITH ITS OWN DNA.
OFTEN, THIS BACTERIAL DNA COMES FROM A DEAD BACTERIUM.
THIS BLENDING OF GENES IS, AT ITS CORE, SEX.
DNA
I'LL NEVER FORGET YOU, BOB...
YOU COULD SAY THEN THAT BACTERIA ARE LITERALLY NECROPHILES.

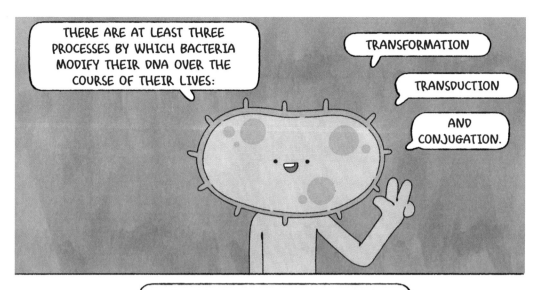

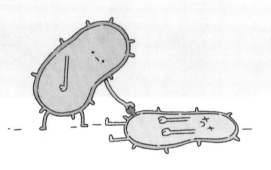

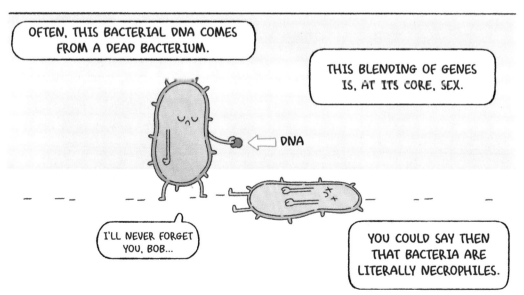

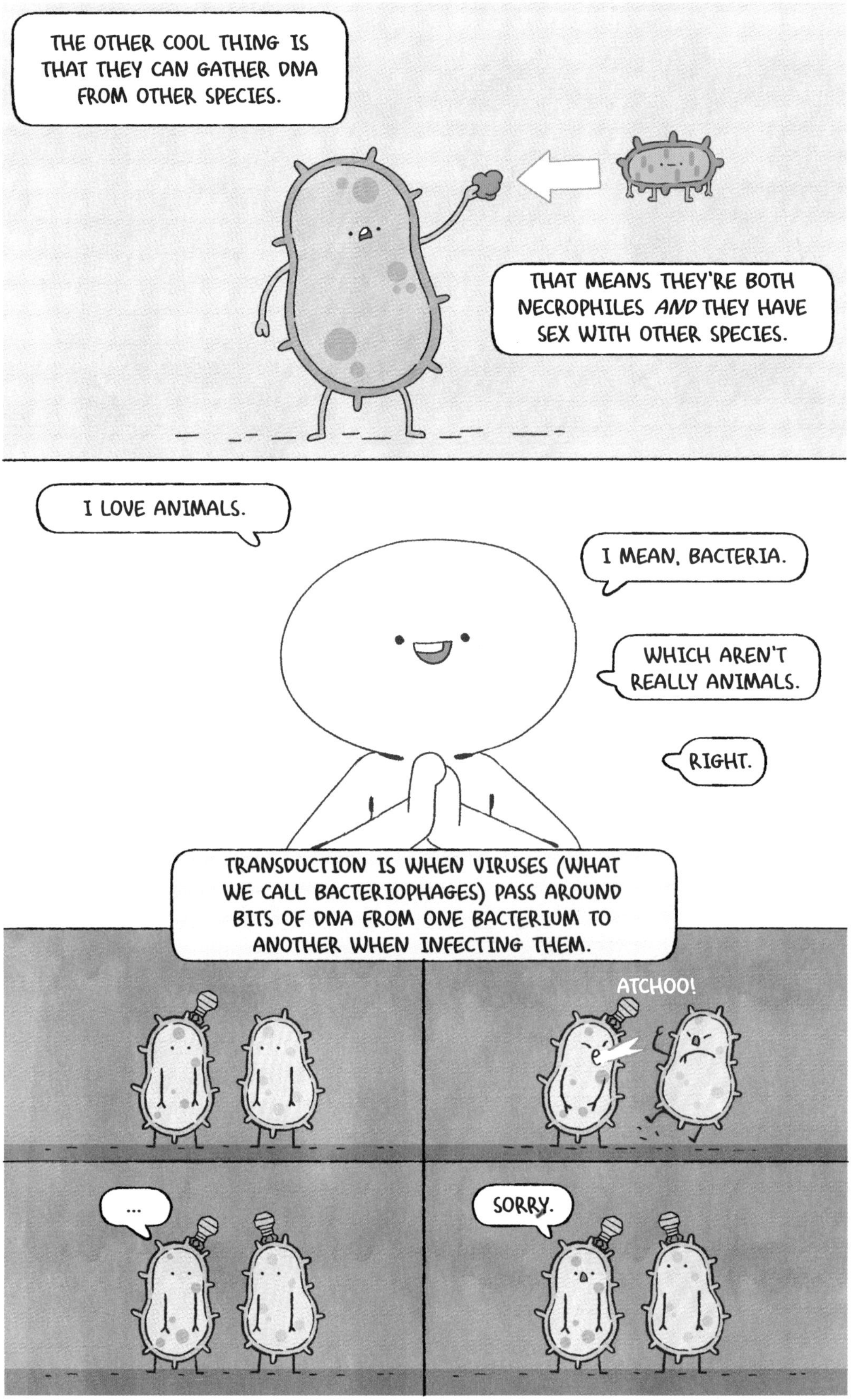
THE OTHER COOL THING IS THAT THEY CAN GATHER DNA FROM OTHER SPECIES.
THAT MEANS THEY'RE BOTH NECROPHILES *AND* THEY HAVE SEX WITH OTHER SPECIES.
I LOVE ANIMALS.
I MEAN, BACTERIA.
WHICH AREN'T REALLY ANIMALS.
RIGHT.
TRANSDUCTION IS WHEN VIRUSES (WHAT WE CALL BACTERIOPHAGES) PASS AROUND BITS OF DNA FROM ONE BACTERIUM TO ANOTHER WHEN INFECTING THEM.
ATCHOO!
...
SORRY.

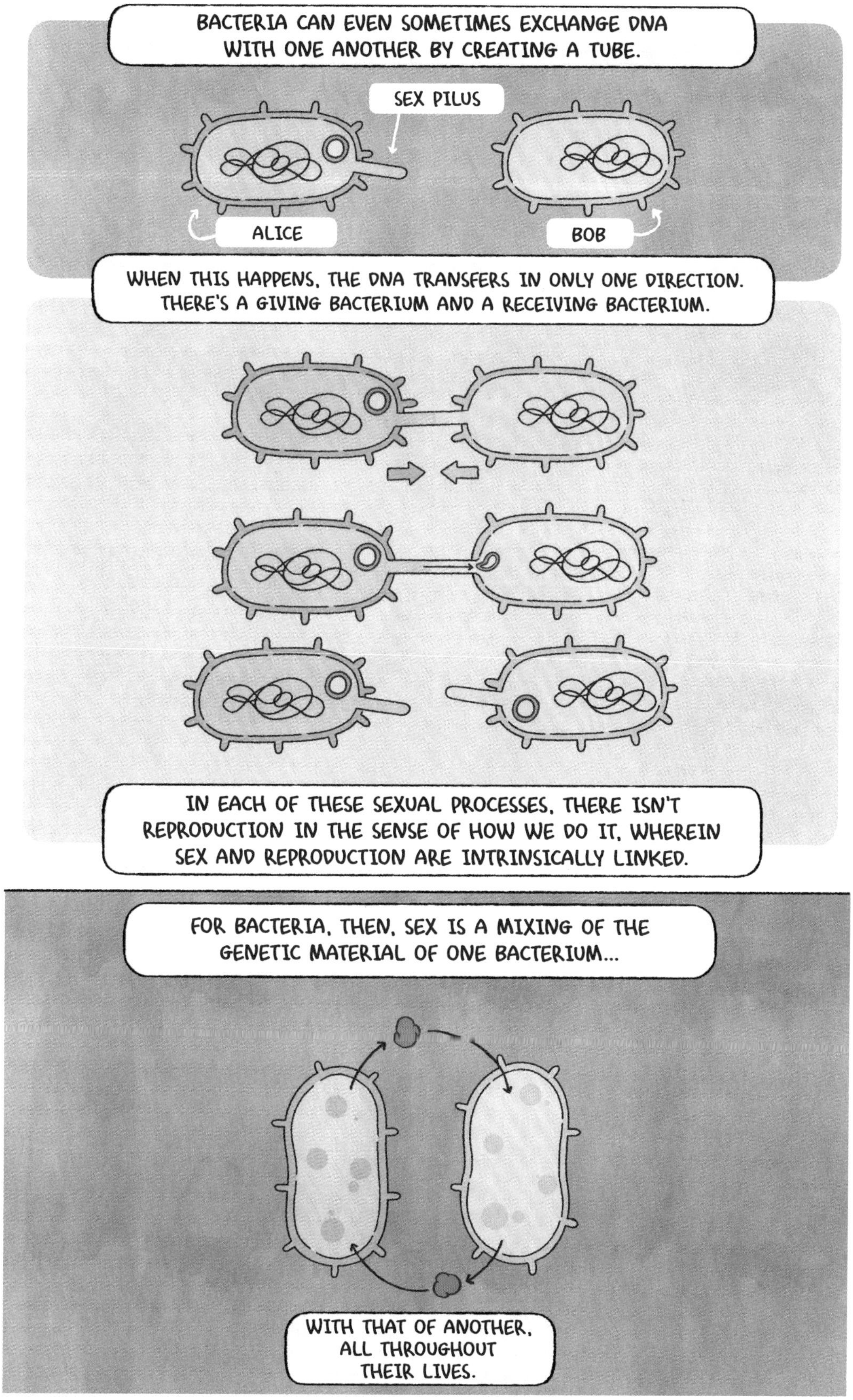

BACTERIA CAN EVEN SOMETIMES EXCHANGE DNA WITH ONE ANOTHER BY CREATING A TUBE.

SEX PILUS

ALICE   BOB

WHEN THIS HAPPENS, THE DNA TRANSFERS IN ONLY ONE DIRECTION. THERE'S A GIVING BACTERIUM AND A RECEIVING BACTERIUM.

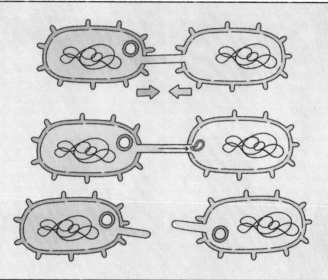

IN EACH OF THESE SEXUAL PROCESSES, THERE ISN'T REPRODUCTION IN THE SENSE OF HOW WE DO IT, WHEREIN SEX AND REPRODUCTION ARE INTRINSICALLY LINKED.

FOR BACTERIA, THEN, SEX IS A MIXING OF THE GENETIC MATERIAL OF ONE BACTERIUM...

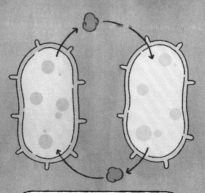

WITH THAT OF ANOTHER, ALL THROUGHOUT THEIR LIVES.

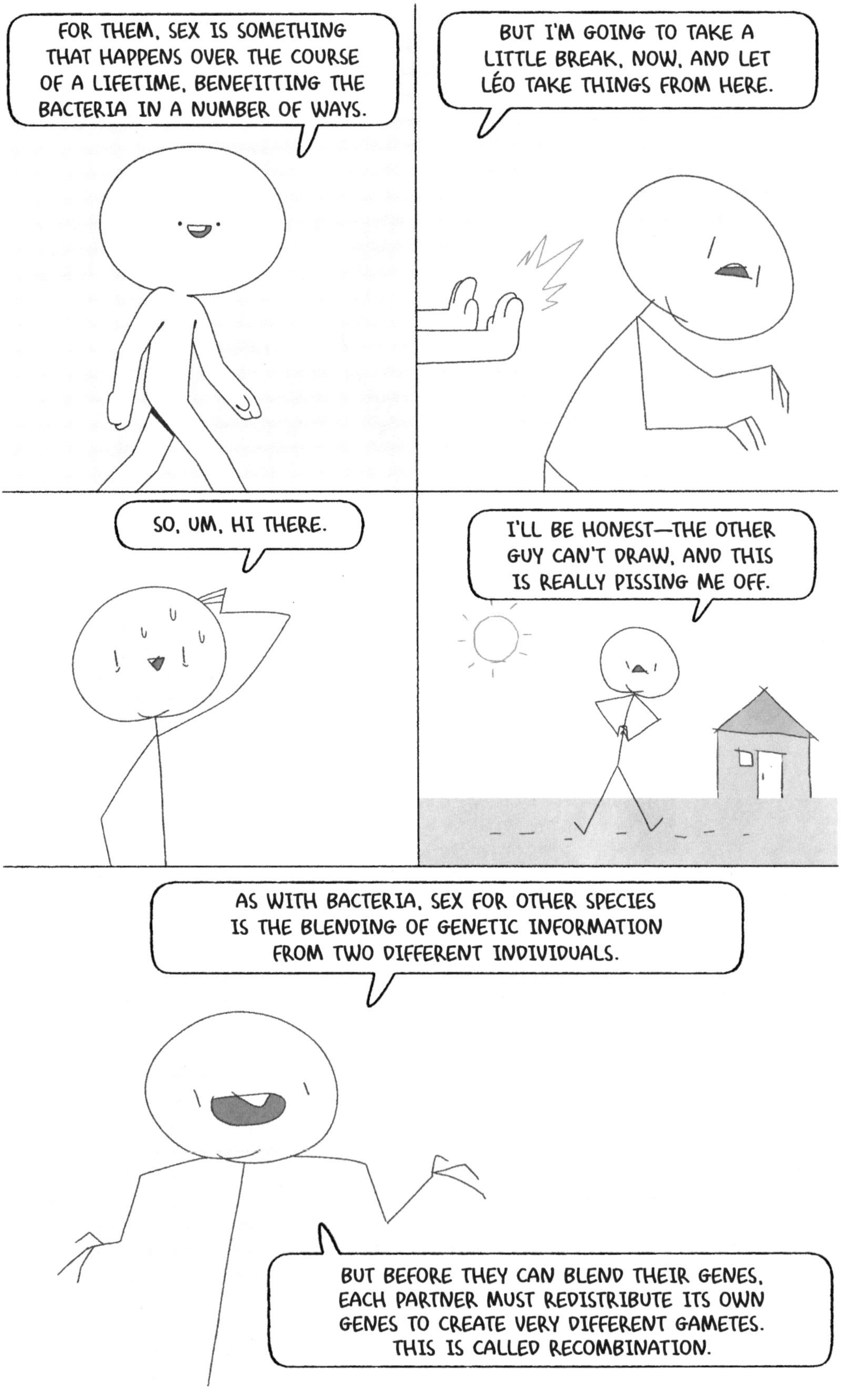
FOR THEM, SEX IS SOMETHING THAT HAPPENS OVER THE COURSE OF A LIFETIME, BENEFITTING THE BACTERIA IN A NUMBER OF WAYS.
BUT I'M GOING TO TAKE A LITTLE BREAK, NOW, AND LET LÉO TAKE THINGS FROM HERE.
SO, UM, HI THERE.
I'LL BE HONEST—THE OTHER GUY CAN'T DRAW, AND THIS IS REALLY PISSING ME OFF.
AS WITH BACTERIA, SEX FOR OTHER SPECIES IS THE BLENDING OF GENETIC INFORMATION FROM TWO DIFFERENT INDIVIDUALS.
BUT BEFORE THEY CAN BLEND THEIR GENES, EACH PARTNER MUST REDISTRIBUTE ITS OWN GENES TO CREATE VERY DIFFERENT GAMETES. THIS IS CALLED RECOMBINATION.

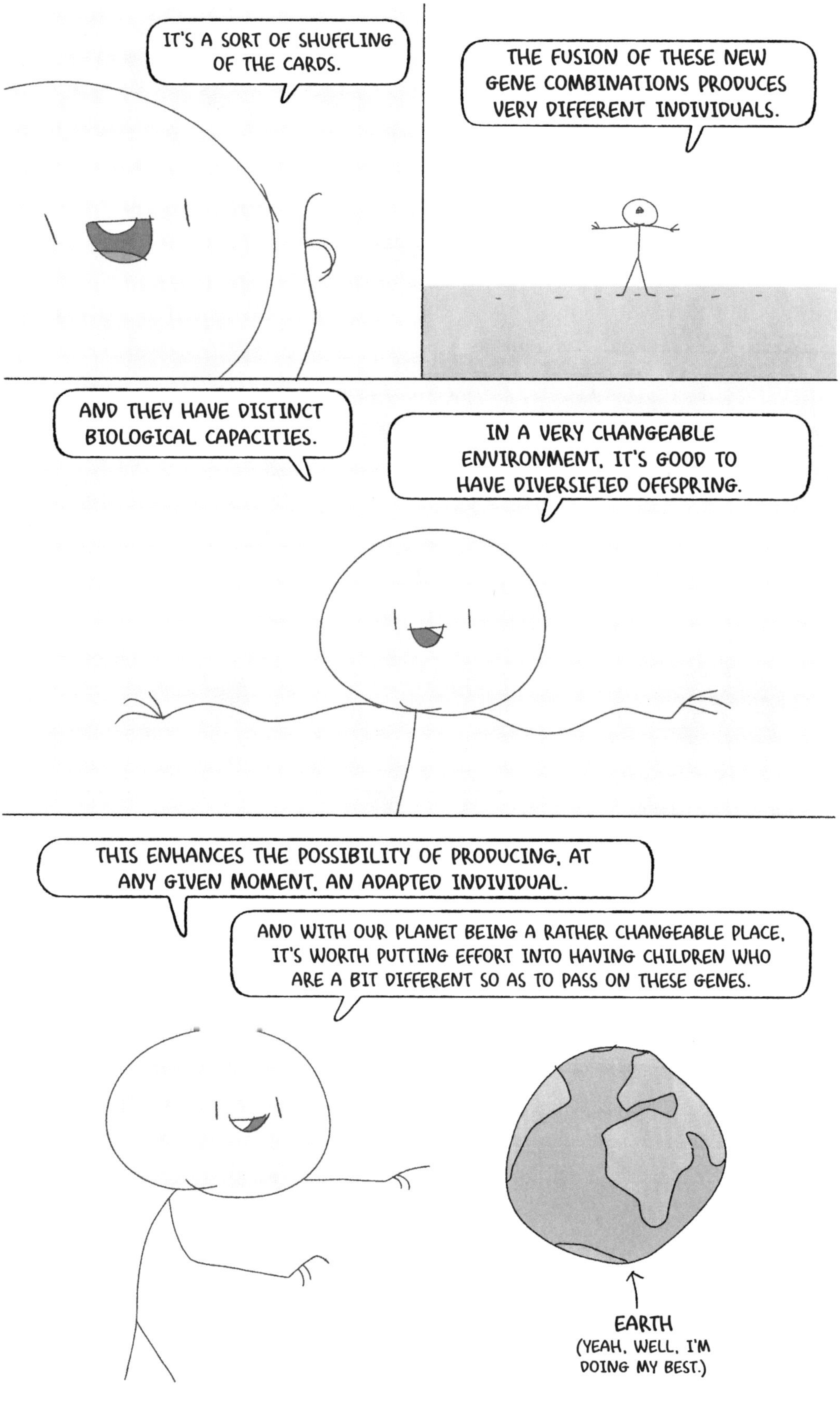
IT'S A SORT OF SHUFFLING OF THE CARDS.
THE FUSION OF THESE NEW GENE COMBINATIONS PRODUCES VERY DIFFERENT INDIVIDUALS.
AND THEY HAVE DISTINCT BIOLOGICAL CAPACITIES.
IN A VERY CHANGEABLE ENVIRONMENT, IT'S GOOD TO HAVE DIVERSIFIED OFFSPRING.
THIS ENHANCES THE POSSIBILITY OF PRODUCING, AT ANY GIVEN MOMENT, AN ADAPTED INDIVIDUAL.
AND WITH OUR PLANET BEING A RATHER CHANGEABLE PLACE, IT'S WORTH PUTTING EFFORT INTO HAVING CHILDREN WHO ARE A BIT DIFFERENT SO AS TO PASS ON THESE GENES.
EARTH
(YEAH, WELL, I'M DOING MY BEST.)

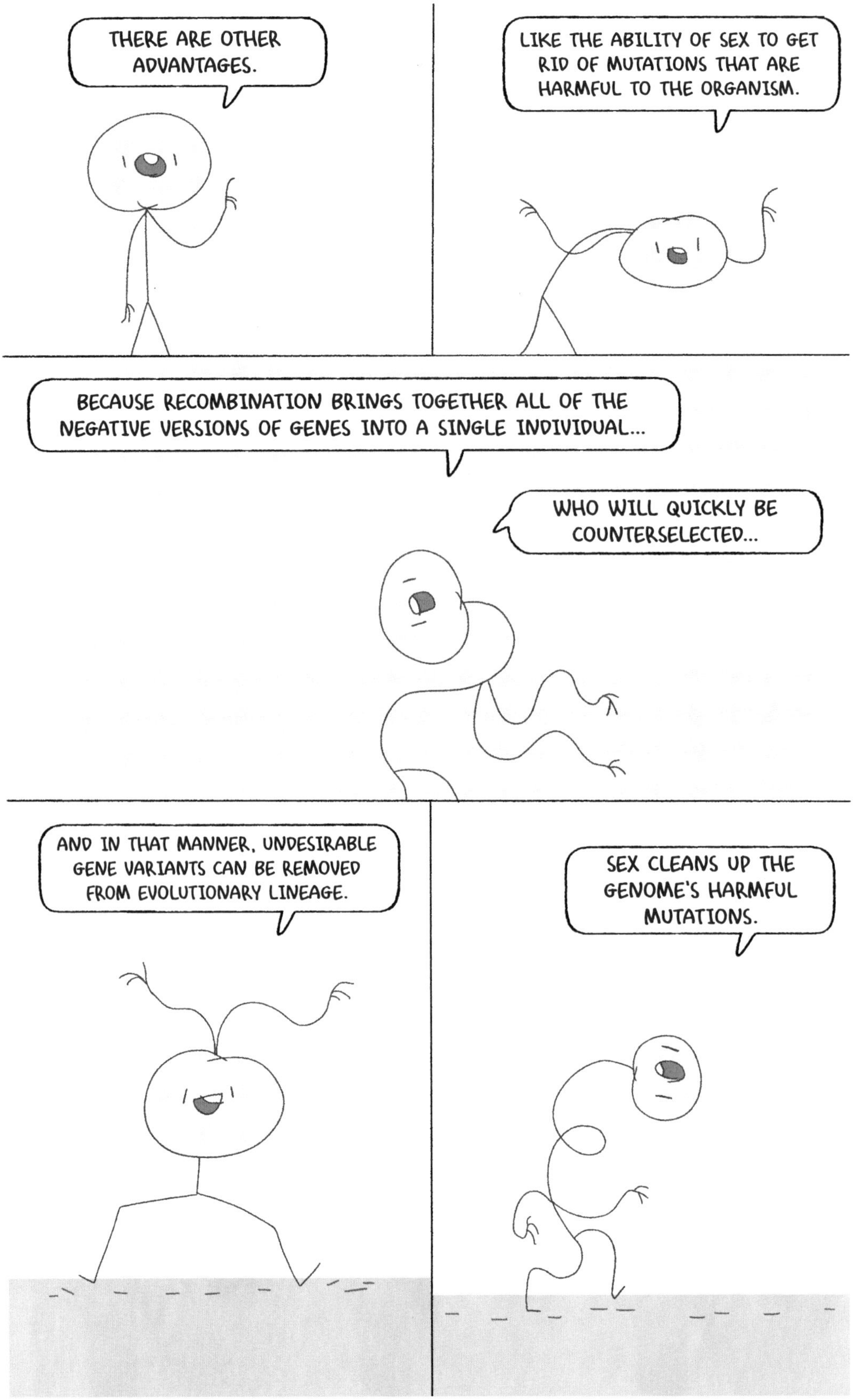
THERE ARE OTHER ADVANTAGES.
LIKE THE ABILITY OF SEX TO GET RID OF MUTATIONS THAT ARE HARMFUL TO THE ORGANISM.
BECAUSE RECOMBINATION BRINGS TOGETHER ALL OF THE NEGATIVE VERSIONS OF GENES INTO A SINGLE INDIVIDUAL...
WHO WILL QUICKLY BE COUNTERSELECTED...
AND IN THAT MANNER, UNDESIRABLE GENE VARIANTS CAN BE REMOVED FROM EVOLUTIONARY LINEAGE.
SEX CLEANS UP THE GENOME'S HARMFUL MUTATIONS.

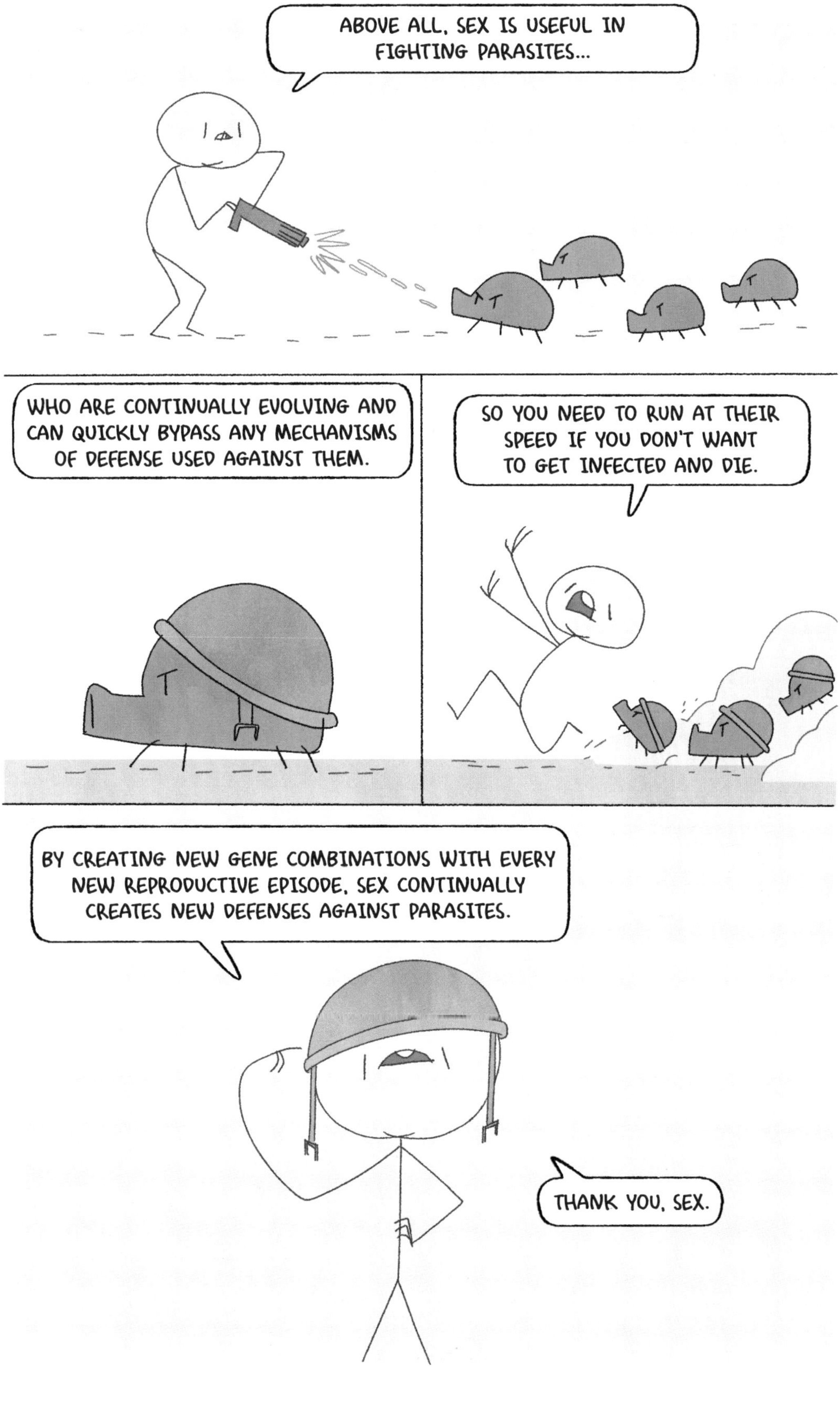
ABOVE ALL, SEX IS USEFUL IN FIGHTING PARASITES...
WHO ARE CONTINUALLY EVOLVING AND CAN QUICKLY BYPASS ANY MECHANISMS OF DEFENSE USED AGAINST THEM.
SO YOU NEED TO RUN AT THEIR SPEED IF YOU DON'T WANT TO GET INFECTED AND DIE.
BY CREATING NEW GENE COMBINATIONS WITH EVERY NEW REPRODUCTIVE EPISODE, SEX CONTINUALLY CREATES NEW DEFENSES AGAINST PARASITES.
THANK YOU, SEX.

WE LOOK TO ALL OF THESE HYPOTHESES TO EXPLAIN THE PARADOX OF SEX. BUT IN TRUTH, IT CONTINUES TO BE RATHER MYSTERIOUS.

TO BETTER UNDERSTAND, WE CAN STUDY THE SPECIES THAT MIX SEXUAL REPRODUCTION WITH CLONING.

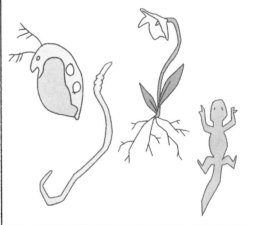

AMONG THESE, WE HAVE WATER FLEAS, NEMATODES, AND A NUMBER OF PLANTS AND LIZARDS.

BUT THE MOST INTERESTING CASE IS THAT OF THE BDELLOIDS, WHICH HAVE COMPLETELY LOST THE ABILITY TO HAVE SEX.

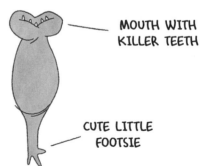

MOUTH WITH KILLER TEETH

CUTE LITTLE FOOTSIE

THESE ANIMALS MEASURE 500 MICRONS AND ARE SHAPED LIKE A SACK.

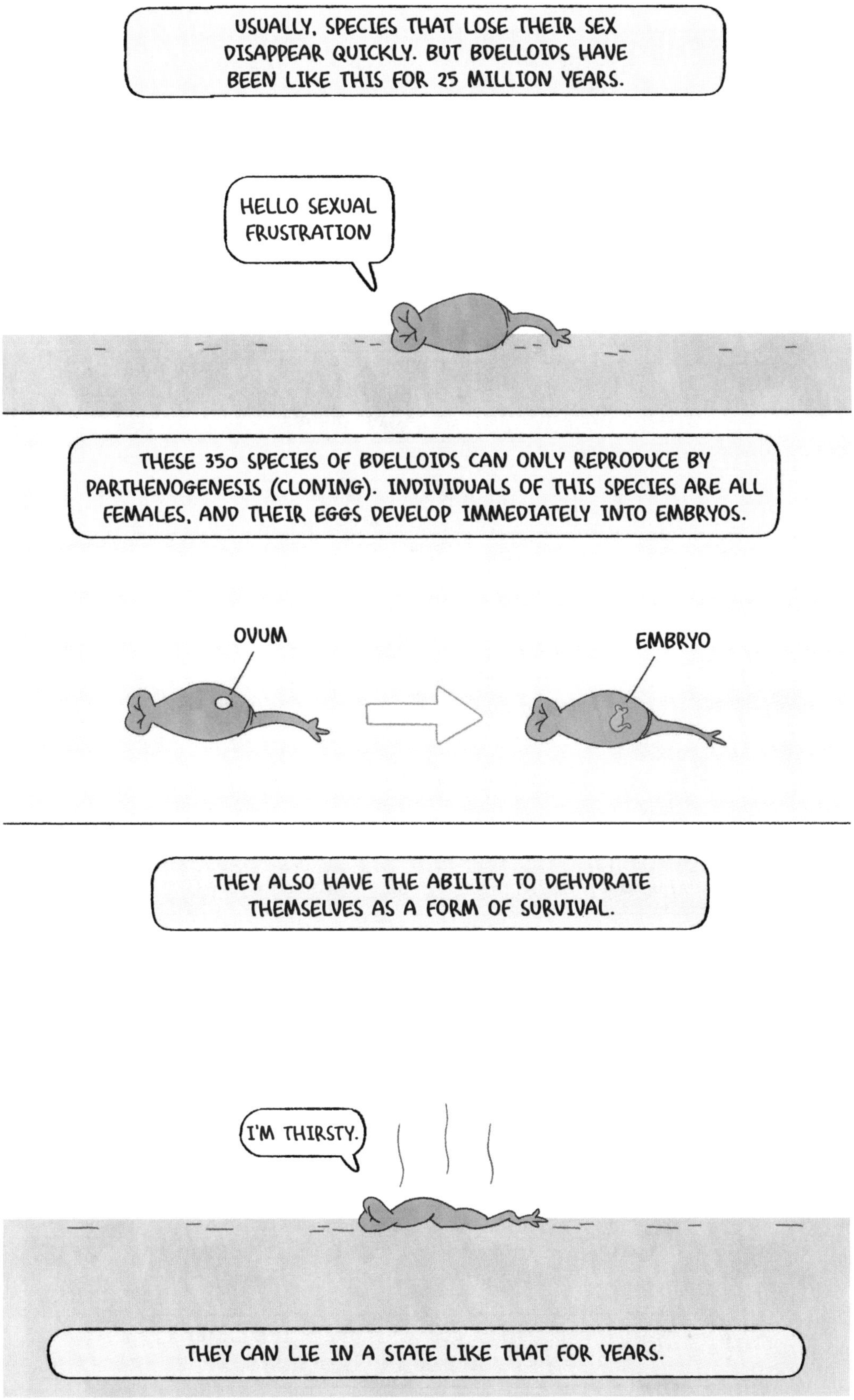
USUALLY, SPECIES THAT LOSE THEIR SEX DISAPPEAR QUICKLY. BUT BDELLOIDS HAVE BEEN LIKE THIS FOR 25 MILLION YEARS.
HELLO SEXUAL FRUSTRATION
THESE 350 SPECIES OF BDELLOIDS CAN ONLY REPRODUCE BY PARTHENOGENESIS (CLONING). INDIVIDUALS OF THIS SPECIES ARE ALL FEMALES, AND THEIR EGGS DEVELOP IMMEDIATELY INTO EMBRYOS.
OVUM
EMBRYO
THEY ALSO HAVE THE ABILITY TO DEHYDRATE THEMSELVES AS A FORM OF SURVIVAL.
I'M THIRSTY.
THEY CAN LIE IN A STATE LIKE THAT FOR YEARS.

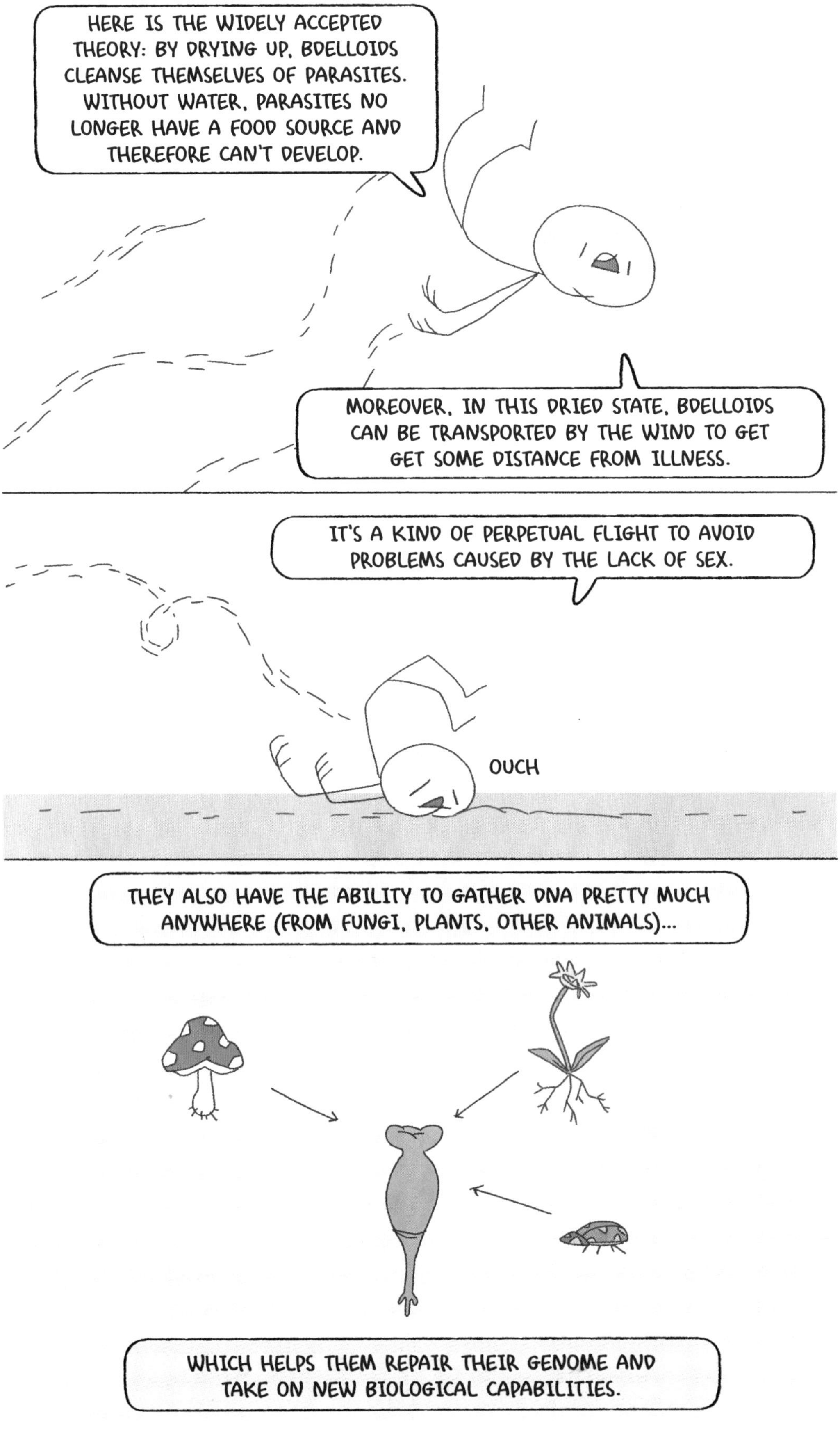
HERE IS THE WIDELY ACCEPTED THEORY: BY DRYING UP, BDELLOIDS CLEANSE THEMSELVES OF PARASITES. WITHOUT WATER, PARASITES NO LONGER HAVE A FOOD SOURCE AND THEREFORE CAN'T DEVELOP.
MOREOVER, IN THIS DRIED STATE, BDELLOIDS CAN BE TRANSPORTED BY THE WIND TO GET GET SOME DISTANCE FROM ILLNESS.
IT'S A KIND OF PERPETUAL FLIGHT TO AVOID PROBLEMS CAUSED BY THE LACK OF SEX.
OUCH
THEY ALSO HAVE THE ABILITY TO GATHER DNA PRETTY MUCH ANYWHERE (FROM FUNGI, PLANTS, OTHER ANIMALS)...
WHICH HELPS THEM REPAIR THEIR GENOME AND TAKE ON NEW BIOLOGICAL CAPABILITIES.

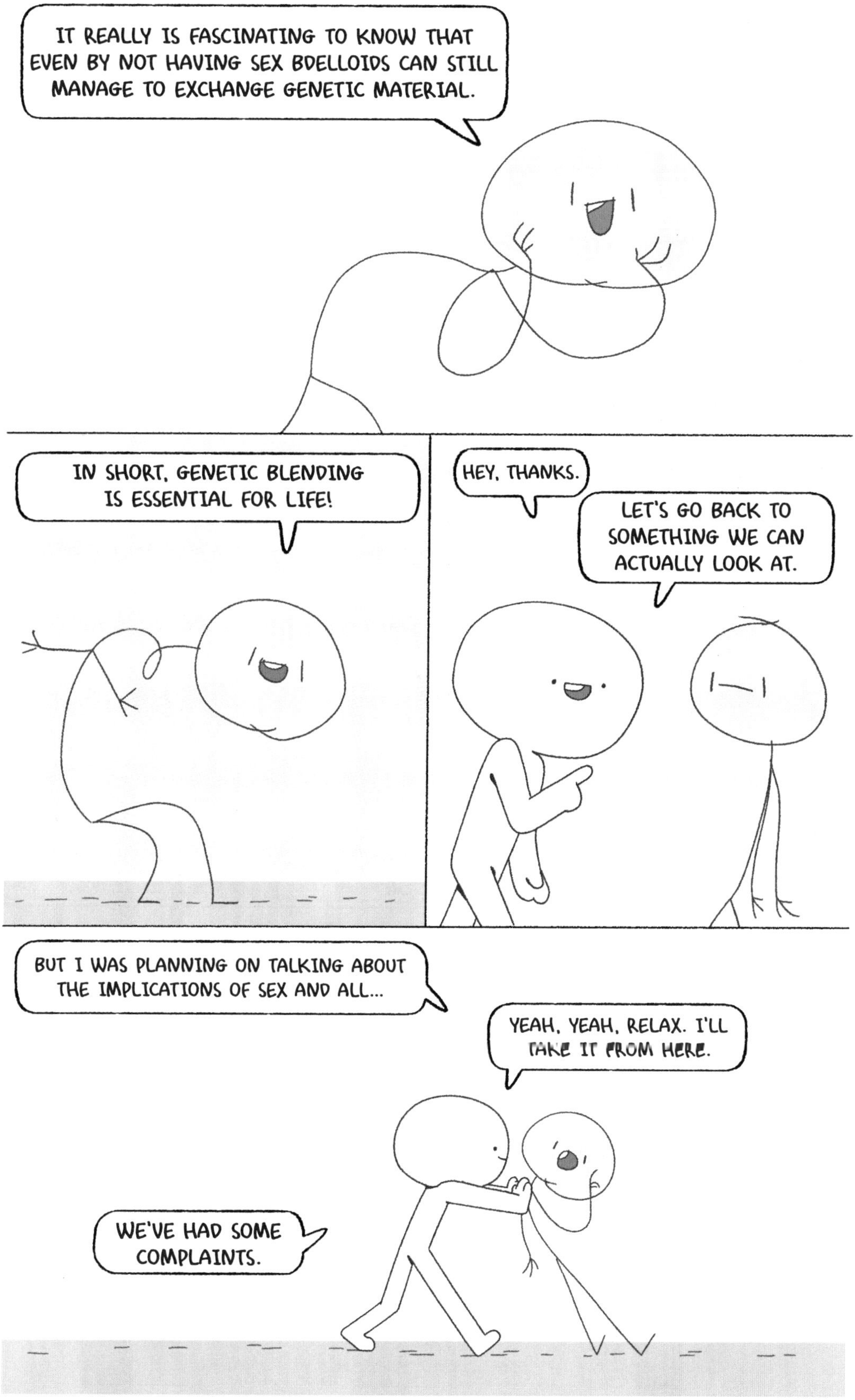
IT REALLY IS FASCINATING TO KNOW THAT EVEN BY NOT HAVING SEX BDELLOIDS CAN STILL MANAGE TO EXCHANGE GENETIC MATERIAL.
IN SHORT, GENETIC BLENDING IS ESSENTIAL FOR LIFE!
HEY, THANKS.
LET'S GO BACK TO SOMETHING WE CAN ACTUALLY LOOK AT.
BUT I WAS PLANNING ON TALKING ABOUT THE IMPLICATIONS OF SEX AND ALL...
YEAH, YEAH, RELAX. I'LL TAKE IT FROM HERE.
WE'VE HAD SOME COMPLAINTS.

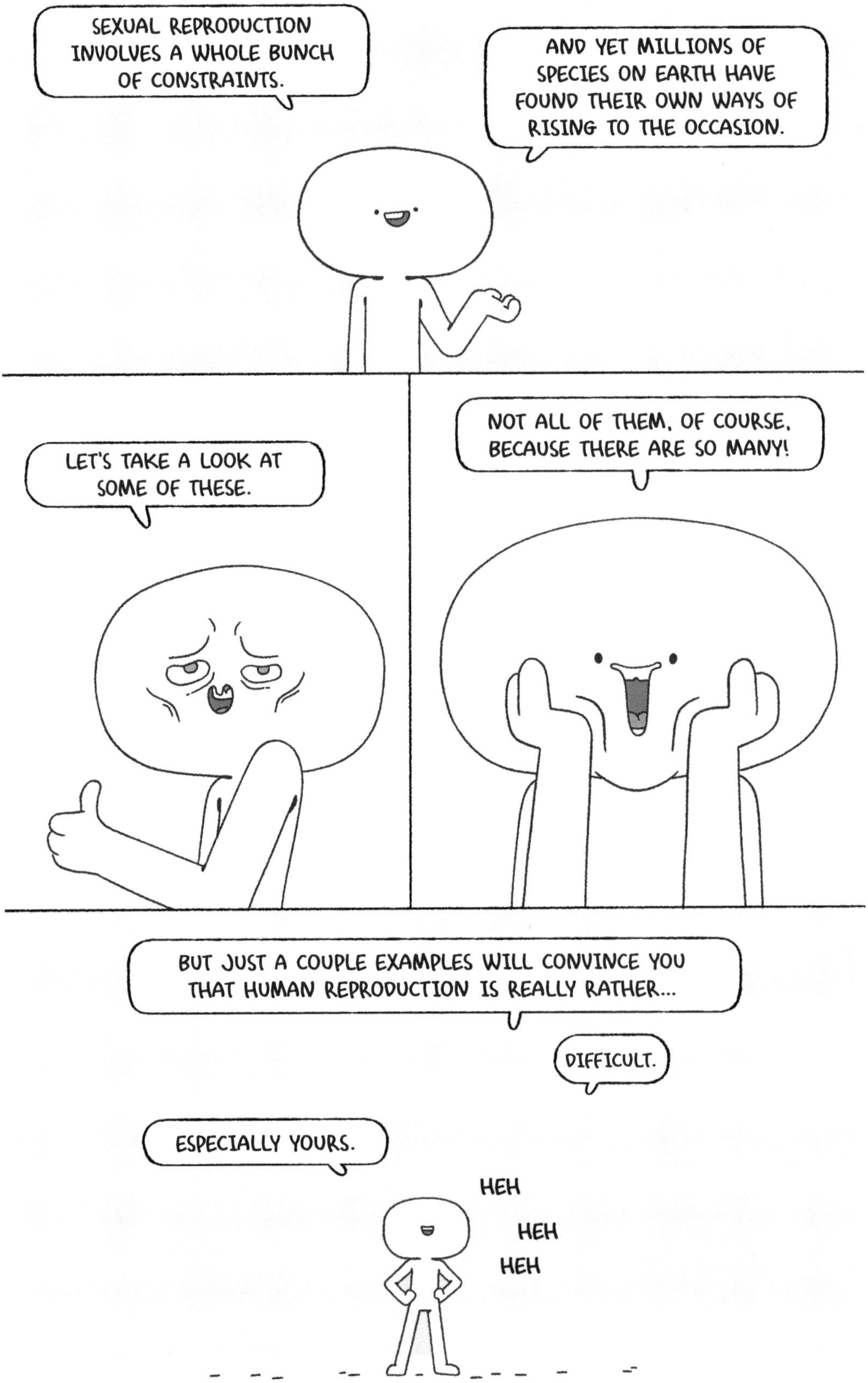

SEXUAL REPRODUCTION INVOLVES A WHOLE BUNCH OF CONSTRAINTS.
AND YET MILLIONS OF SPECIES ON EARTH HAVE FOUND THEIR OWN WAYS OF RISING TO THE OCCASION.
LET'S TAKE A LOOK AT SOME OF THESE.
NOT ALL OF THEM, OF COURSE, BECAUSE THERE ARE SO MANY!
BUT JUST A COUPLE EXAMPLES WILL CONVINCE YOU THAT HUMAN REPRODUCTION IS REALLY RATHER...
DIFFICULT.
ESPECIALLY YOURS.
HEH
HEH
HEH

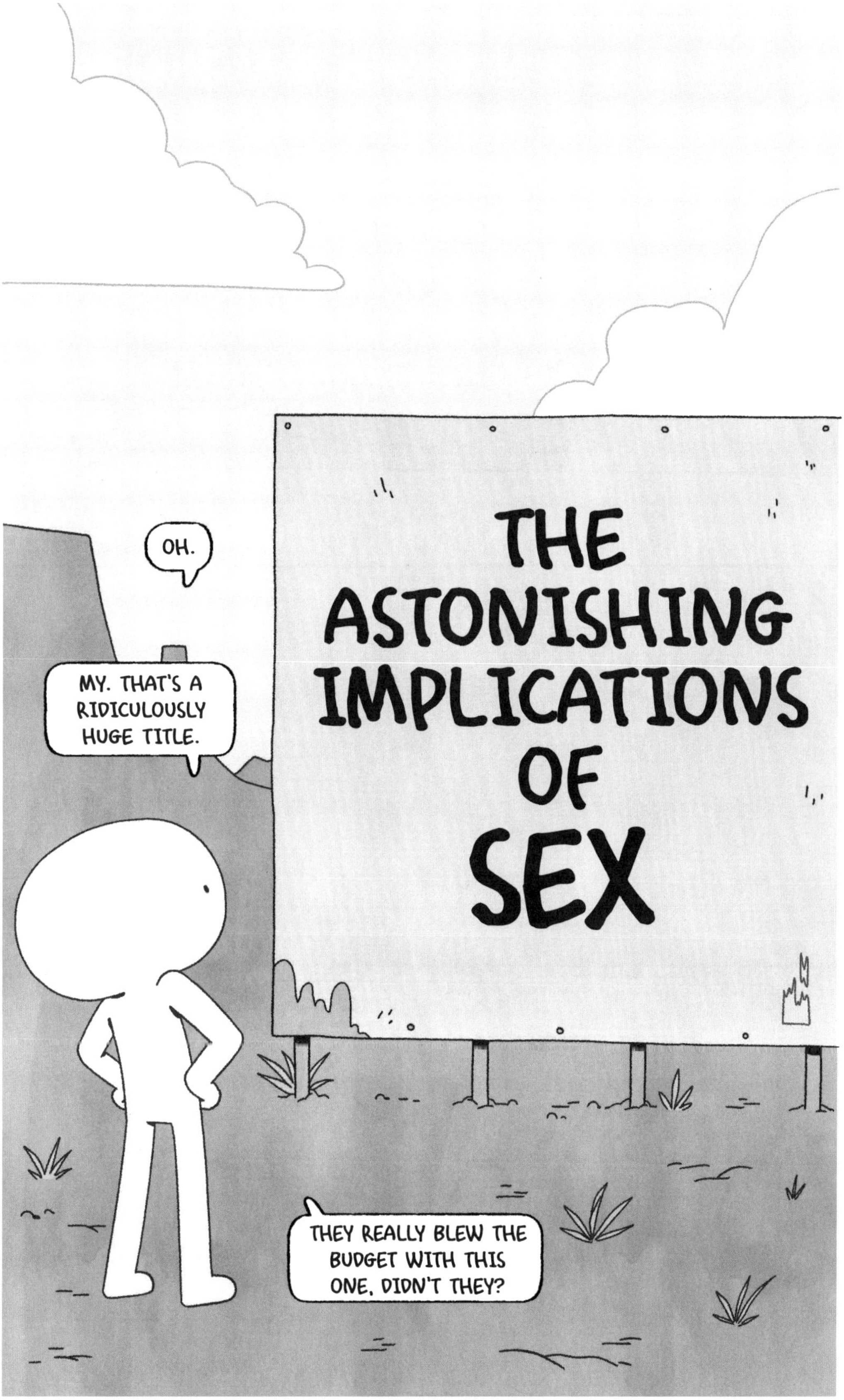
OH.
MY. THAT'S A RIDICULOUSLY HUGE TITLE.
THE ASTONISHING IMPLICATIONS OF SEX
THEY REALLY BLEW THE BUDGET WITH THIS ONE, DIDN'T THEY?

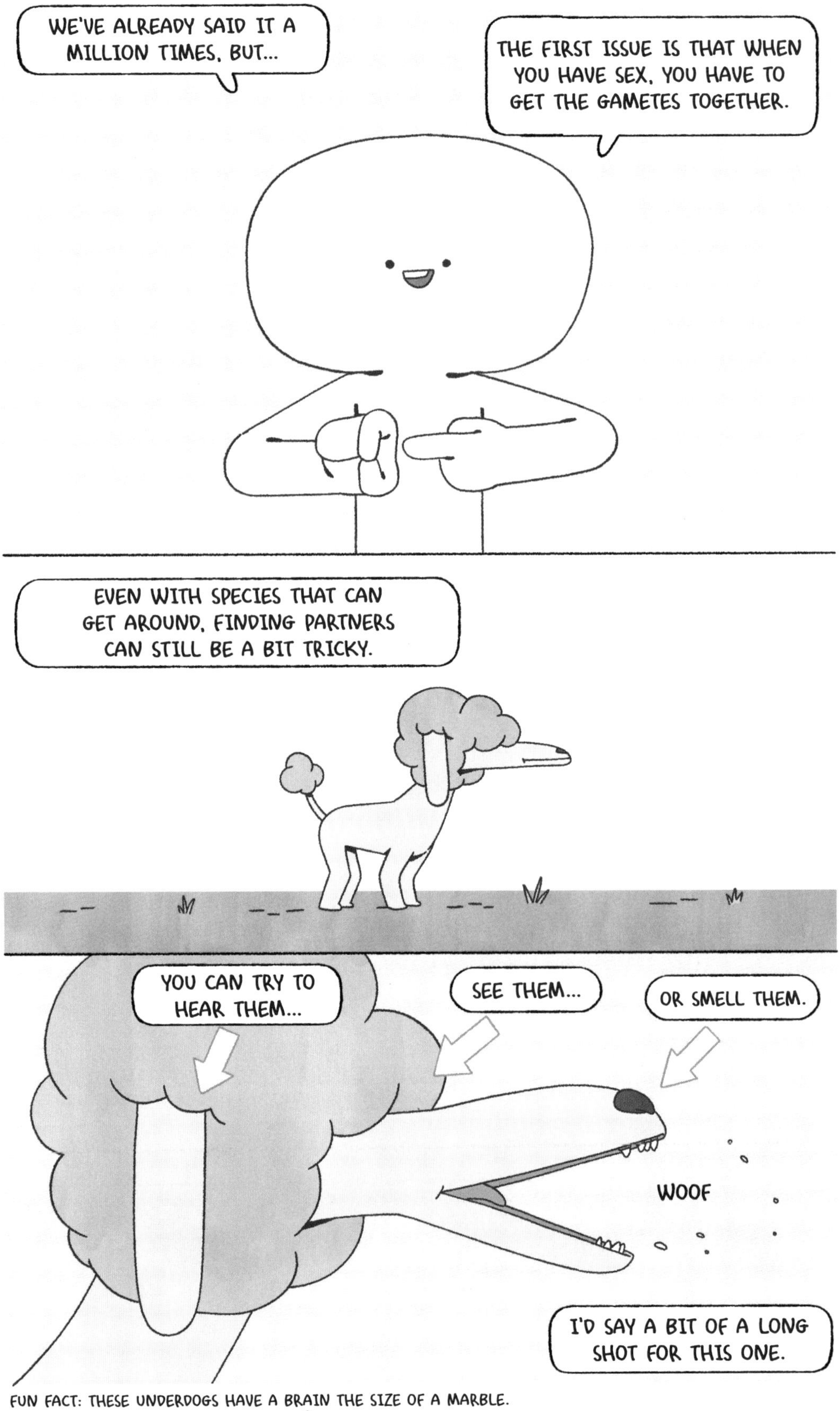
WE'VE ALREADY SAID IT A MILLION TIMES, BUT...
THE FIRST ISSUE IS THAT WHEN YOU HAVE SEX, YOU HAVE TO GET THE GAMETES TOGETHER.
EVEN WITH SPECIES THAT CAN GET AROUND, FINDING PARTNERS CAN STILL BE A BIT TRICKY.
YOU CAN TRY TO HEAR THEM...
SEE THEM...
OR SMELL THEM.
WOOF
I'D SAY A BIT OF A LONG SHOT FOR THIS ONE.

FUN FACT: THESE UNDERDOGS HAVE A BRAIN THE SIZE OF A MARBLE.
FUN FACT 2: THAT STATEMENT IS FALSE. ↑

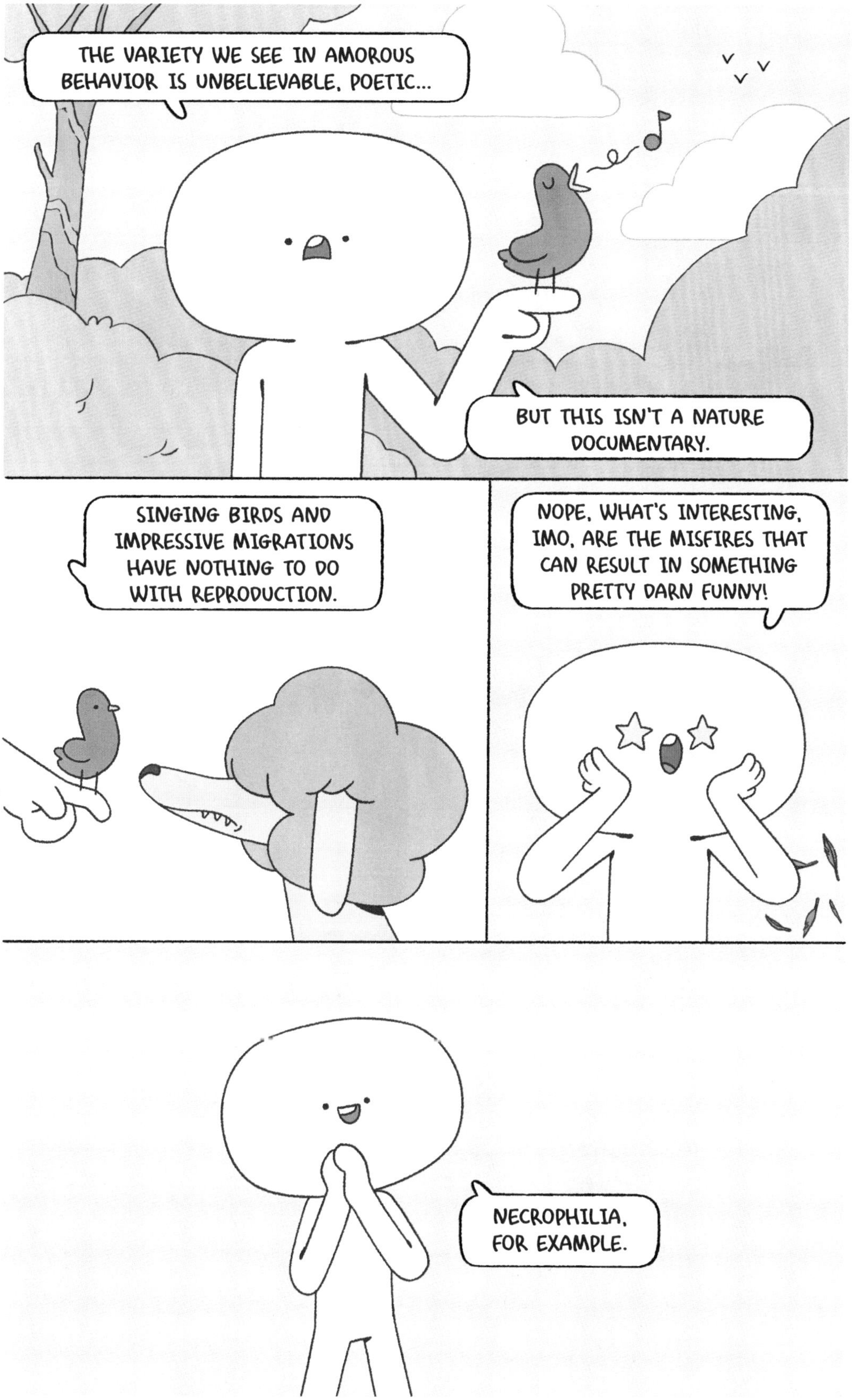
THE VARIETY WE SEE IN AMOROUS BEHAVIOR IS UNBELIEVABLE, POETIC...
BUT THIS ISN'T A NATURE DOCUMENTARY.
SINGING BIRDS AND IMPRESSIVE MIGRATIONS HAVE NOTHING TO DO WITH REPRODUCTION.
NOPE, WHAT'S INTERESTING, IMO, ARE THE MISFIRES THAT CAN RESULT IN SOMETHING PRETTY DARN FUNNY!
NECROPHILIA, FOR EXAMPLE.

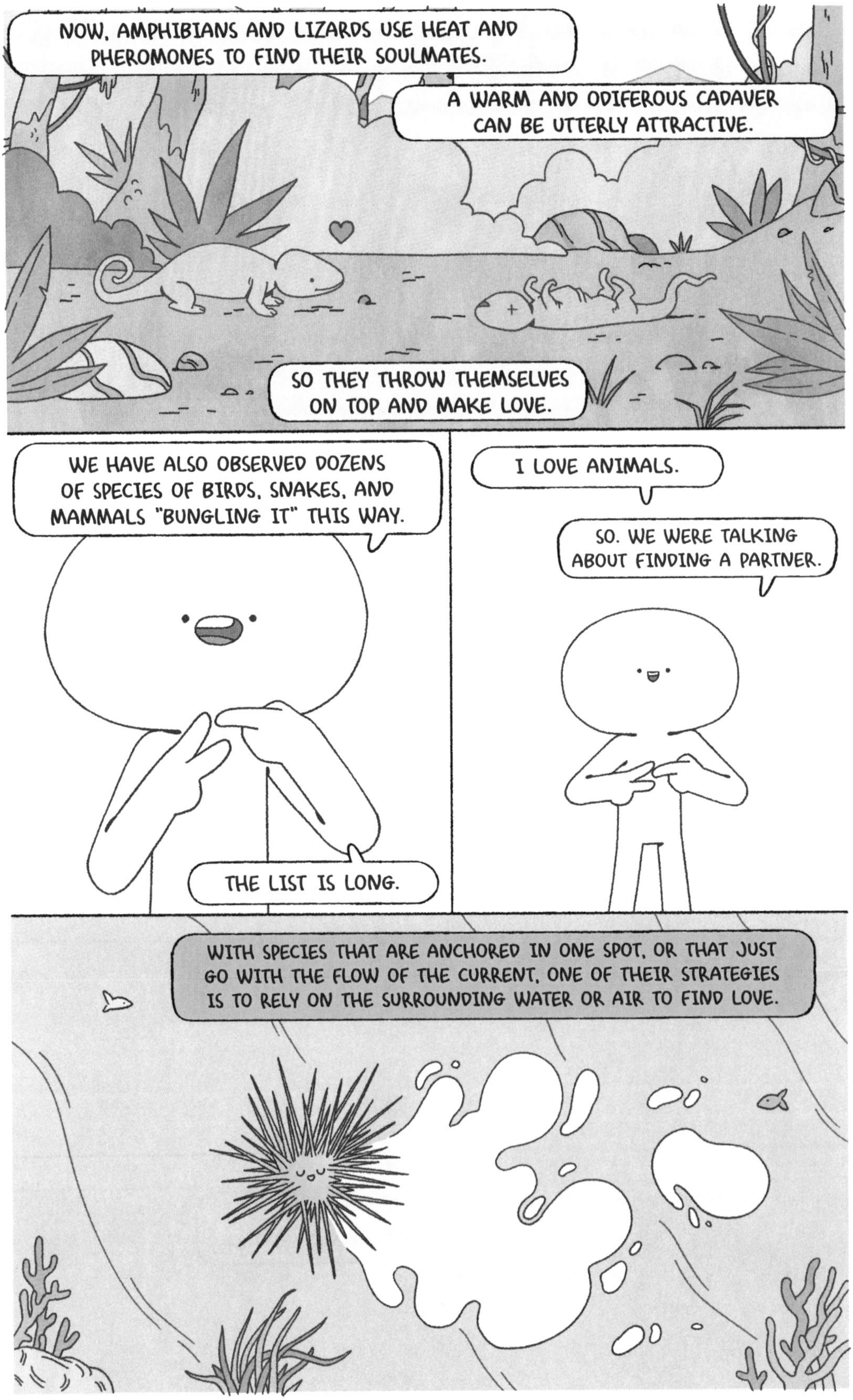
NOW, AMPHIBIANS AND LIZARDS USE HEAT AND PHEROMONES TO FIND THEIR SOULMATES.
A WARM AND ODIFEROUS CADAVER CAN BE UTTERLY ATTRACTIVE.
SO THEY THROW THEMSELVES ON TOP AND MAKE LOVE.
WE HAVE ALSO OBSERVED DOZENS OF SPECIES OF BIRDS, SNAKES, AND MAMMALS "BUNGLING IT" THIS WAY.
THE LIST IS LONG.
I LOVE ANIMALS.
SO. WE WERE TALKING ABOUT FINDING A PARTNER.
WITH SPECIES THAT ARE ANCHORED IN ONE SPOT, OR THAT JUST GO WITH THE FLOW OF THE CURRENT, ONE OF THEIR STRATEGIES IS TO RELY ON THE SURROUNDING WATER OR AIR TO FIND LOVE.

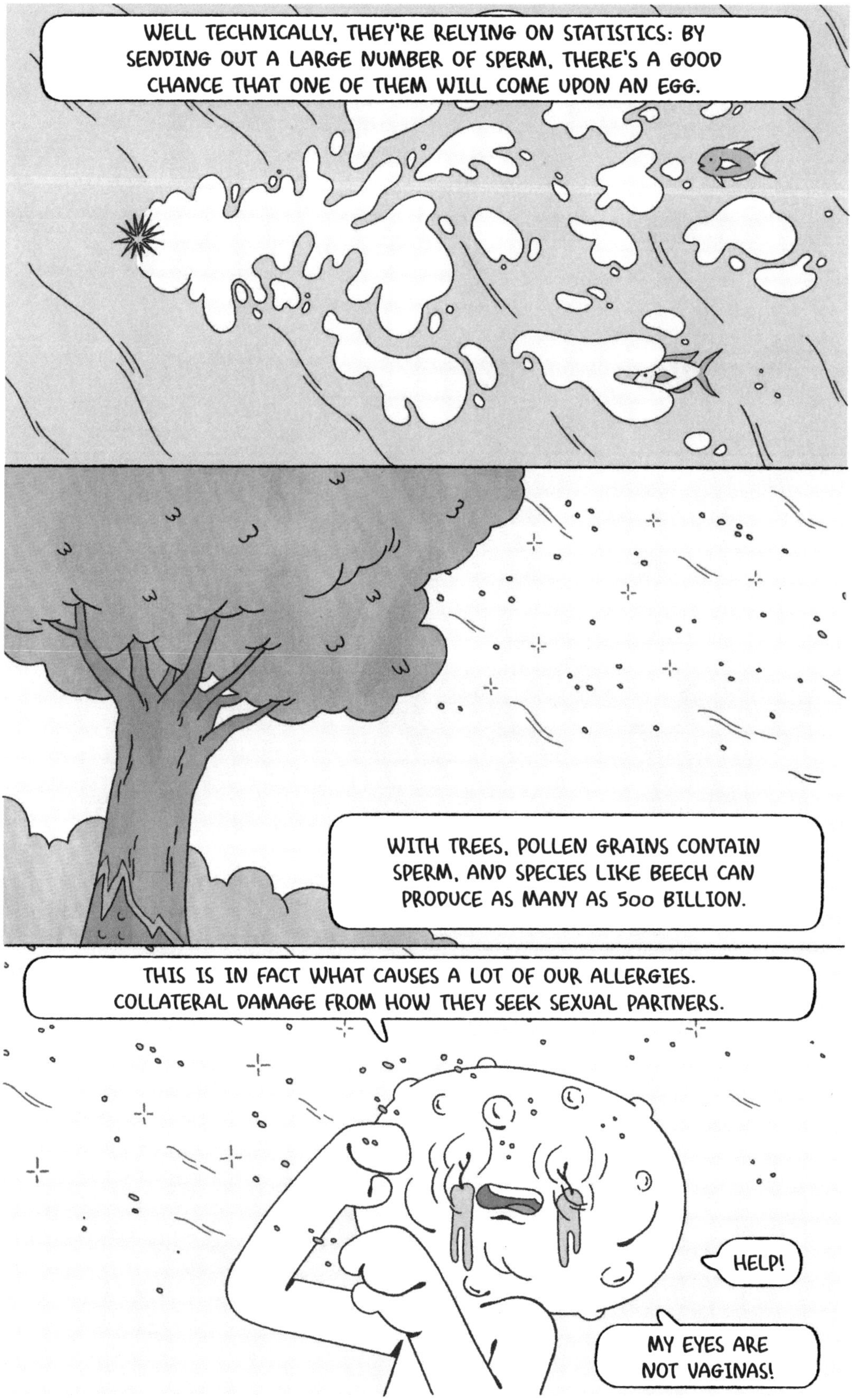
WELL TECHNICALLY, THEY'RE RELYING ON STATISTICS: BY SENDING OUT A LARGE NUMBER OF SPERM, THERE'S A GOOD CHANCE THAT ONE OF THEM WILL COME UPON AN EGG.
WITH TREES, POLLEN GRAINS CONTAIN SPERM, AND SPECIES LIKE BEECH CAN PRODUCE AS MANY AS 500 BILLION.
THIS IS IN FACT WHAT CAUSES A LOT OF OUR ALLERGIES. COLLATERAL DAMAGE FROM HOW THEY SEEK SEXUAL PARTNERS.
HELP!
MY EYES ARE NOT VAGINAS!

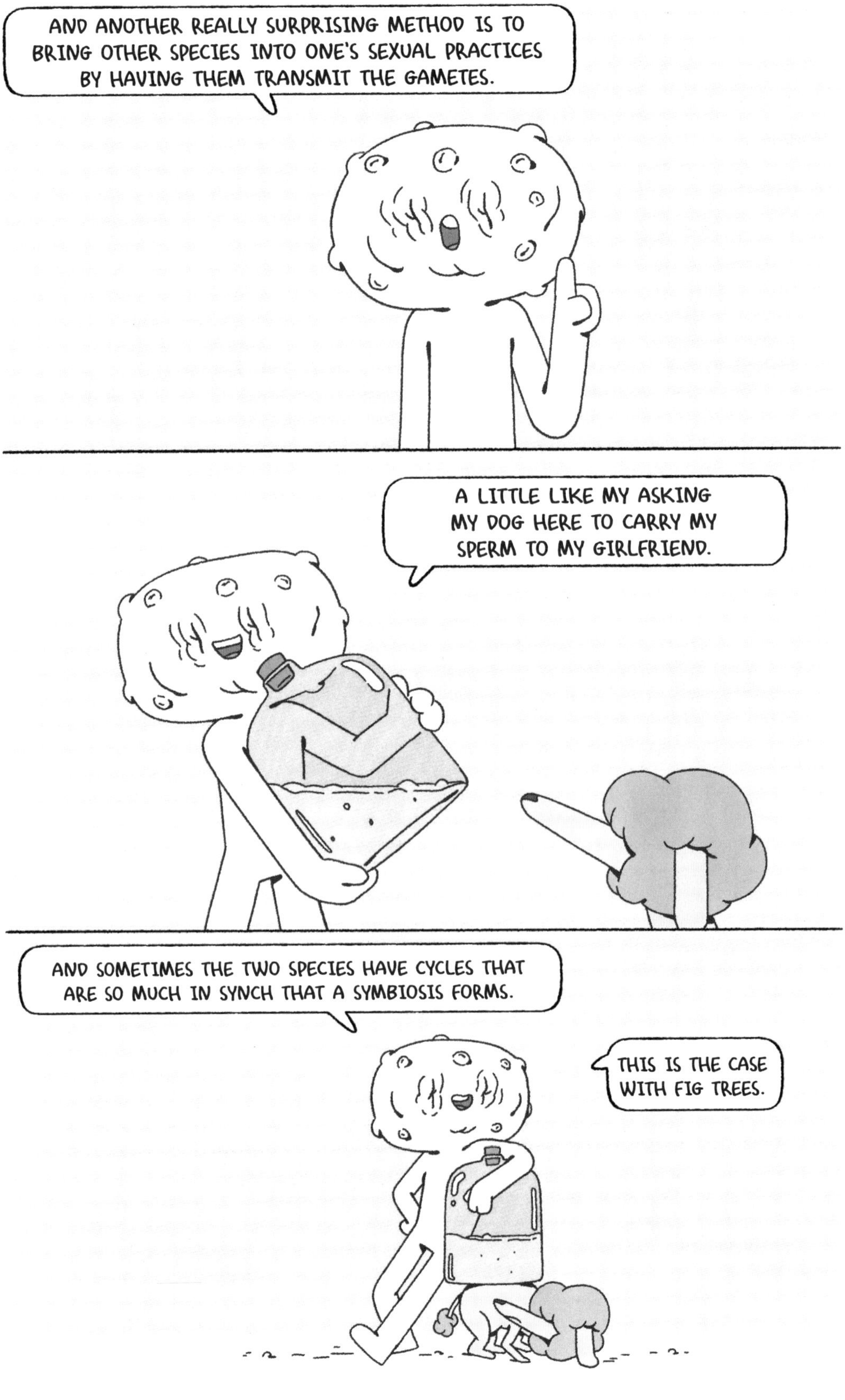
AND ANOTHER REALLY SURPRISING METHOD IS TO BRING OTHER SPECIES INTO ONE'S SEXUAL PRACTICES BY HAVING THEM TRANSMIT THE GAMETES.
A LITTLE LIKE MY ASKING MY DOG HERE TO CARRY MY SPERM TO MY GIRLFRIEND.
AND SOMETIMES THE TWO SPECIES HAVE CYCLES THAT ARE SO MUCH IN SYNCH THAT A SYMBIOSIS FORMS.
THIS IS THE CASE WITH FIG TREES.

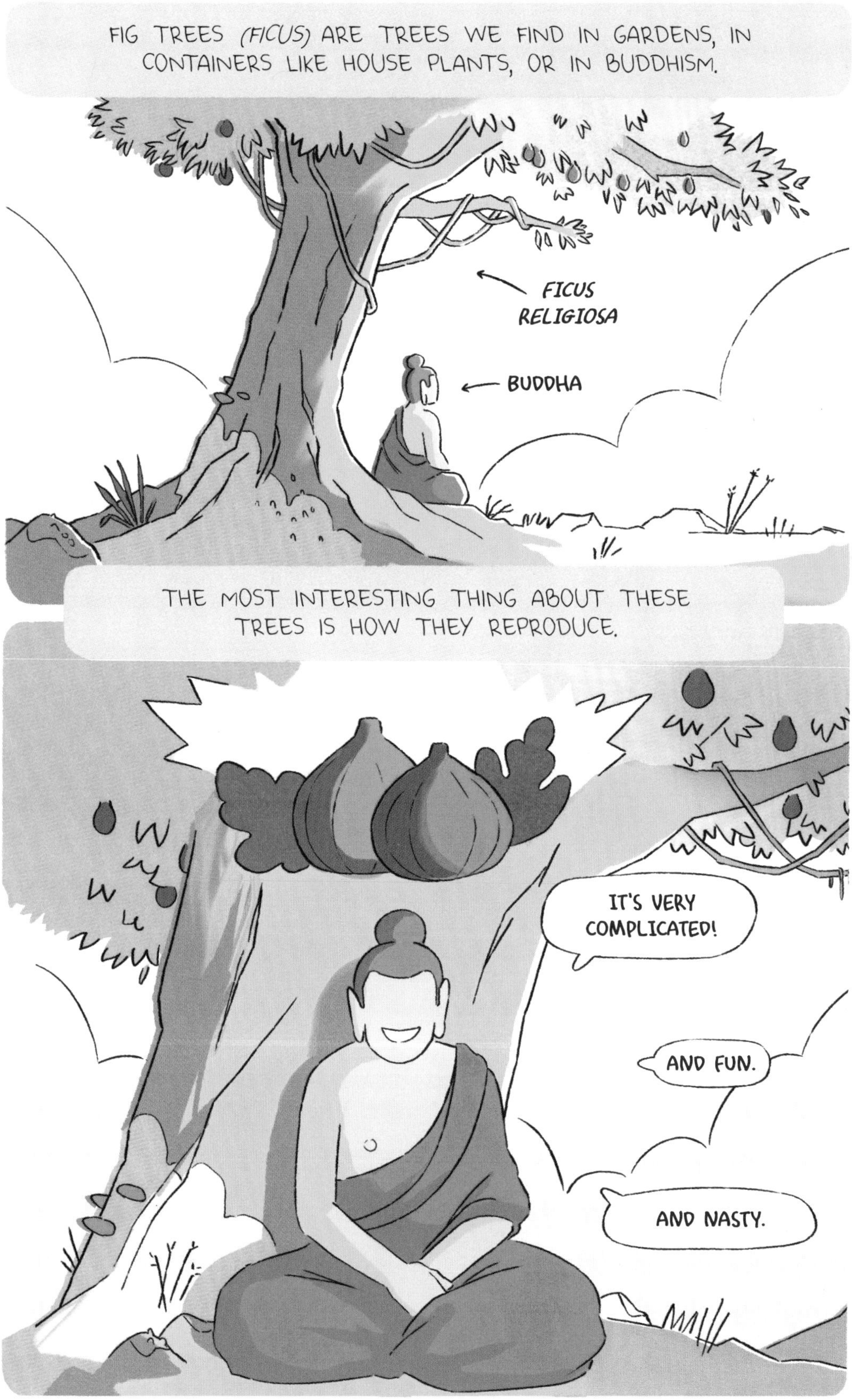
FIG TREES (FICUS) ARE TREES WE FIND IN GARDENS, IN CONTAINERS LIKE HOUSE PLANTS, OR IN BUDDHISM.
FICUS RELIGIOSA
BUDDHA
THE MOST INTERESTING THING ABOUT THESE TREES IS HOW THEY REPRODUCE.
IT'S VERY COMPLICATED!
AND FUN.
AND NASTY.

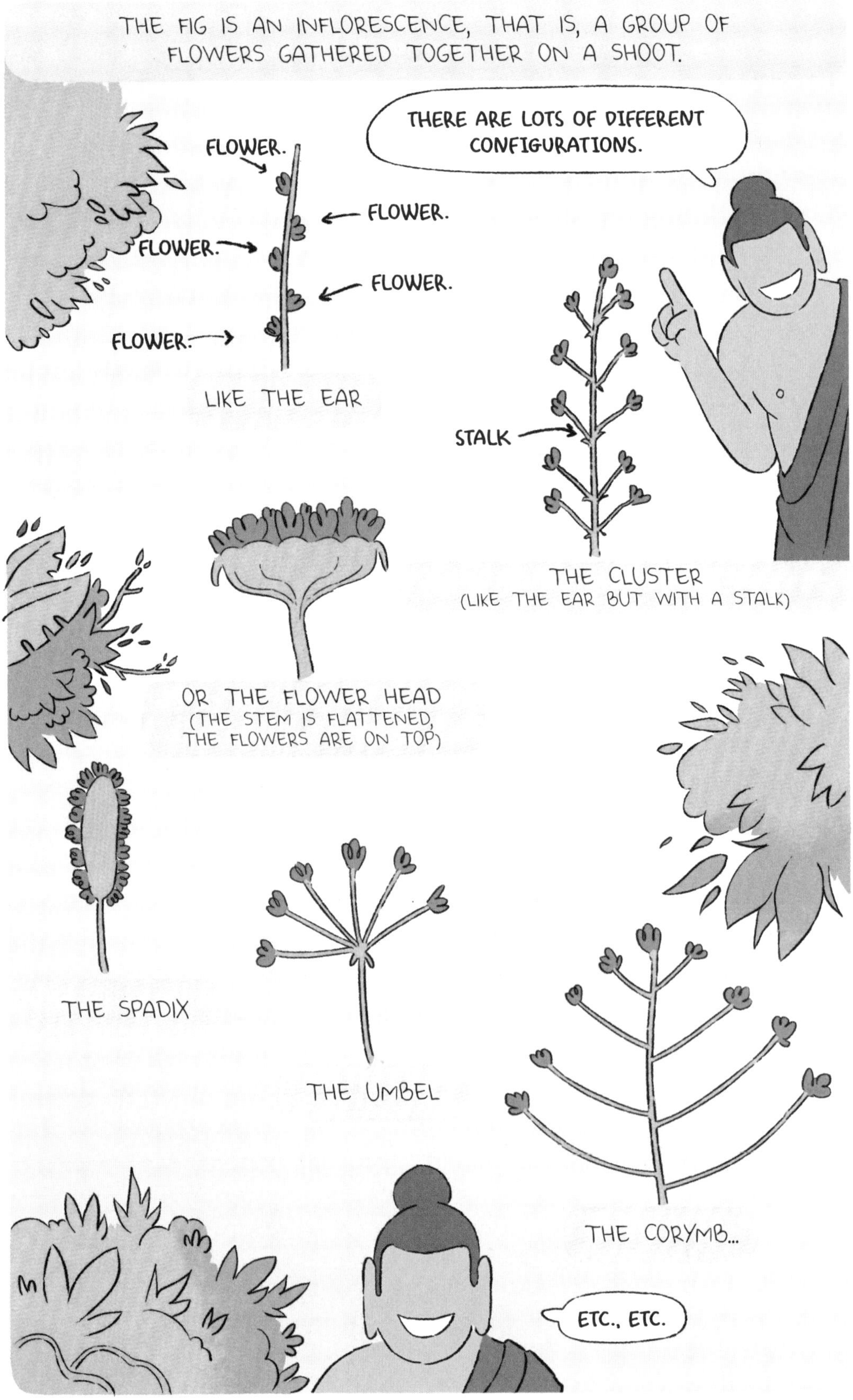
THE FIG IS AN INFLORESCENCE, THAT IS, A GROUP OF FLOWERS GATHERED TOGETHER ON A SHOOT.
THERE ARE LOTS OF DIFFERENT CONFIGURATIONS.
FLOWER.
FLOWER.
FLOWER.
FLOWER.
FLOWER.
LIKE THE EAR
STALK
THE CLUSTER
(LIKE THE EAR BUT WITH A STALK)
OR THE FLOWER HEAD
(THE STEM IS FLATTENED, THE FLOWERS ARE ON TOP)
THE SPADIX
THE UMBEL
THE CORYMB...
ETC., ETC.

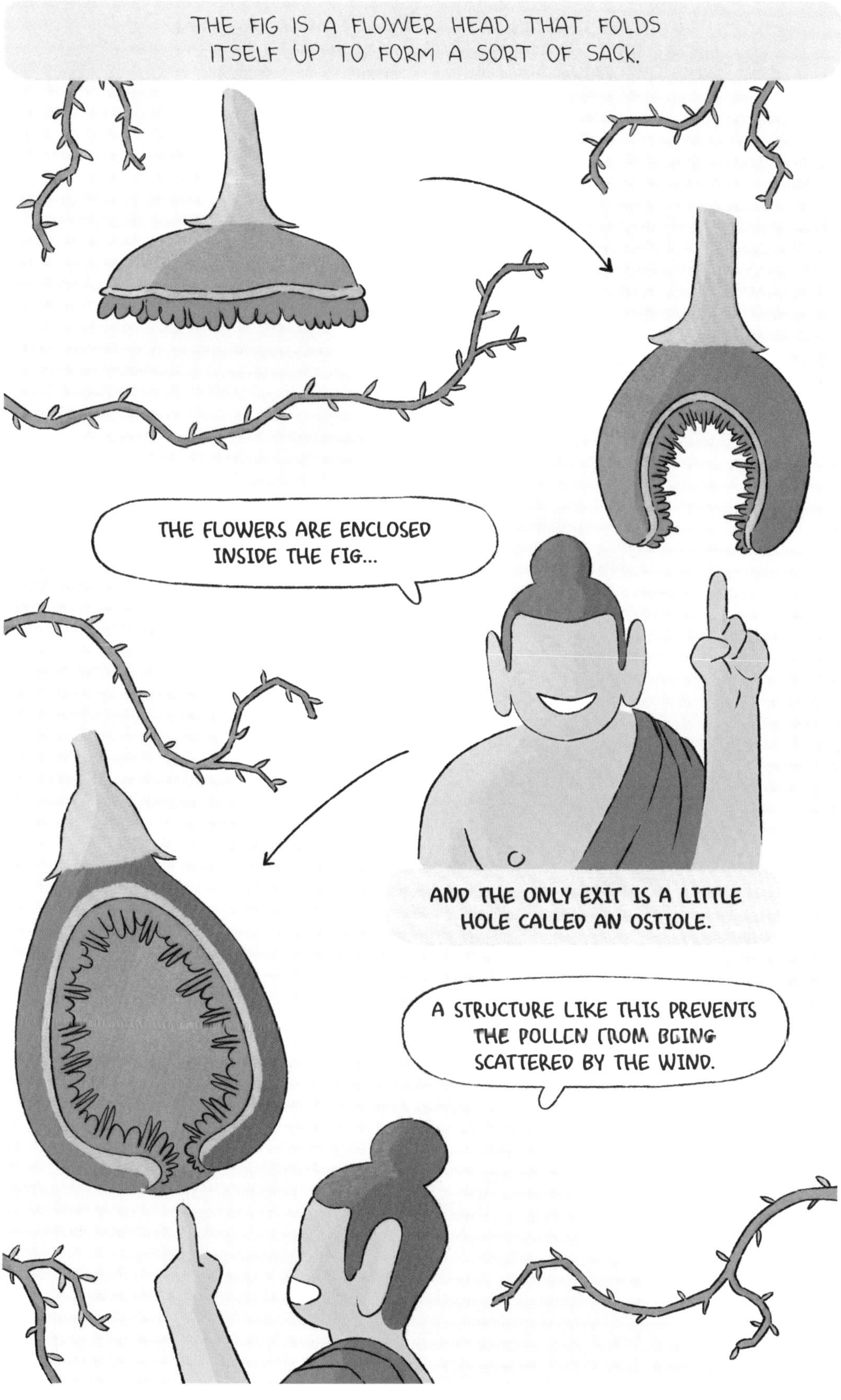
THE FIG IS A FLOWER HEAD THAT FOLDS ITSELF UP TO FORM A SORT OF SACK.
THE FLOWERS ARE ENCLOSED INSIDE THE FIG...
AND THE ONLY EXIT IS A LITTLE HOLE CALLED AN OSTIOLE.
A STRUCTURE LIKE THIS PREVENTS THE POLLEN FROM BEING SCATTERED BY THE WIND.

A VERY SPECIALIZED POLLINATOR WILL TAKE CARE OF FERTILIZATION: A MEMBER OF *AGAONIDAE* CALLED THE FIG WASP.

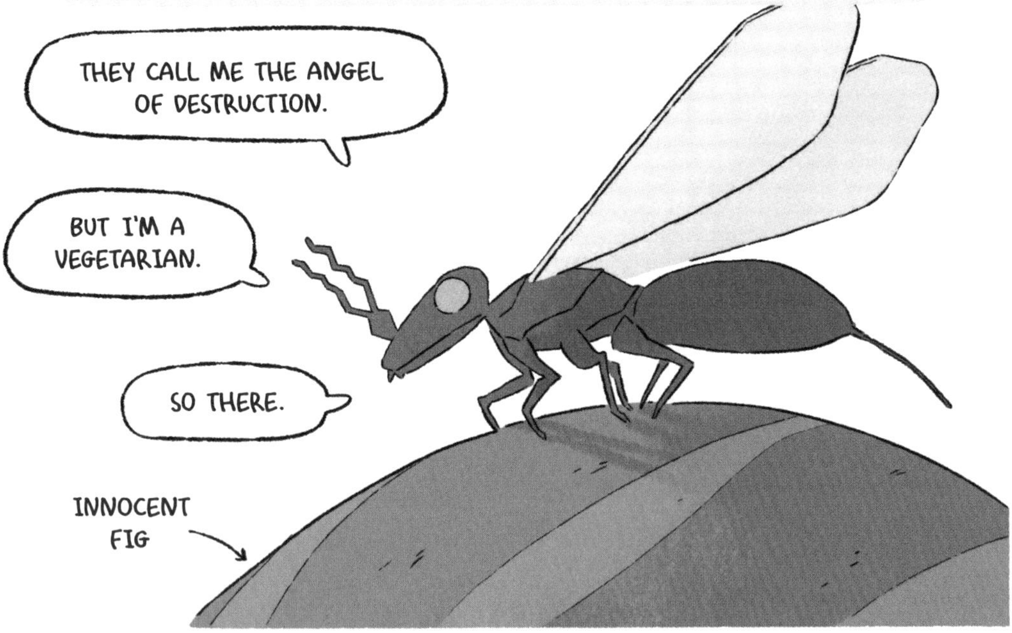

THEY CALL ME THE ANGEL OF DESTRUCTION.
BUT I'M A VEGETARIAN.
SO THERE.
INNOCENT FIG

THERE IS TYPICALLY A SPECIES OF *AGAONIDAE* FOR EACH SPECIES OF FIG TREE, WITH THE SEXUALITY OF EACH SPECIES BEING CLOSELY RELATED.

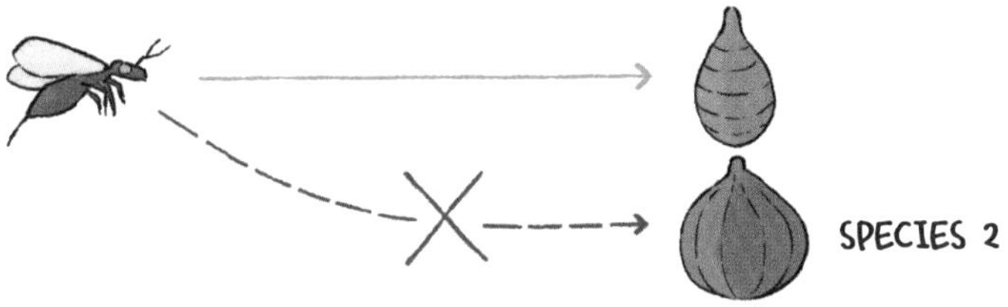

SPECIES 2

THE STORY BEGINS WITH THE ARRIVAL OF A FEMALE WASP AT THE ENTRANCE TO THE FIG.

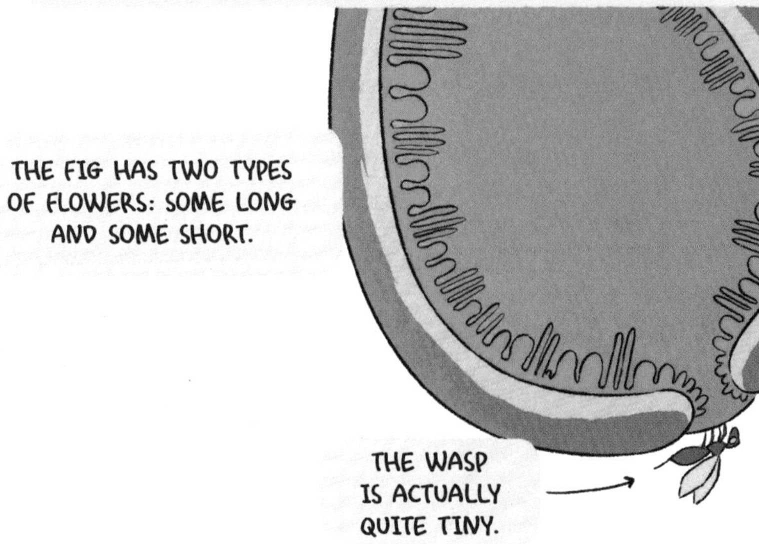

THE FIG HAS TWO TYPES OF FLOWERS: SOME LONG AND SOME SHORT.
THE WASP IS ACTUALLY QUITE TINY.

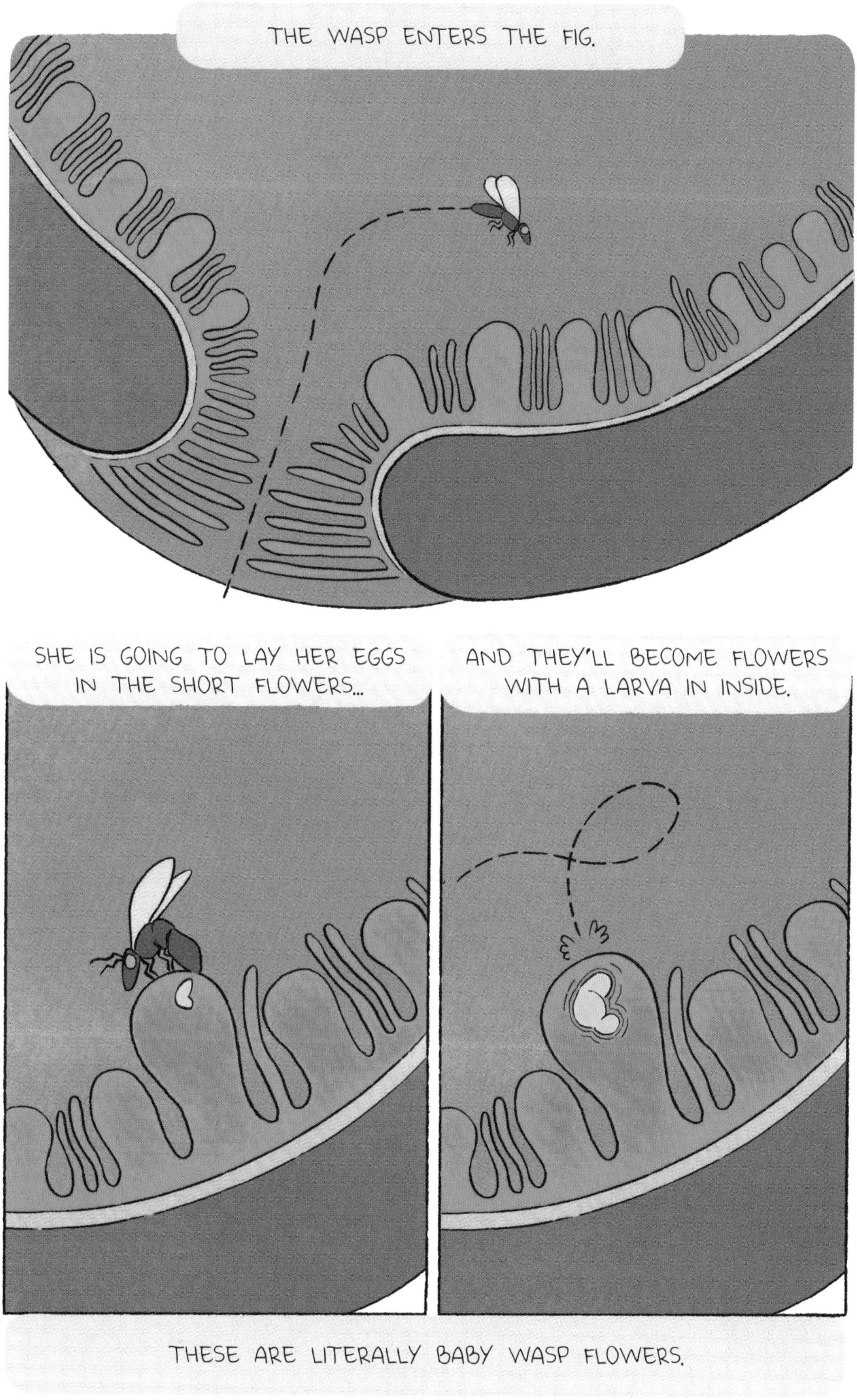
THE WASP ENTERS THE FIG.
SHE IS GOING TO LAY HER EGGS IN THE SHORT FLOWERS...
AND THEY'LL BECOME FLOWERS WITH A LARVA IN INSIDE.
THESE ARE LITERALLY BABY WASP FLOWERS.

WHILE SHE'S LAYING EGGS, THE WASP WILL BRUSH UP AGAINST THE LONG FLOWERS.

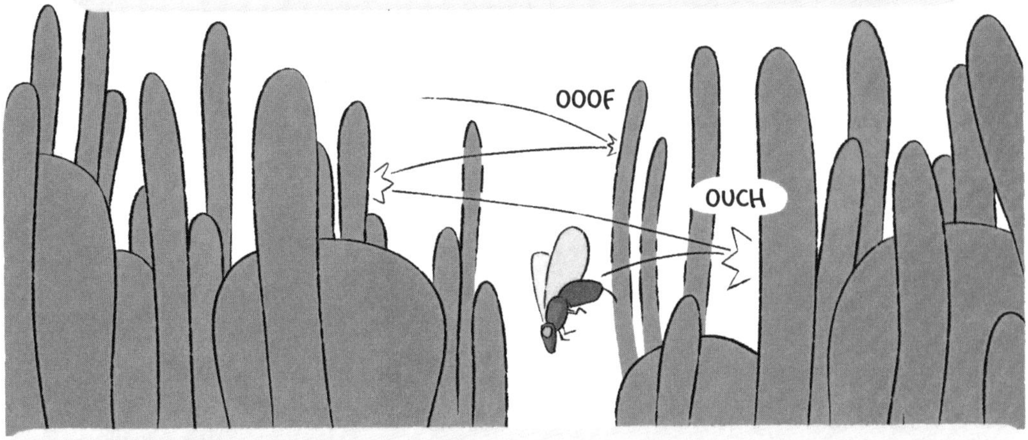
OOOF
OUCH

THE POLLEN SHE'S CARRYING WILL POLLINATE THESE FLOWERS.

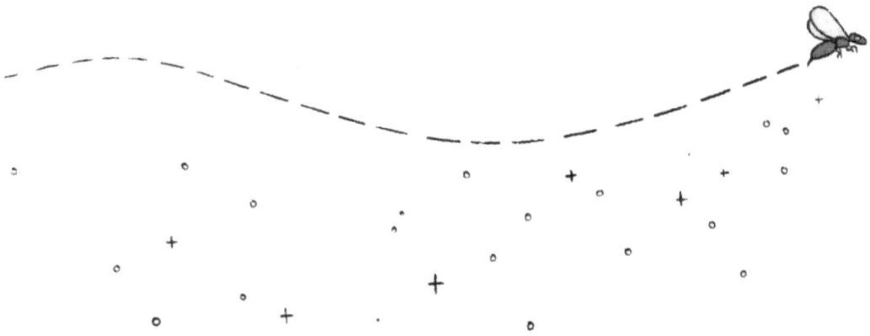

THEN SHE COMES BACK OUT, LEAVING BEHIND HER THE LARVAE GROWING IN THE FLOWERS, AS WELL AS THE POLLINATED FLOWERS.

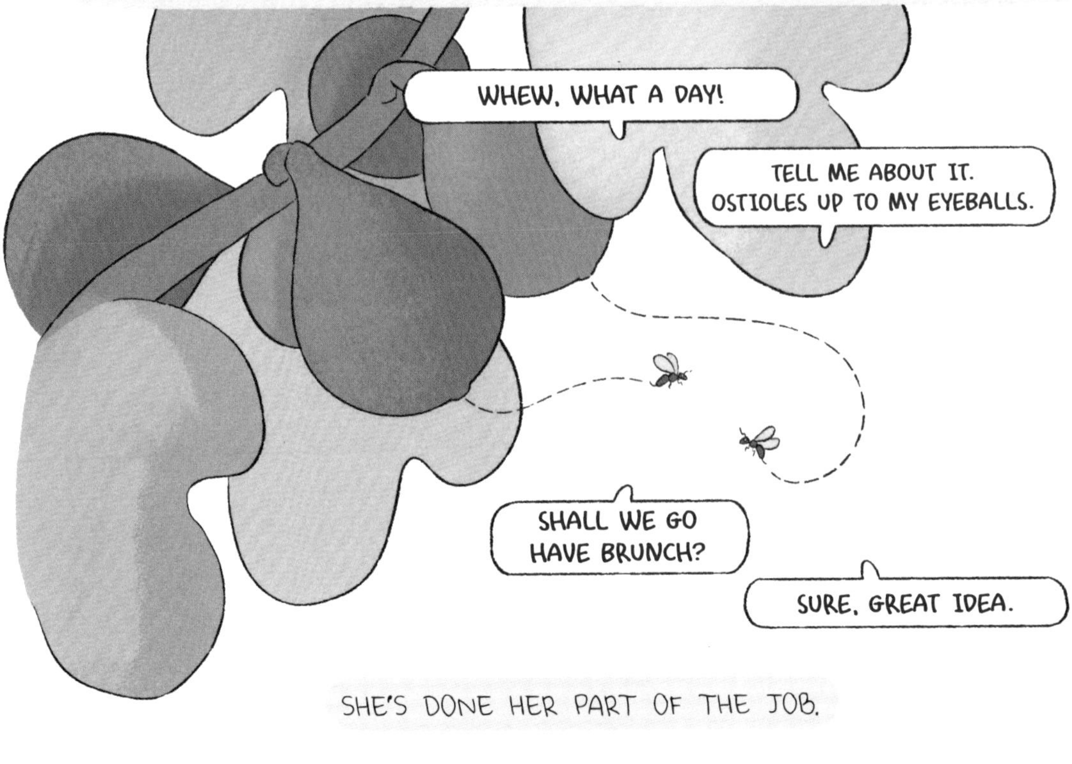
WHEW. WHAT A DAY!
TELL ME ABOUT IT. OSTIOLES UP TO MY EYEBALLS.
SHALL WE GO HAVE BRUNCH?
SURE. GREAT IDEA.

SHE'S DONE HER PART OF THE JOB.

THE MALE LARVAE LEAVE THEIR FLOWERS FIRST. THEY'RE SOMEWHAT DEFORMED, WITH TINY FEET, ANTENNAE, AND NO WINGS OR EYES.

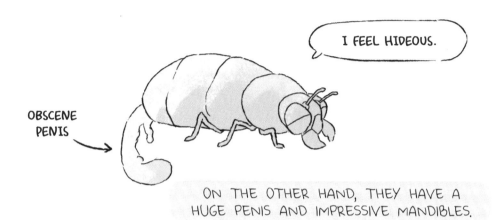

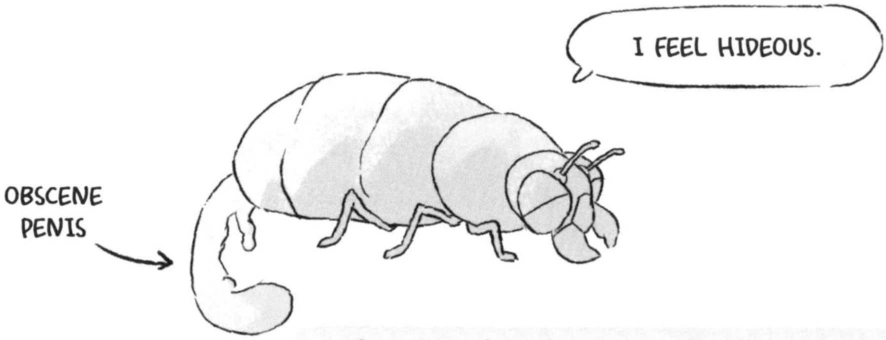
I FEEL HIDEOUS.
OBSCENE PENIS

ON THE OTHER HAND, THEY HAVE A HUGE PENIS AND IMPRESSIVE MANDIBLES.

THEY GO IN SEARCH OF FLOWERS CONTAINING THE FEMALES AND USE THEIR MANDIBLES TO MAKE A LITTLE OPENING. THEY FERTILIZE THEM VIA THIS FRUITY GLORY HOLE.

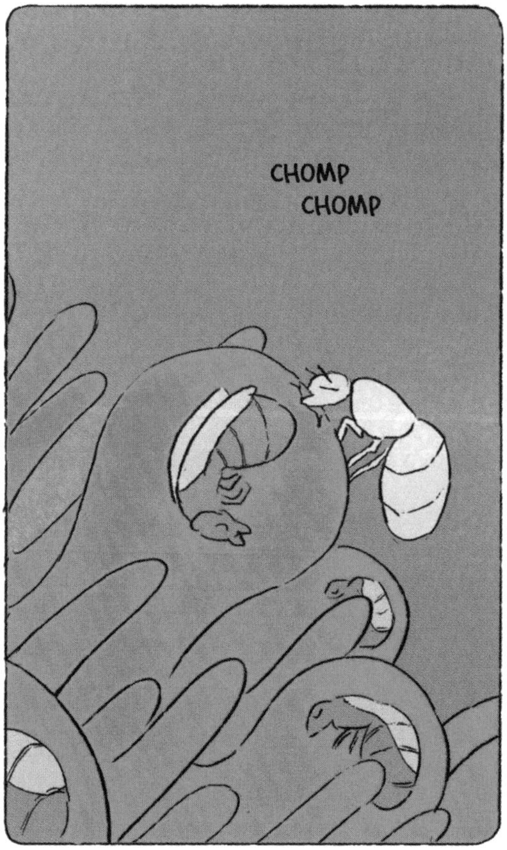
CHOMP
CHOMP

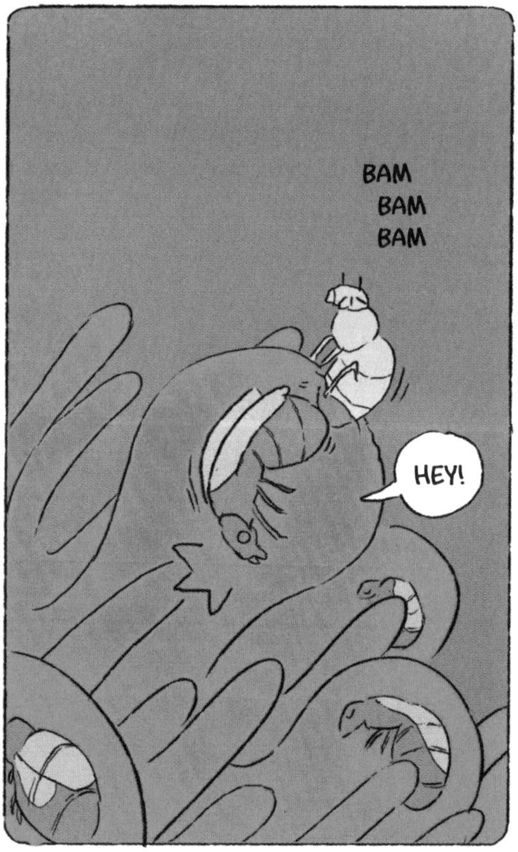
BAM
BAM
BAM
HEY!

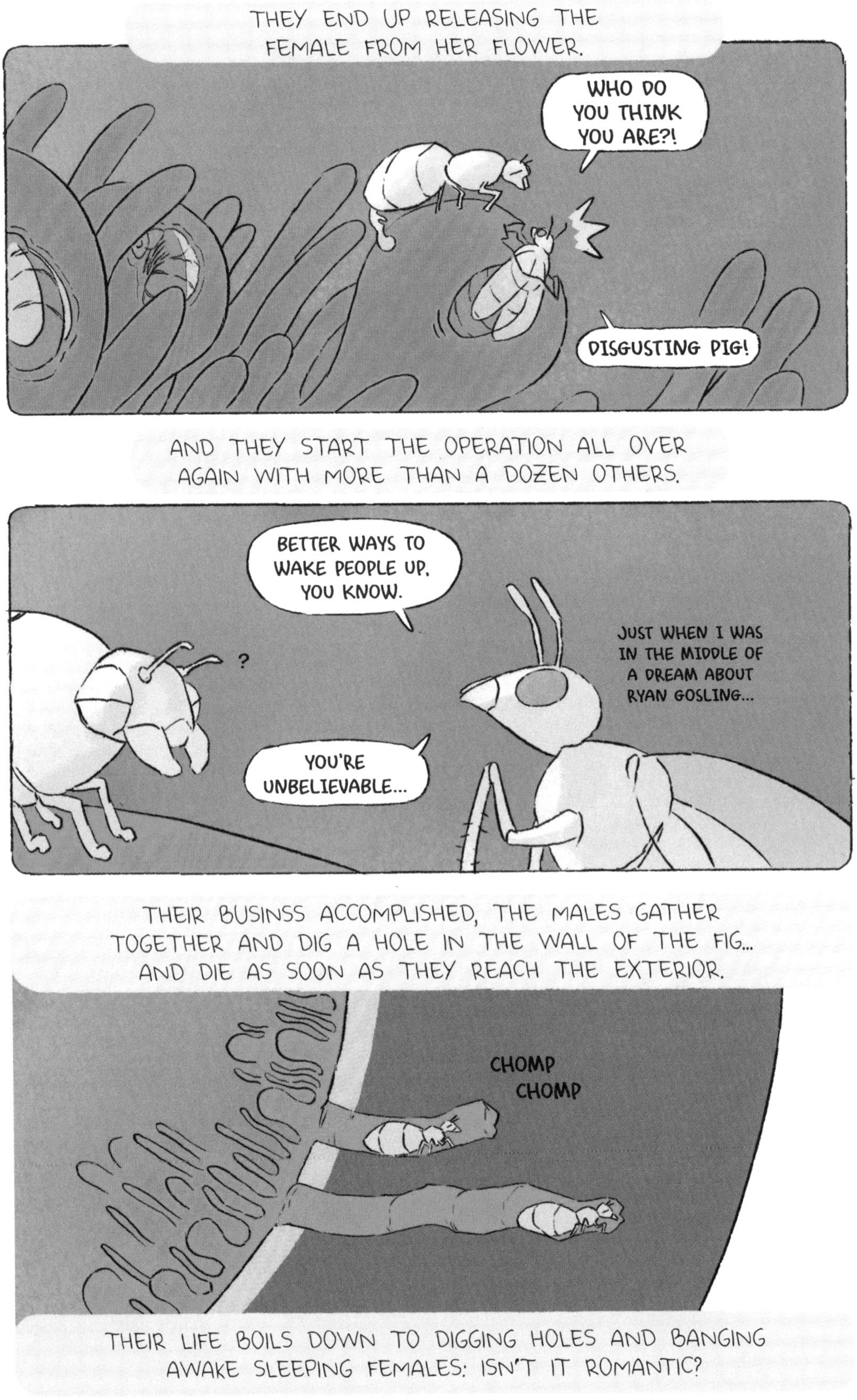
THEY END UP RELEASING THE FEMALE FROM HER FLOWER.
WHO DO YOU THINK YOU ARE?!
DISGUSTING PIG!
AND THEY START THE OPERATION ALL OVER AGAIN WITH MORE THAN A DOZEN OTHERS.
BETTER WAYS TO WAKE PEOPLE UP, YOU KNOW.
?
JUST WHEN I WAS IN THE MIDDLE OF A DREAM ABOUT RYAN GOSLING...
YOU'RE UNBELIEVABLE...
THEIR BUSINSS ACCOMPLISHED, THE MALES GATHER TOGETHER AND DIG A HOLE IN THE WALL OF THE FIG... AND DIE AS SOON AS THEY REACH THE EXTERIOR.
CHOMP
CHOMP
THEIR LIFE BOILS DOWN TO DIGGING HOLES AND BANGING AWAKE SLEEPING FEMALES: ISN'T IT ROMANTIC?

AS FOR THE FEMALES, NOW FREE AND FERTILIZED, THEY GATHER UP POLLEN AND STORE IT IN A SPECIAL POCKET IN THEIR THORAX.

THEN THEY'LL LEAVE USING THE TUNNELS DUG BY THE MALES, AND THEY'LL GO IN SEARCH OF A FIG IN WHICH TO LAY THEIR OWN EGGS...

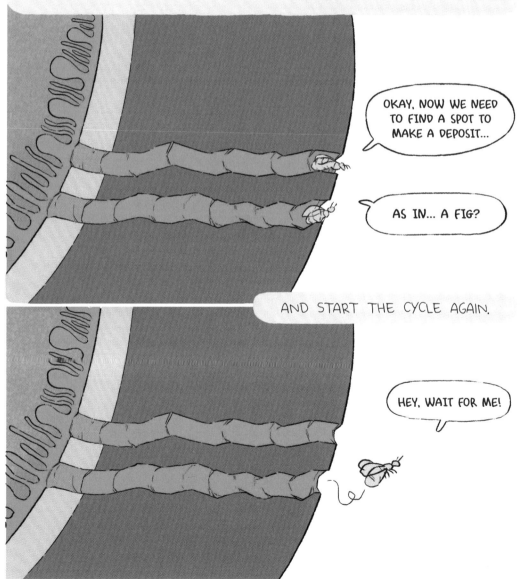

OKAY, NOW WE NEED TO FIND A SPOT TO MAKE A DEPOSIT...

AS IN... A FIG?

AND START THE CYCLE AGAIN.

HEY, WAIT FOR ME!

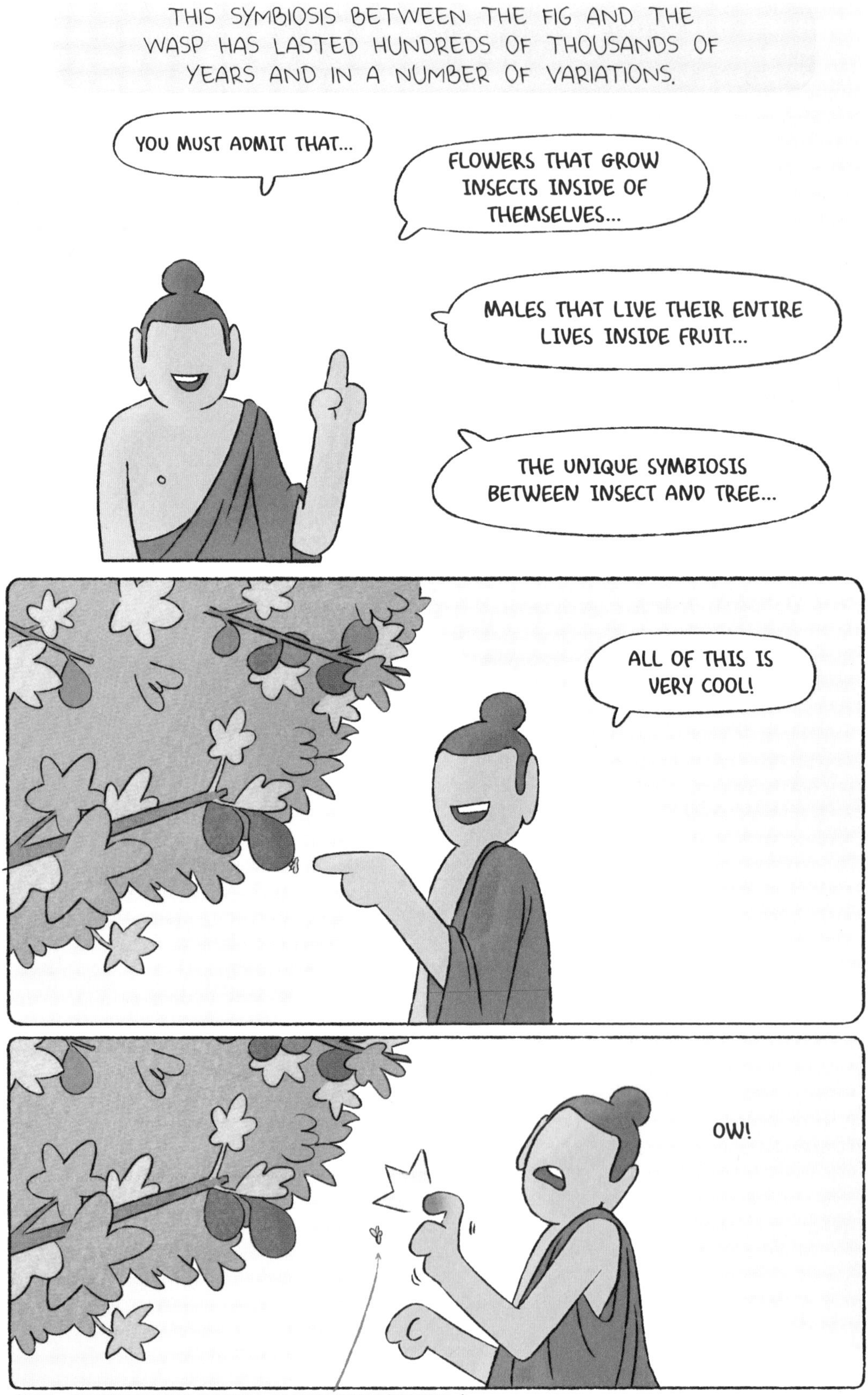
THIS SYMBIOSIS BETWEEN THE FIG AND THE WASP HAS LASTED HUNDREDS OF THOUSANDS OF YEARS AND IN A NUMBER OF VARIATIONS.
YOU MUST ADMIT THAT...
FLOWERS THAT GROW INSECTS INSIDE OF THEMSELVES...
MALES THAT LIVE THEIR ENTIRE LIVES INSIDE FRUIT...
THE UNIQUE SYMBIOSIS BETWEEN INSECT AND TREE...
ALL OF THIS IS VERY COOL!
OW!
(NON-*AGAONIDAE* WASP BECAUSE *AGAONIDAE* DON'T STING...)

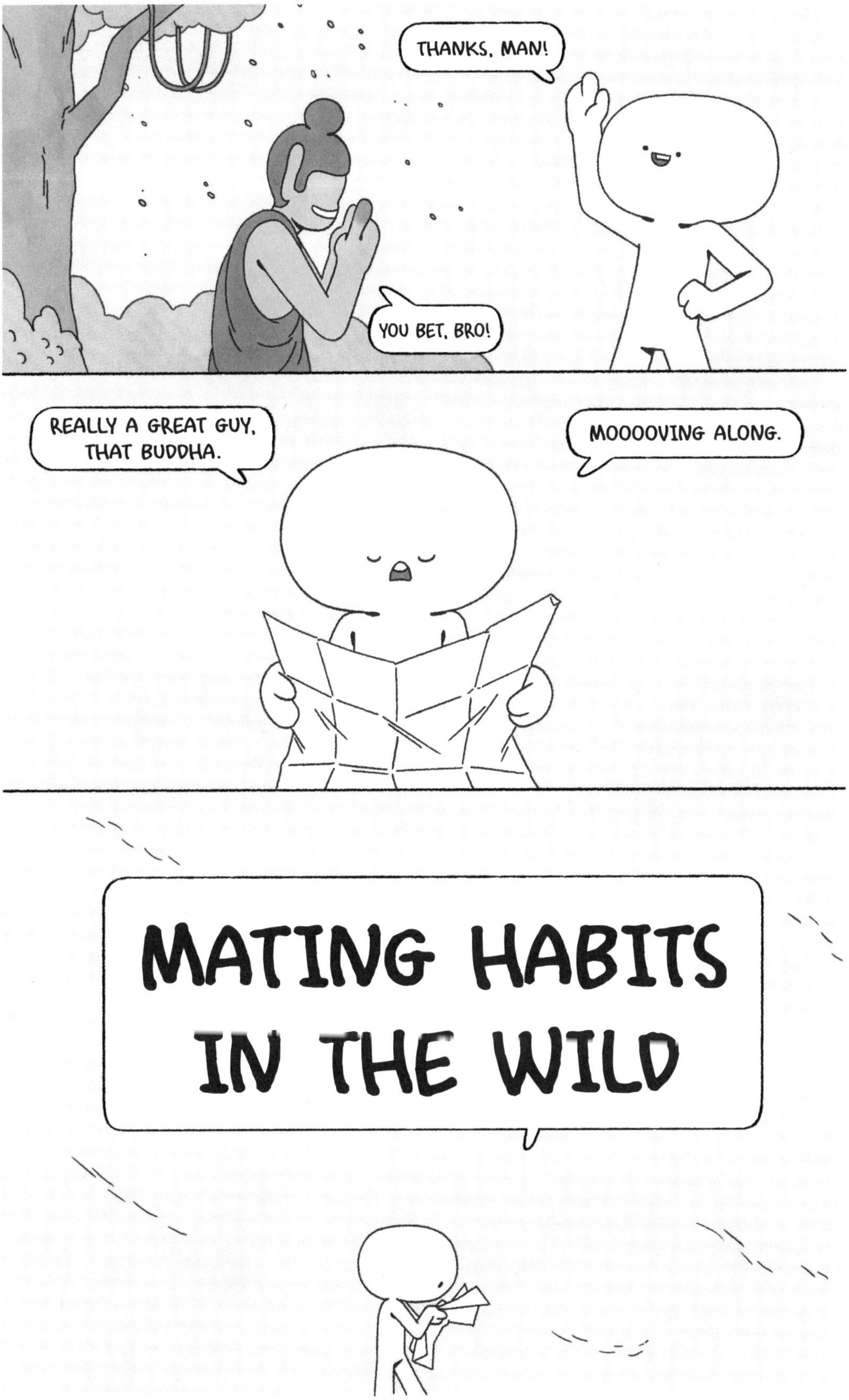
THANKS, MAN!
YOU BET, BRO!
REALLY A GREAT GUY, THAT BUDDHA.
MOOOOVING ALONG.
MATING HABITS IN THE WILD

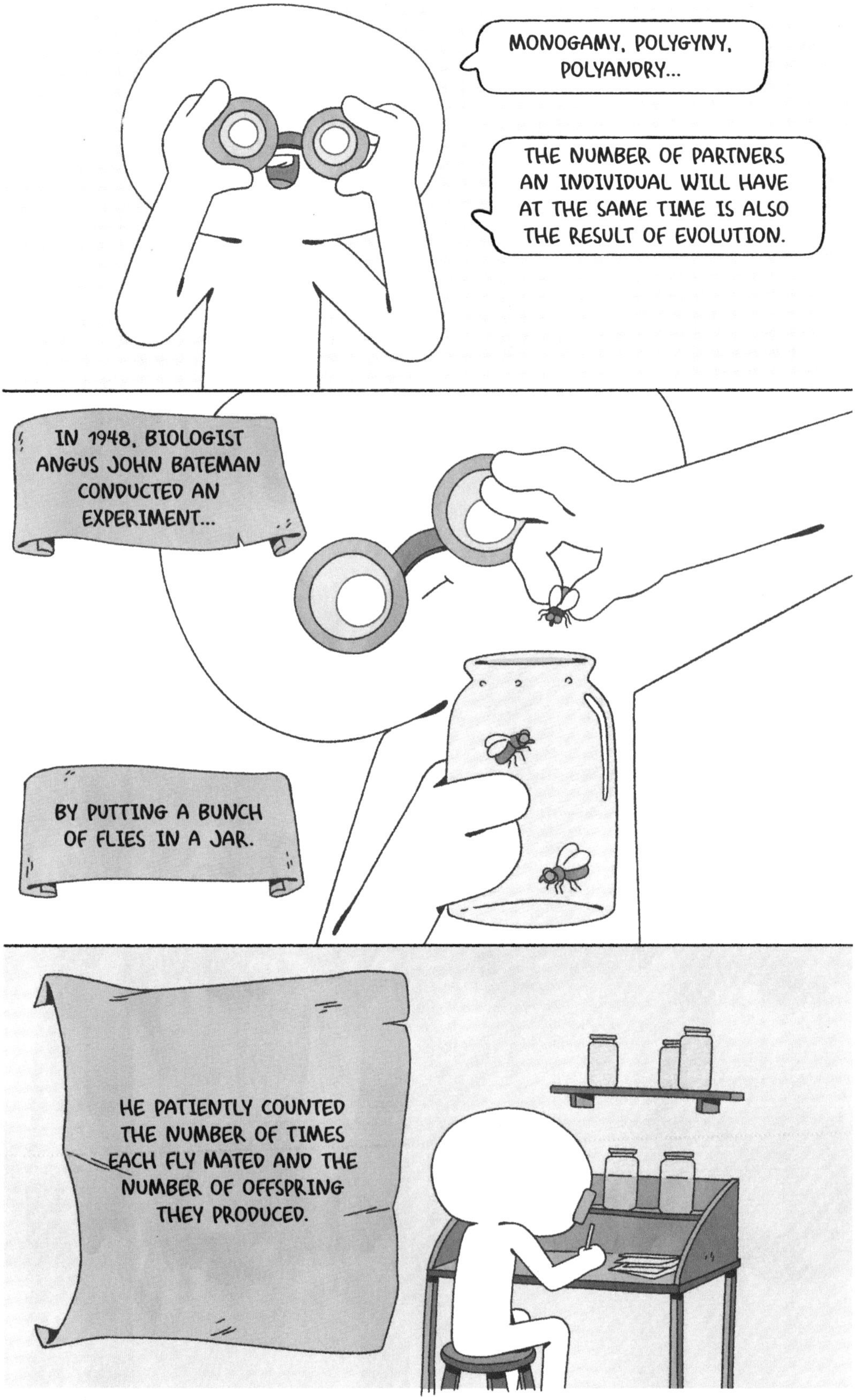
MONOGAMY, POLYGYNY, POLYANDRY...
THE NUMBER OF PARTNERS AN INDIVIDUAL WILL HAVE AT THE SAME TIME IS ALSO THE RESULT OF EVOLUTION.
IN 1948, BIOLOGIST ANGUS JOHN BATEMAN CONDUCTED AN EXPERIMENT...
BY PUTTING A BUNCH OF FLIES IN A JAR.
HE PATIENTLY COUNTED THE NUMBER OF TIMES EACH FLY MATED AND THE NUMBER OF OFFSPRING THEY PRODUCED.

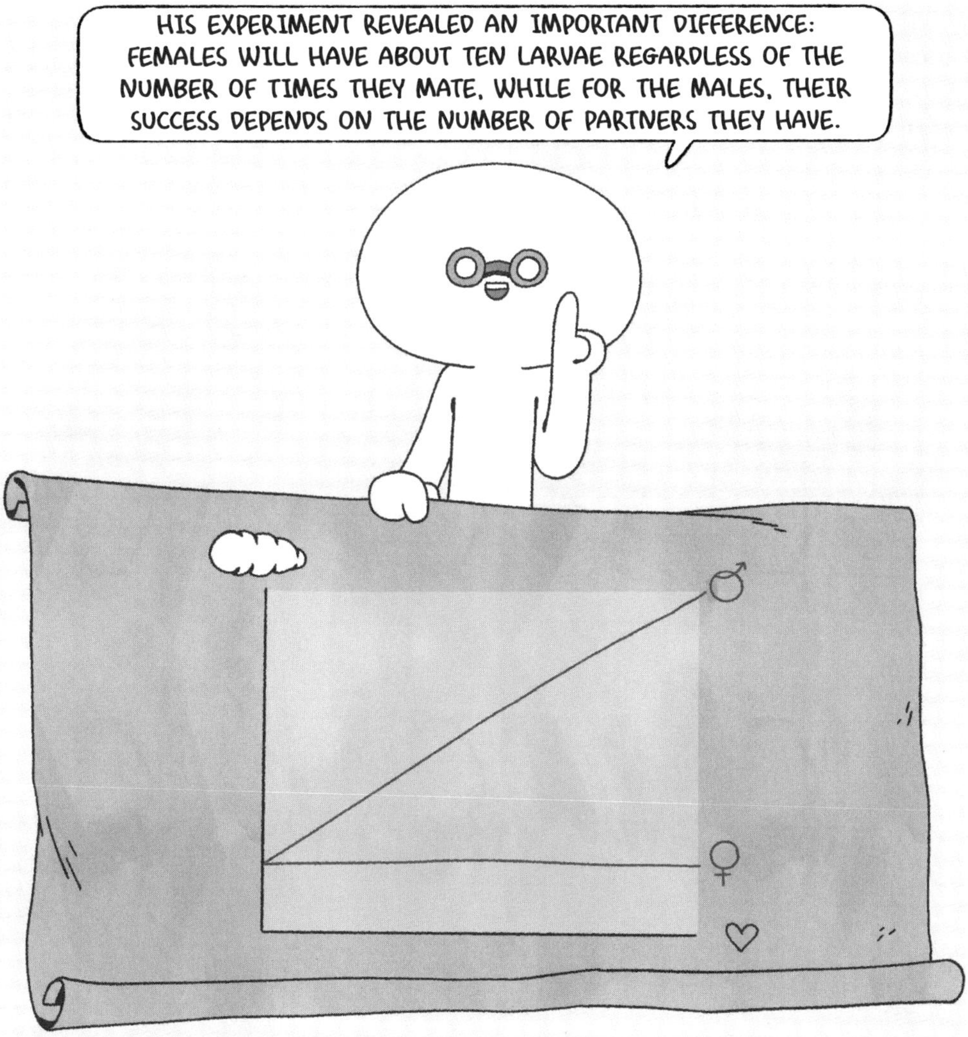
HIS EXPERIMENT REVEALED AN IMPORTANT DIFFERENCE: FEMALES WILL HAVE ABOUT TEN LARVAE REGARDLESS OF THE NUMBER OF TIMES THEY MATE, WHILE FOR THE MALES, THEIR SUCCESS DEPENDS ON THE NUMBER OF PARTNERS THEY HAVE.
♂
♀
♡

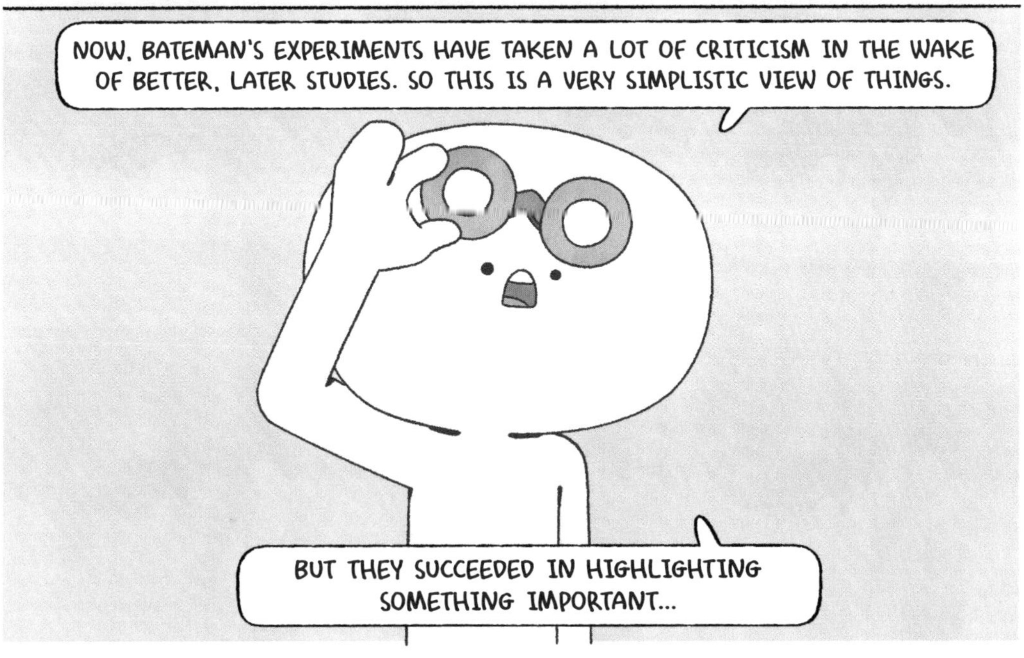
NOW, BATEMAN'S EXPERIMENTS HAVE TAKEN A LOT OF CRITICISM IN THE WAKE OF BETTER, LATER STUDIES. SO THIS IS A VERY SIMPLISTIC VIEW OF THINGS.
BUT THEY SUCCEEDED IN HIGHLIGHTING SOMETHING IMPORTANT...

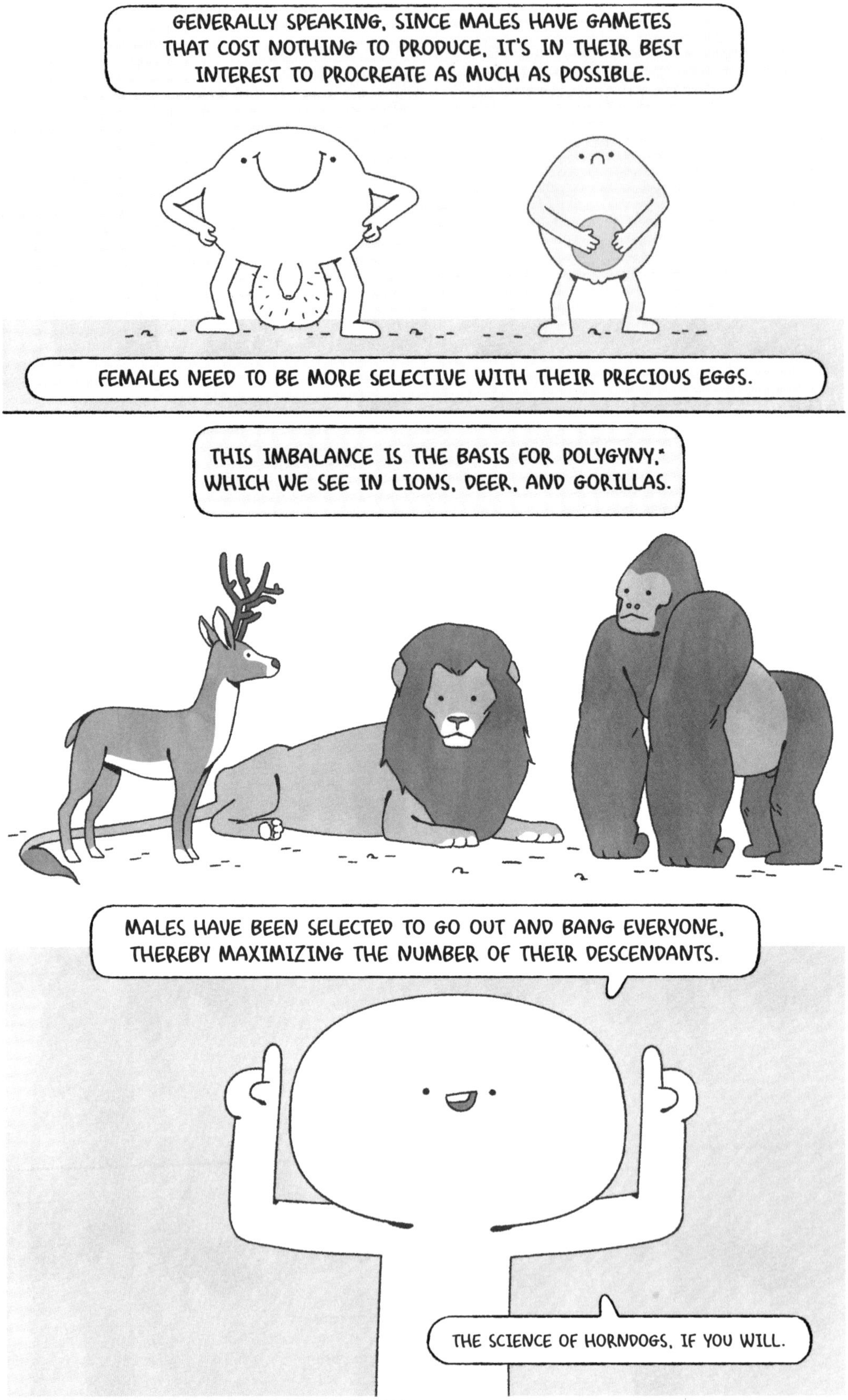
GENERALLY SPEAKING, SINCE MALES HAVE GAMETES THAT COST NOTHING TO PRODUCE, IT'S IN THEIR BEST INTEREST TO PROCREATE AS MUCH AS POSSIBLE.
FEMALES NEED TO BE MORE SELECTIVE WITH THEIR PRECIOUS EGGS.
THIS IMBALANCE IS THE BASIS FOR POLYGYNY,* WHICH WE SEE IN LIONS, DEER, AND GORILLAS.
MALES HAVE BEEN SELECTED TO GO OUT AND BANG EVERYONE, THEREBY MAXIMIZING THE NUMBER OF THEIR DESCENDANTS.
THE SCIENCE OF HORNDOGS, IF YOU WILL.

*WHEN A MALE REPRODUCES WITH MULTIPLE FEMALES.

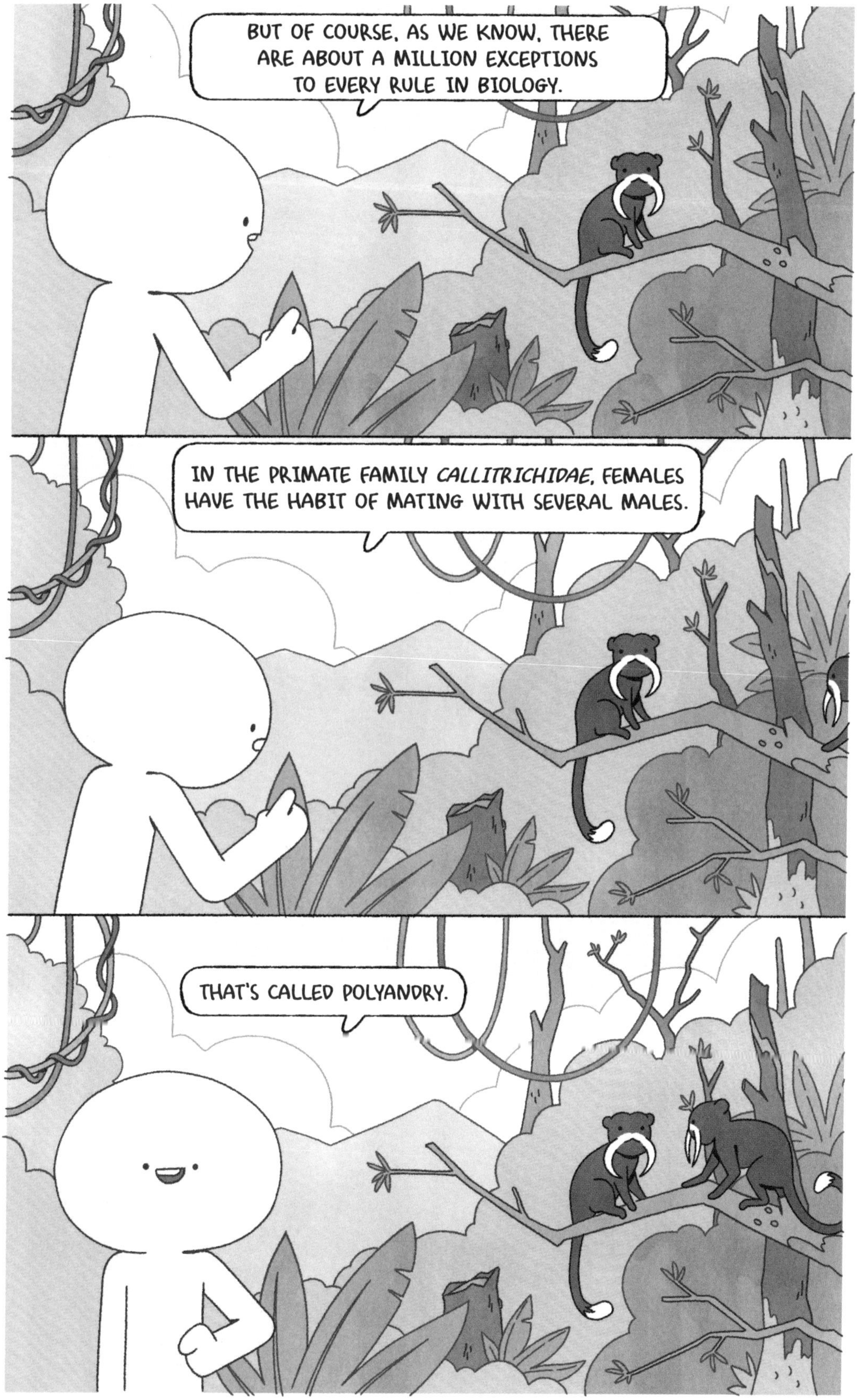
BUT OF COURSE, AS WE KNOW, THERE ARE ABOUT A MILLION EXCEPTIONS TO EVERY RULE IN BIOLOGY.
IN THE PRIMATE FAMILY *CALLITRICHIDAE*, FEMALES HAVE THE HABIT OF MATING WITH SEVERAL MALES.
THAT'S CALLED POLYANDRY.

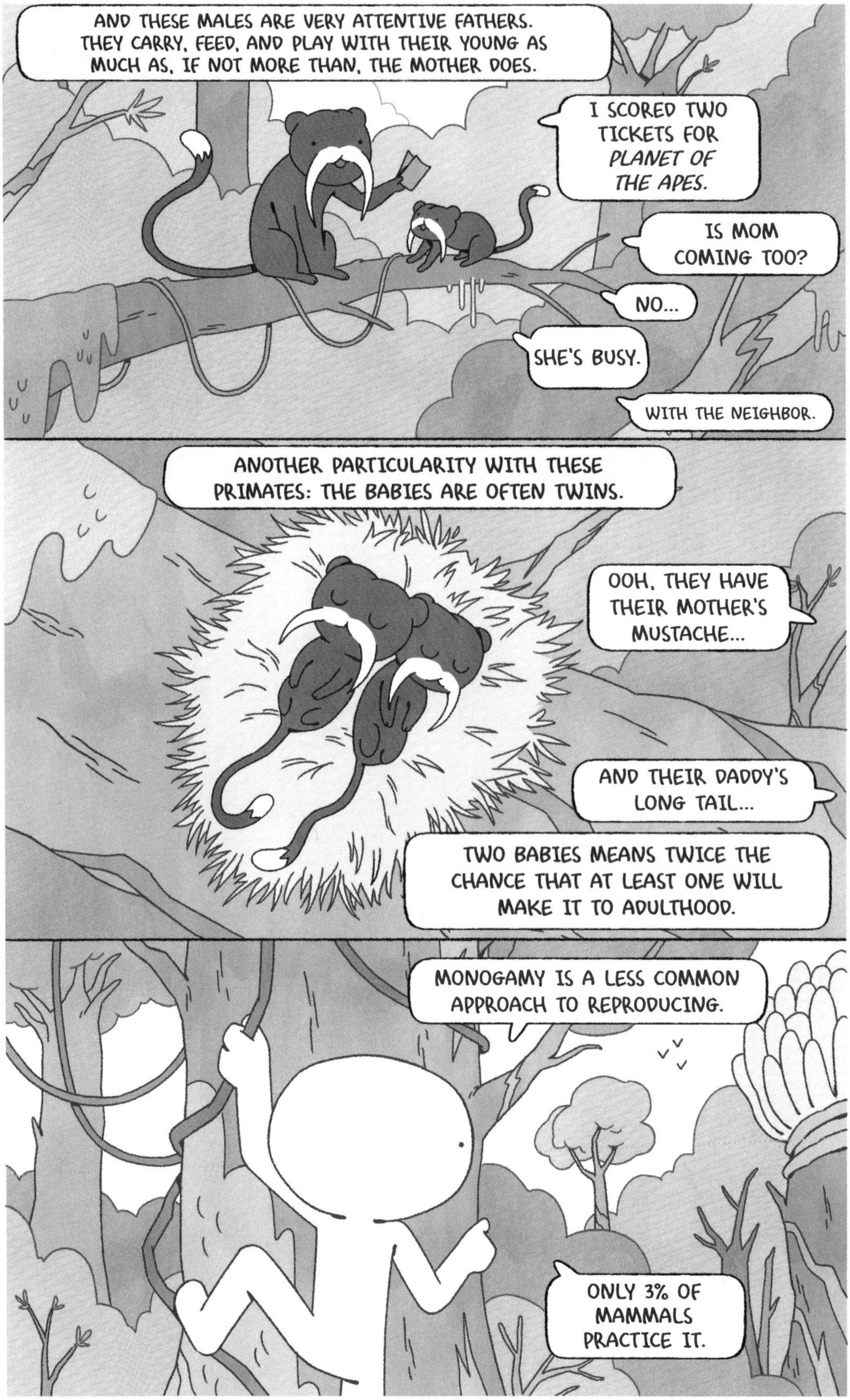
AND THESE MALES ARE VERY ATTENTIVE FATHERS. THEY CARRY, FEED, AND PLAY WITH THEIR YOUNG AS MUCH AS, IF NOT MORE THAN, THE MOTHER DOES.
I SCORED TWO TICKETS FOR *PLANET OF THE APES.*
IS MOM COMING TOO?
NO...
SHE'S BUSY.
WITH THE NEIGHBOR.
ANOTHER PARTICULARITY WITH THESE PRIMATES: THE BABIES ARE OFTEN TWINS.
OOH, THEY HAVE THEIR MOTHER'S MUSTACHE...
AND THEIR DADDY'S LONG TAIL...
TWO BABIES MEANS TWICE THE CHANCE THAT AT LEAST ONE WILL MAKE IT TO ADULTHOOD.
MONOGAMY IS A LESS COMMON APPROACH TO REPRODUCING.
ONLY 3% OF MAMMALS PRACTICE IT.

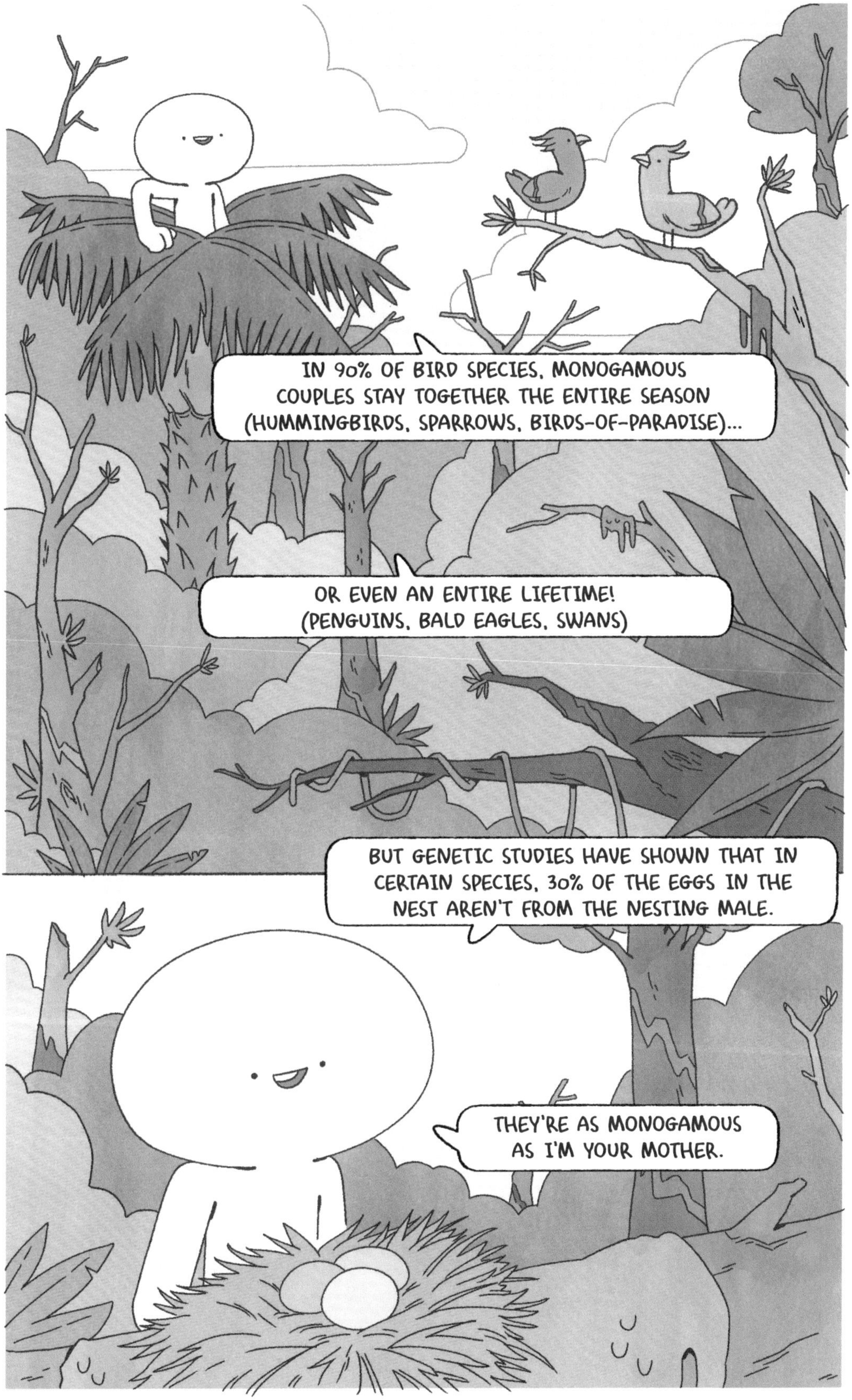
IN 90% OF BIRD SPECIES, MONOGAMOUS COUPLES STAY TOGETHER THE ENTIRE SEASON (HUMMINGBIRDS, SPARROWS, BIRDS-OF-PARADISE)...
OR EVEN AN ENTIRE LIFETIME! (PENGUINS, BALD EAGLES, SWANS)
BUT GENETIC STUDIES HAVE SHOWN THAT IN CERTAIN SPECIES, 30% OF THE EGGS IN THE NEST AREN'T FROM THE NESTING MALE.
THEY'RE AS MONOGAMOUS AS I'M YOUR MOTHER.

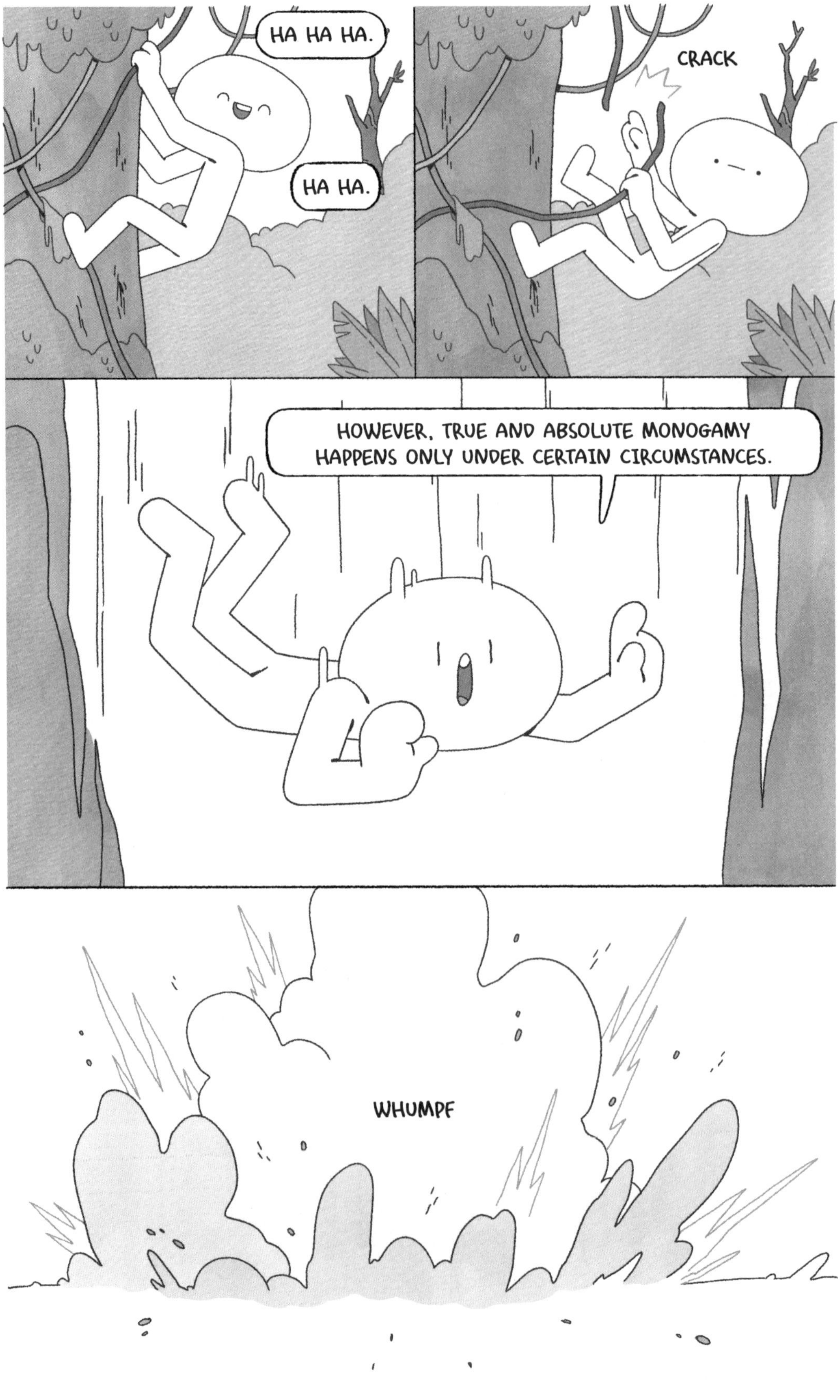
HA HA HA.
HA HA.
CRACK
HOWEVER, TRUE AND ABSOLUTE MONOGAMY HAPPENS ONLY UNDER CERTAIN CIRCUMSTANCES.
WHUMPF

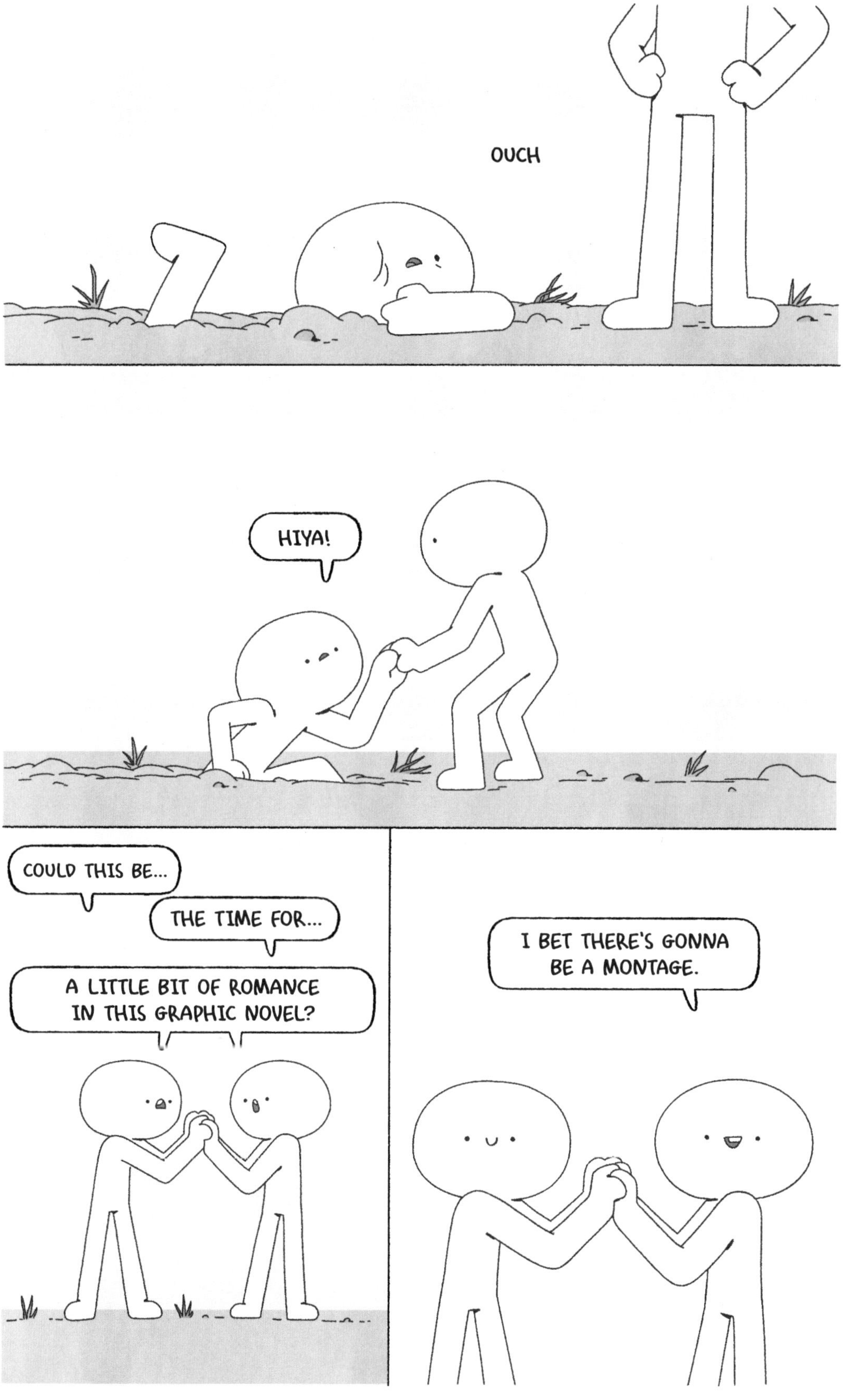
OUCH
HIYA!
COULD THIS BE...
THE TIME FOR...
A LITTLE BIT OF ROMANCE IN THIS GRAPHIC NOVEL?
I BET THERE'S GONNA BE A MONTAGE.

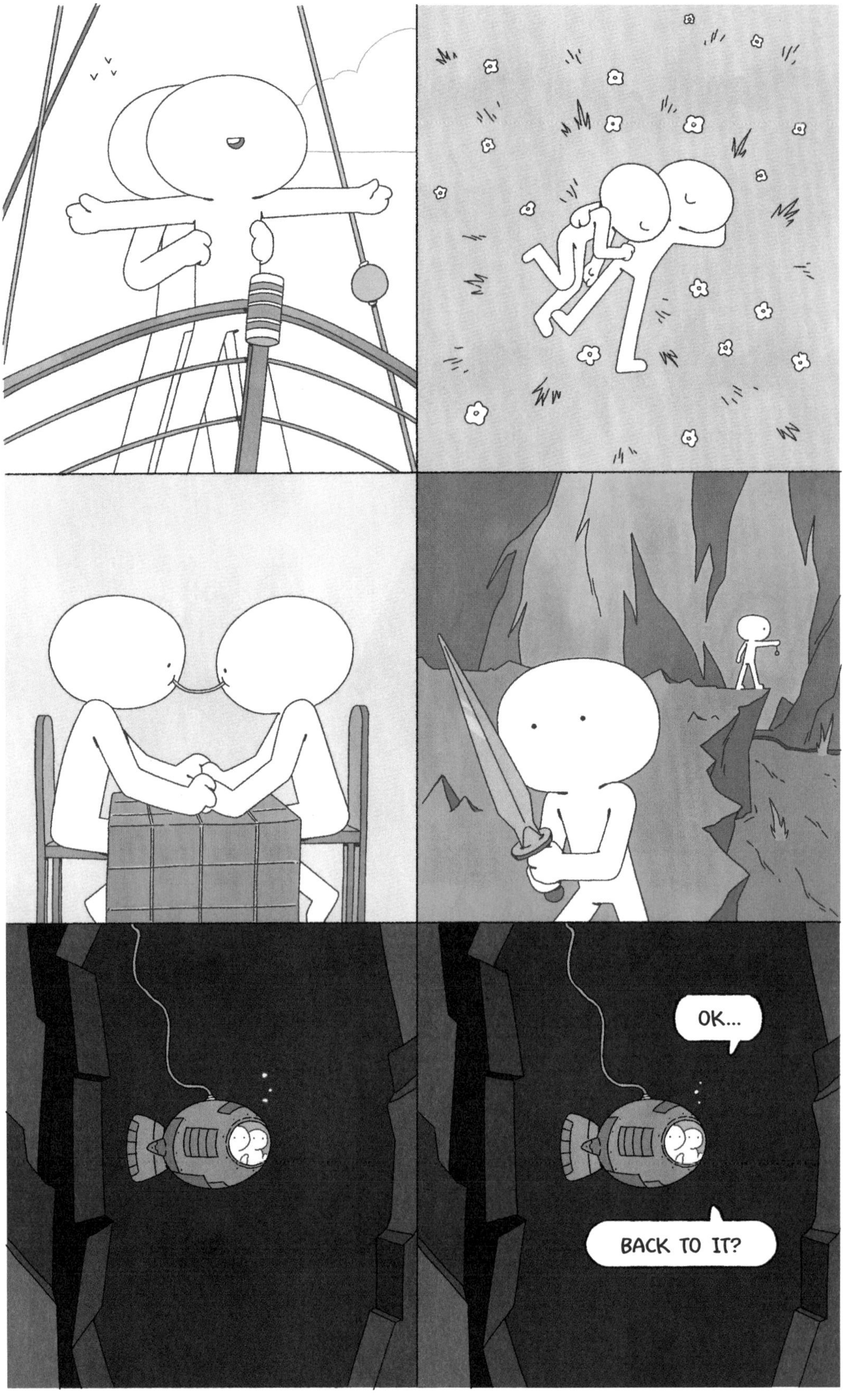
OK...
BACK TO IT?

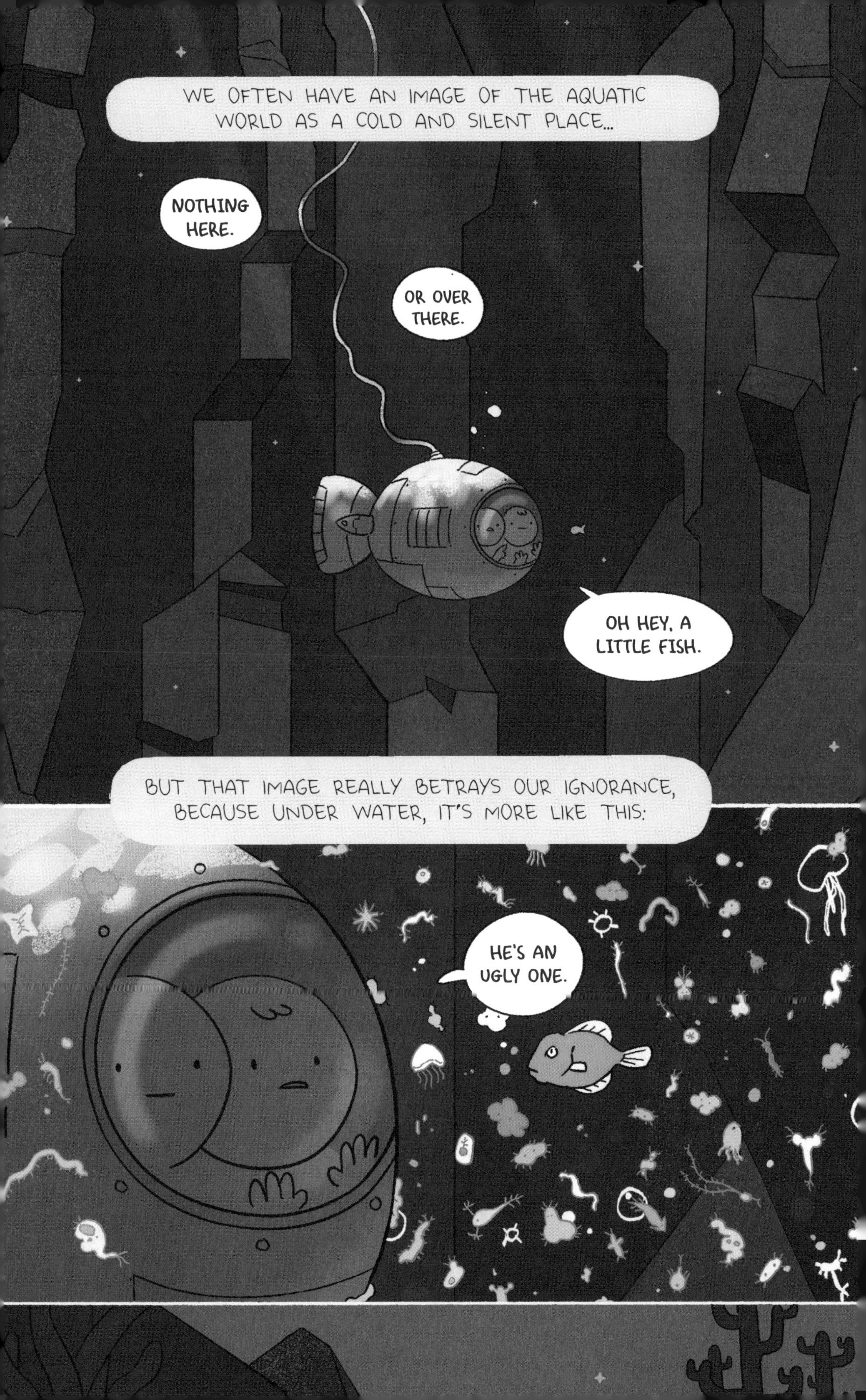
WE OFTEN HAVE AN IMAGE OF THE AQUATIC WORLD AS A COLD AND SILENT PLACE...
NOTHING HERE.
OR OVER THERE.
OH HEY, A LITTLE FISH.
BUT THAT IMAGE REALLY BETRAYS OUR IGNORANCE, BECAUSE UNDER WATER, IT'S MORE LIKE THIS:
HE'S AN UGLY ONE.

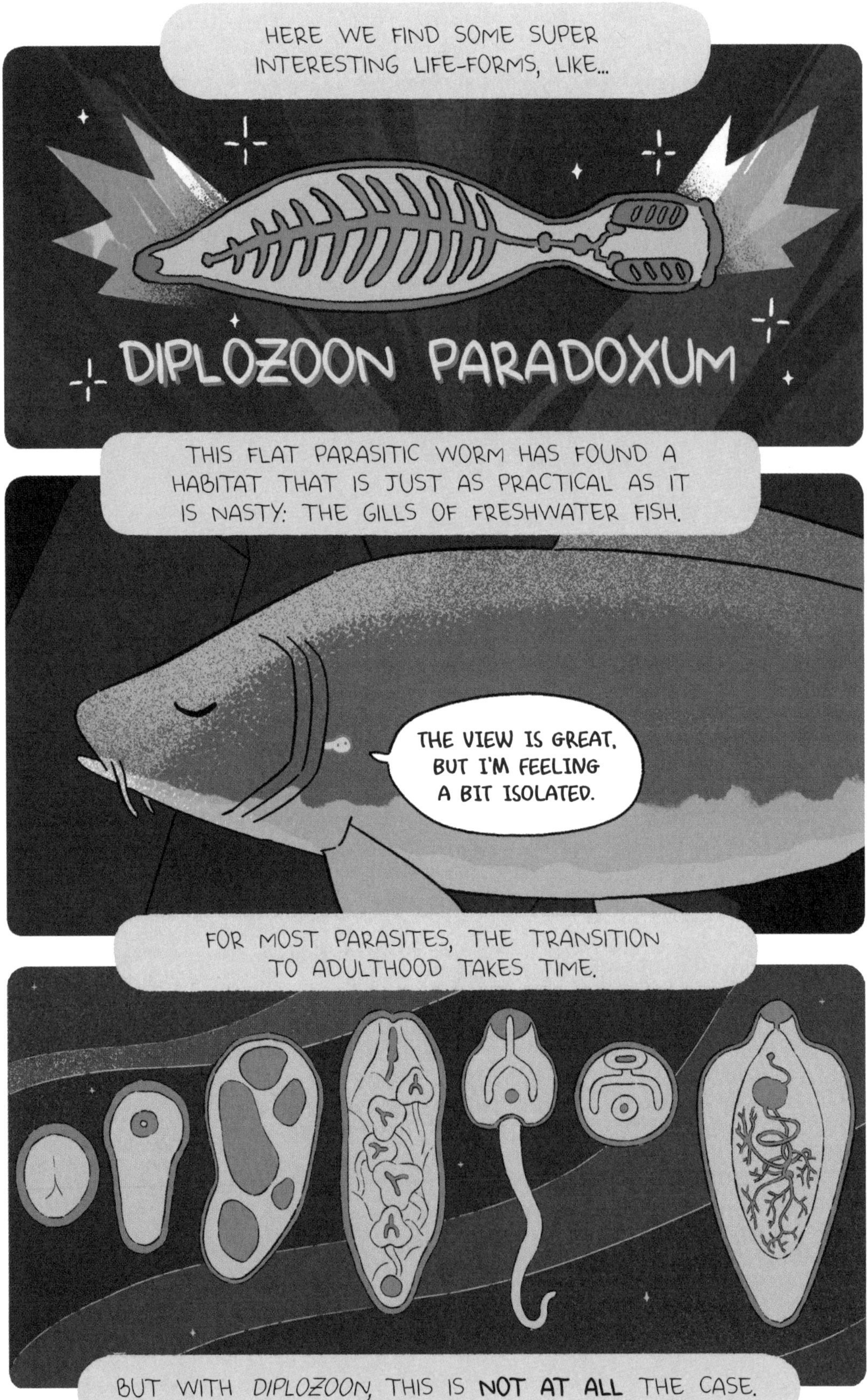
HERE WE FIND SOME SUPER INTERESTING LIFE-FORMS, LIKE...
DIPLOZOON PARADOXUM
THIS FLAT PARASITIC WORM HAS FOUND A HABITAT THAT IS JUST AS PRACTICAL AS IT IS NASTY: THE GILLS OF FRESHWATER FISH.
THE VIEW IS GREAT, BUT I'M FEELING A BIT ISOLATED.
FOR MOST PARASITES, THE TRANSITION TO ADULTHOOD TAKES TIME.
BUT WITH *DIPLOZOON*, THIS IS **NOT AT ALL** THE CASE.

DIPLOZOON CAN'T BECOME ADULTS UNLESS THEY MERGE WITH ANOTHER *DIPLOZOON* LARVA.

DIPLOZOON
ADULT

UM, HI...
HI...
YOU WANNA MAKE A MALE OR A FEMALE?
DOES IT MATTER?
NAH.

IT'S CONVENIENT BEING HERMAPHRODITIC.

THE WORMS ATTACH, ONE TO THE OTHER, AND THEN LITERALLY FUSE.

YOU IN THERE?

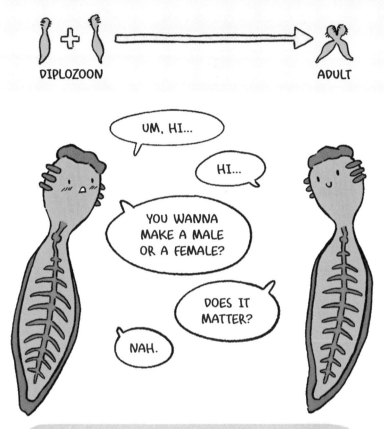

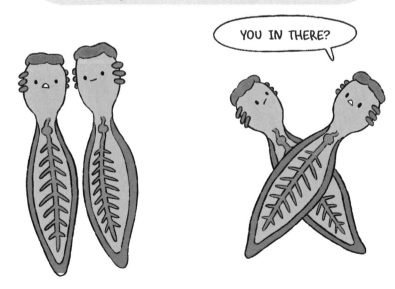

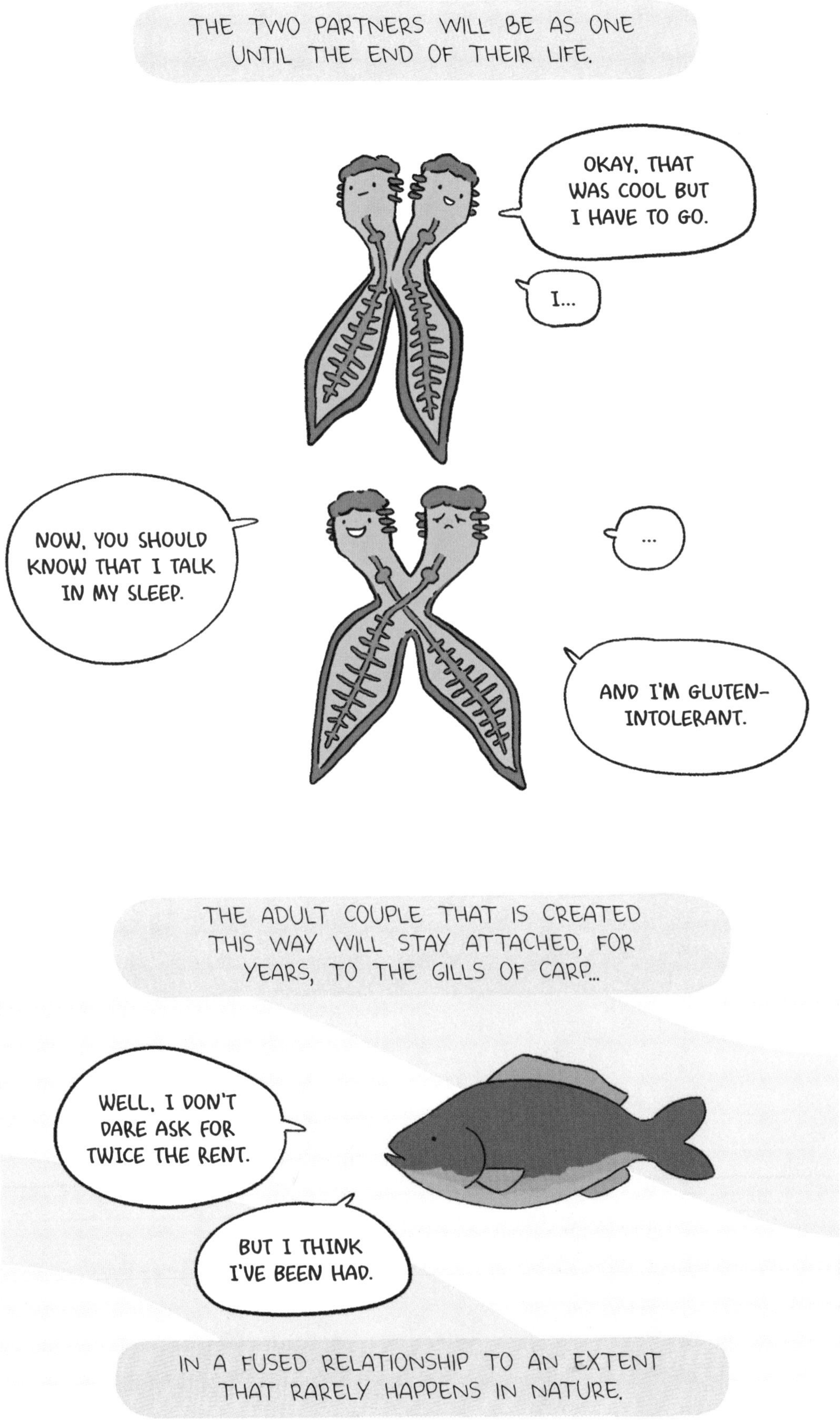
THE TWO PARTNERS WILL BE AS ONE UNTIL THE END OF THEIR LIFE.
OKAY, THAT WAS COOL BUT I HAVE TO GO.
I...
NOW, YOU SHOULD KNOW THAT I TALK IN MY SLEEP.
...
AND I'M GLUTEN-INTOLERANT.
THE ADULT COUPLE THAT IS CREATED THIS WAY WILL STAY ATTACHED, FOR YEARS, TO THE GILLS OF CARP...
WELL, I DON'T DARE ASK FOR TWICE THE RENT.
BUT I THINK I'VE BEEN HAD.
IN A FUSED RELATIONSHIP TO AN EXTENT THAT RARELY HAPPENS IN NATURE.

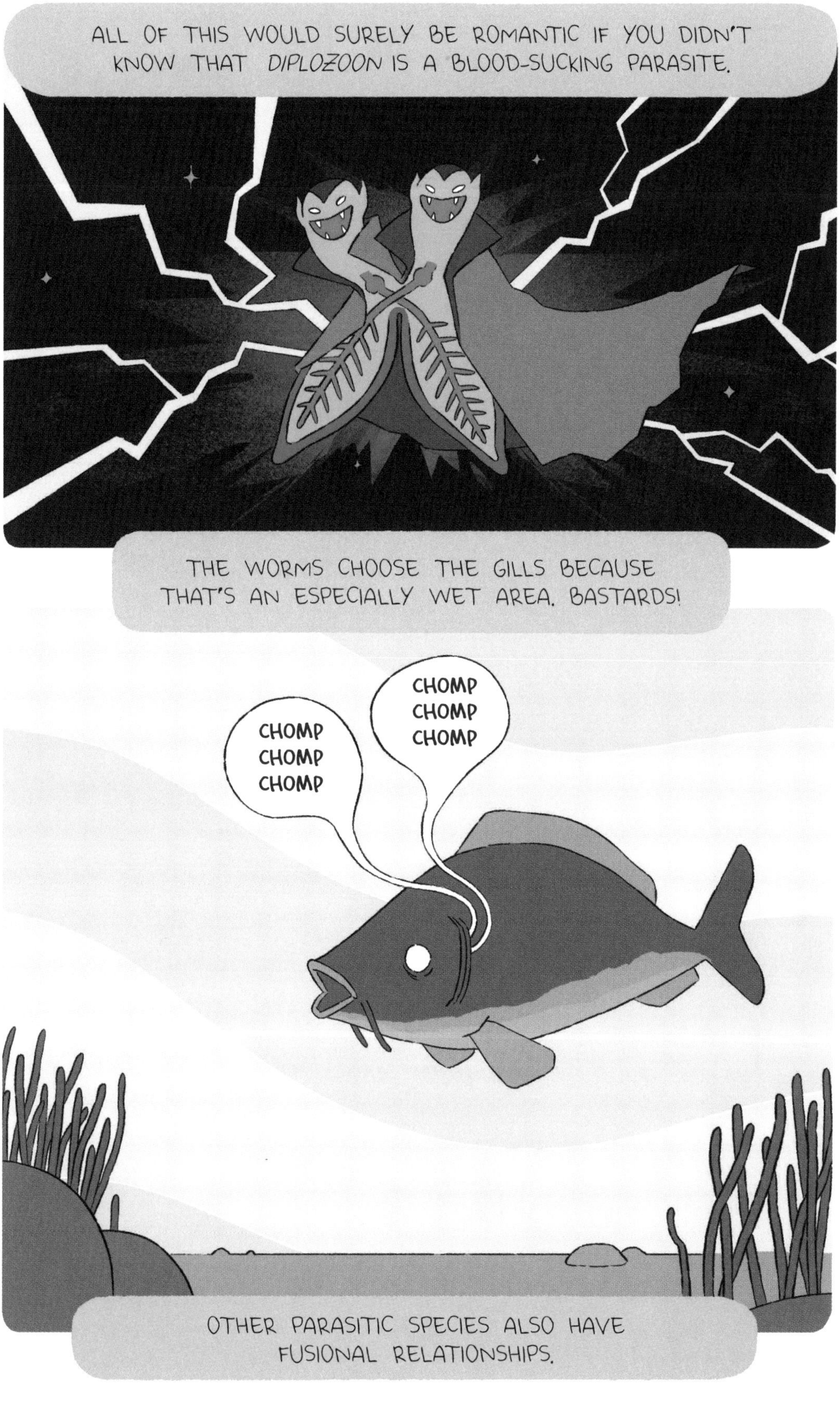
ALL OF THIS WOULD SURELY BE ROMANTIC IF YOU DIDN'T KNOW THAT *DIPLOZOON* IS A BLOOD-SUCKING PARASITE.
THE WORMS CHOOSE THE GILLS BECAUSE THAT'S AN ESPECIALLY WET AREA. BASTARDS!
CHOMP CHOMP CHOMP
CHOMP CHOMP CHOMP
OTHER PARASITIC SPECIES ALSO HAVE FUSIONAL RELATIONSHIPS.

FOR EXAMPLE, IN *SCHISTOSOMA* THE MALE HOSTS THE FEMALE IN HIS BODY, IN A CANAL MADE FOR THAT PURPOSE.

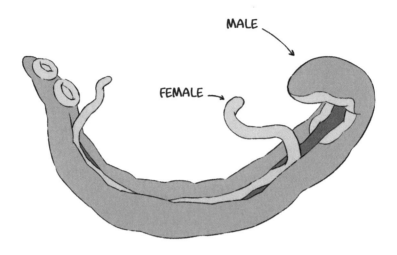

THE FRUIT OF THEIR LOVE IS THE PROLIFIC PRODUCTION OF EGGS (UP TO SEVERAL HUNDREDS EACH DAY), WHICH CAUSE A DEADLY ILLNESS IN HUMANS—SCHISTOSOMIASIS.

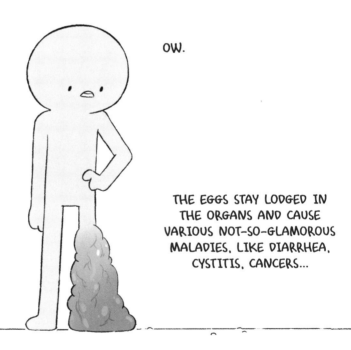

OW.

THE EGGS STAY LODGED IN THE ORGANS AND CAUSE VARIOUS NOT-SO-GLAMOROUS MALADIES, LIKE DIARRHEA, CYSTITIS, CANCERS...

180 MILLION PEOPLE ARE INFECTED WORLDWIDE, AND NEARLY 300,000 DIE EACH YEAR BECAUSE THESE LITTLE FUCKERS ARE CRAZY FOR EACH OTHER <3 <3

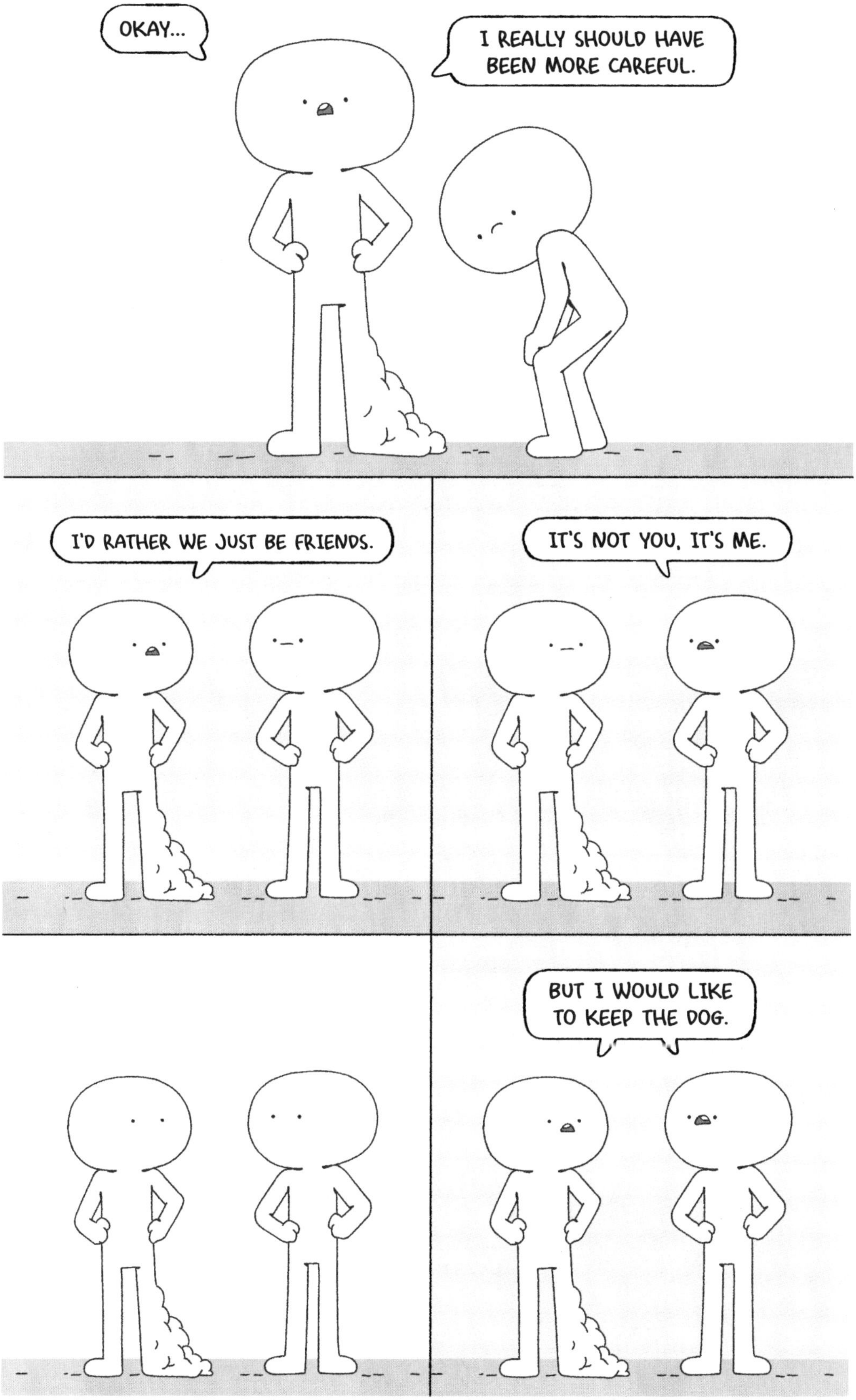
OKAY...
I REALLY SHOULD HAVE BEEN MORE CAREFUL.
I'D RATHER WE JUST BE FRIENDS.
IT'S NOT YOU, IT'S ME.
BUT I WOULD LIKE TO KEEP THE DOG.

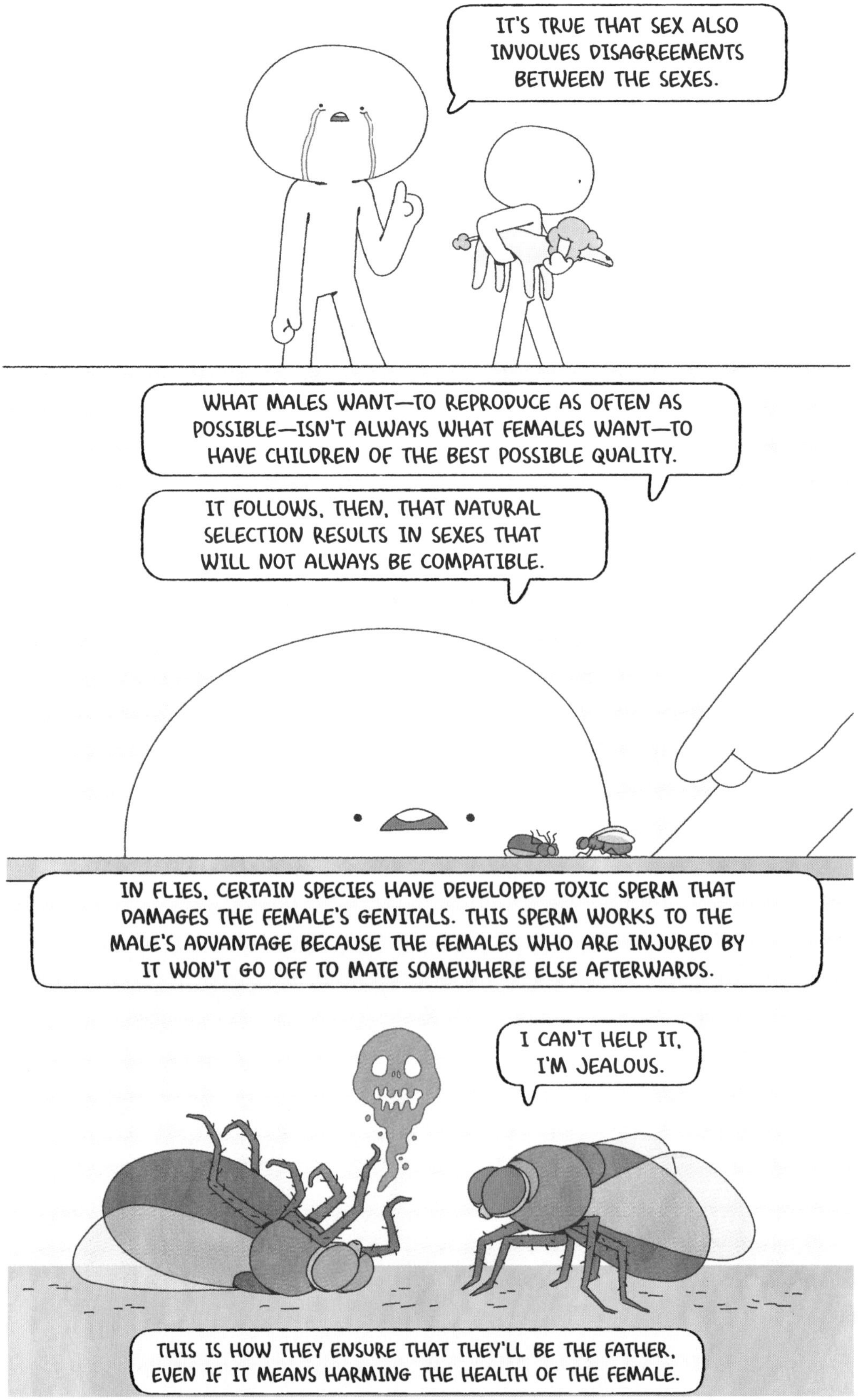
IT'S TRUE THAT SEX ALSO INVOLVES DISAGREEMENTS BETWEEN THE SEXES.
WHAT MALES WANT—TO REPRODUCE AS OFTEN AS POSSIBLE—ISN'T ALWAYS WHAT FEMALES WANT—TO HAVE CHILDREN OF THE BEST POSSIBLE QUALITY.
IT FOLLOWS, THEN, THAT NATURAL SELECTION RESULTS IN SEXES THAT WILL NOT ALWAYS BE COMPATIBLE.
IN FLIES, CERTAIN SPECIES HAVE DEVELOPED TOXIC SPERM THAT DAMAGES THE FEMALE'S GENITALS. THIS SPERM WORKS TO THE MALE'S ADVANTAGE BECAUSE THE FEMALES WHO ARE INJURED BY IT WON'T GO OFF TO MATE SOMEWHERE ELSE AFTERWARDS.
I CAN'T HELP IT, I'M JEALOUS.
THIS IS HOW THEY ENSURE THAT THEY'LL BE THE FATHER, EVEN IF IT MEANS HARMING THE HEALTH OF THE FEMALE.

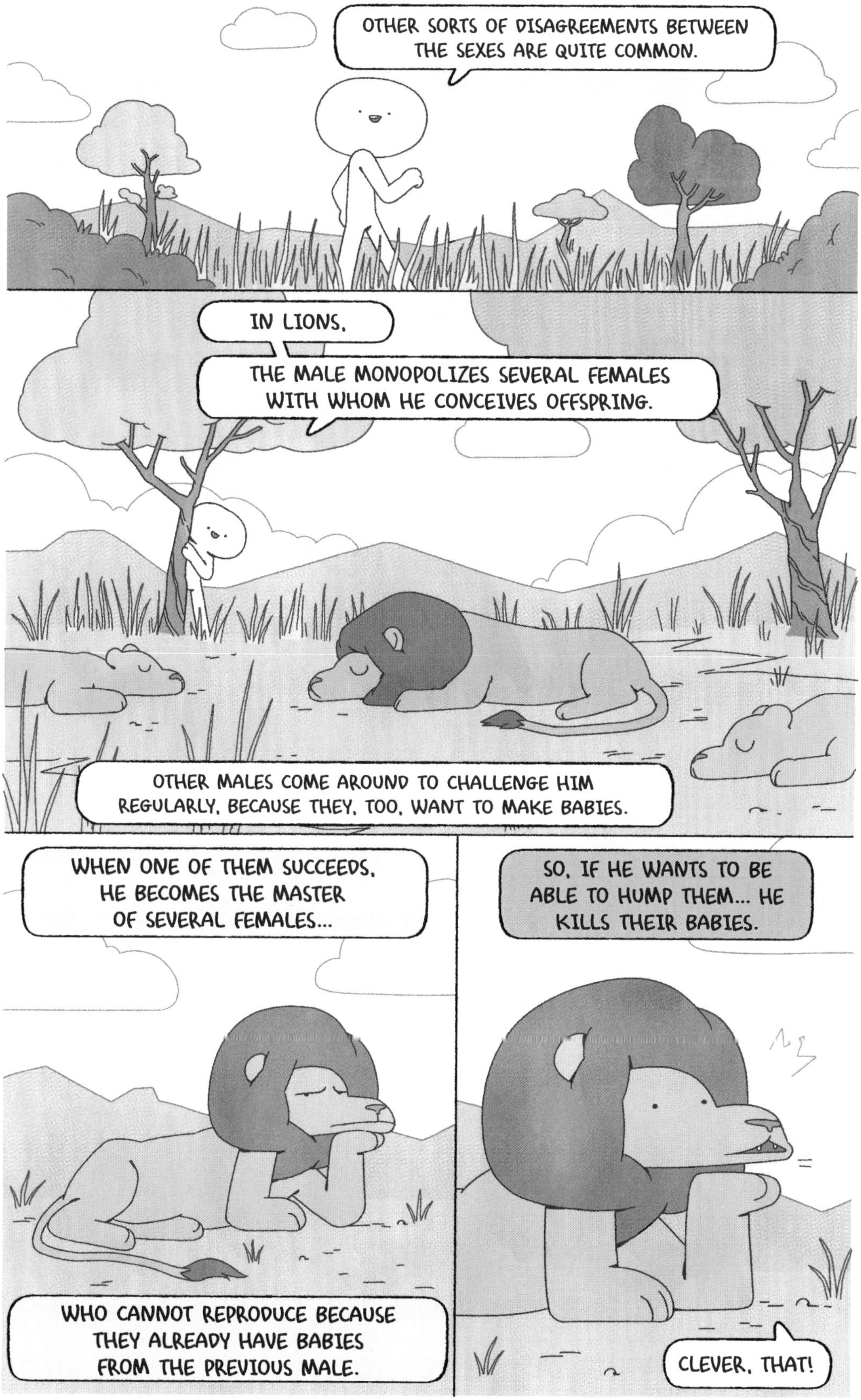
OTHER SORTS OF DISAGREEMENTS BETWEEN THE SEXES ARE QUITE COMMON.
IN LIONS,
THE MALE MONOPOLIZES SEVERAL FEMALES WITH WHOM HE CONCEIVES OFFSPRING.
OTHER MALES COME AROUND TO CHALLENGE HIM REGULARLY, BECAUSE THEY, TOO, WANT TO MAKE BABIES.
WHEN ONE OF THEM SUCCEEDS, HE BECOMES THE MASTER OF SEVERAL FEMALES...
WHO CANNOT REPRODUCE BECAUSE THEY ALREADY HAVE BABIES FROM THE PREVIOUS MALE.
SO, IF HE WANTS TO BE ABLE TO HUMP THEM... HE KILLS THEIR BABIES.
CLEVER, THAT!

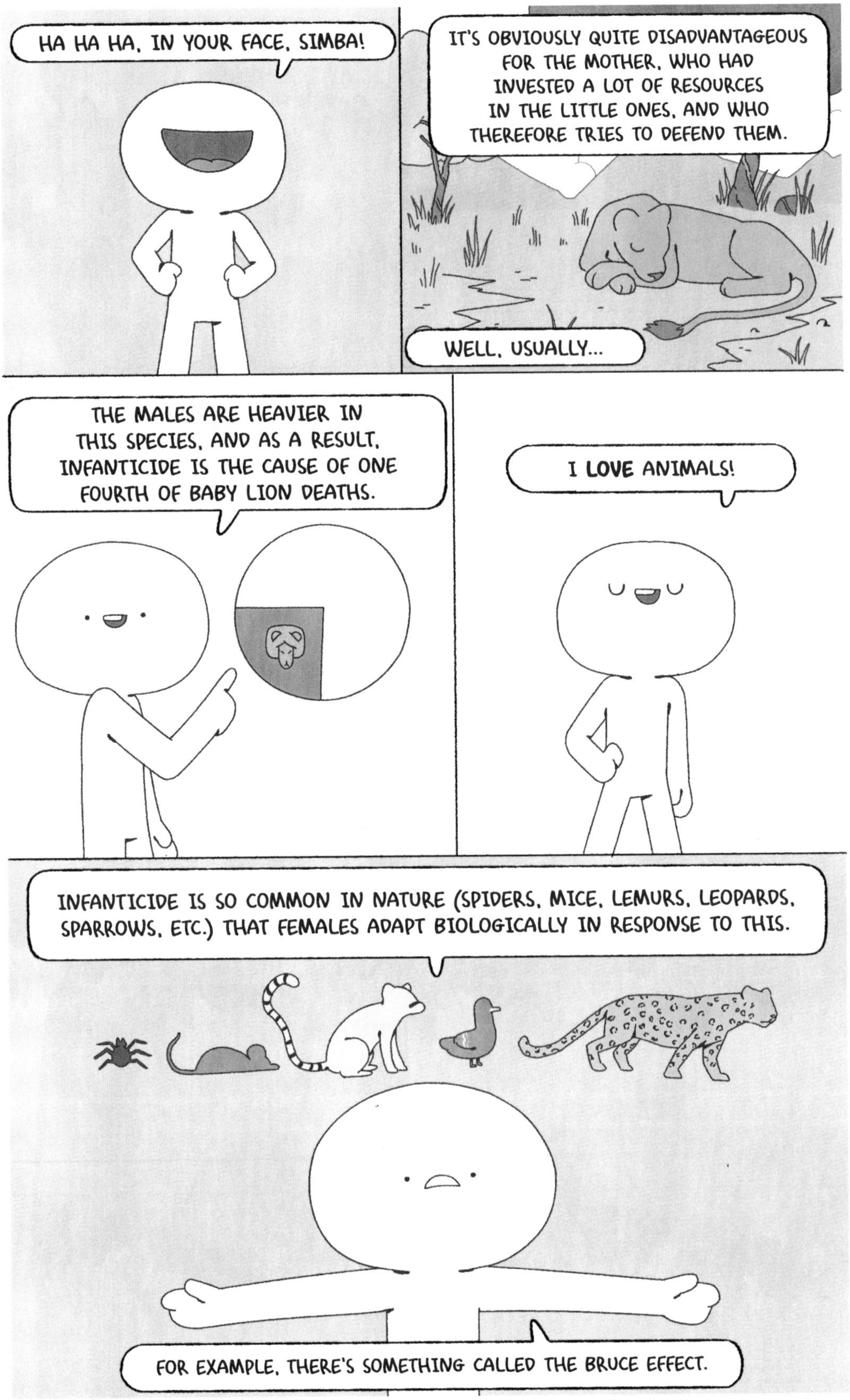
HA HA HA, IN YOUR FACE, SIMBA!
IT'S OBVIOUSLY QUITE DISADVANTAGEOUS FOR THE MOTHER, WHO HAD INVESTED A LOT OF RESOURCES IN THE LITTLE ONES, AND WHO THEREFORE TRIES TO DEFEND THEM.
WELL, USUALLY...
THE MALES ARE HEAVIER IN THIS SPECIES, AND AS A RESULT, INFANTICIDE IS THE CAUSE OF ONE FOURTH OF BABY LION DEATHS.
I **LOVE** ANIMALS!
INFANTICIDE IS SO COMMON IN NATURE (SPIDERS, MICE, LEMURS, LEOPARDS, SPARROWS, ETC.) THAT FEMALES ADAPT BIOLOGICALLY IN RESPONSE TO THIS.
FOR EXAMPLE, THERE'S SOMETHING CALLED THE BRUCE EFFECT.

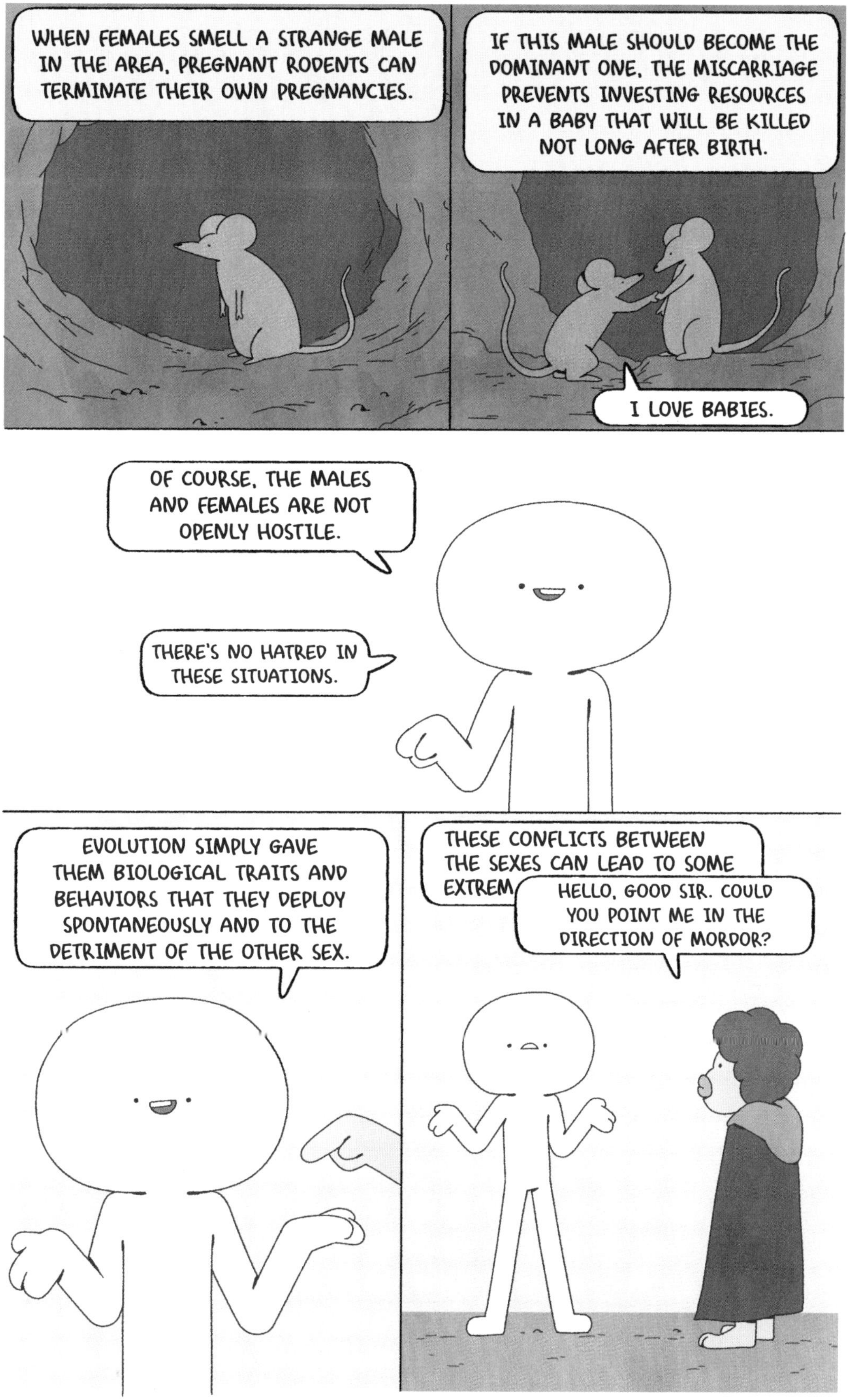
WHEN FEMALES SMELL A STRANGE MALE IN THE AREA, PREGNANT RODENTS CAN TERMINATE THEIR OWN PREGNANCIES.
IF THIS MALE SHOULD BECOME THE DOMINANT ONE, THE MISCARRIAGE PREVENTS INVESTING RESOURCES IN A BABY THAT WILL BE KILLED NOT LONG AFTER BIRTH.
I LOVE BABIES.
OF COURSE, THE MALES AND FEMALES ARE NOT OPENLY HOSTILE.
THERE'S NO HATRED IN THESE SITUATIONS.
EVOLUTION SIMPLY GAVE THEM BIOLOGICAL TRAITS AND BEHAVIORS THAT THEY DEPLOY SPONTANEOUSLY AND TO THE DETRIMENT OF THE OTHER SEX.
THESE CONFLICTS BETWEEN THE SEXES CAN LEAD TO SOME EXTREM
HELLO, GOOD SIR. COULD YOU POINT ME IN THE DIRECTION OF MORDOR?

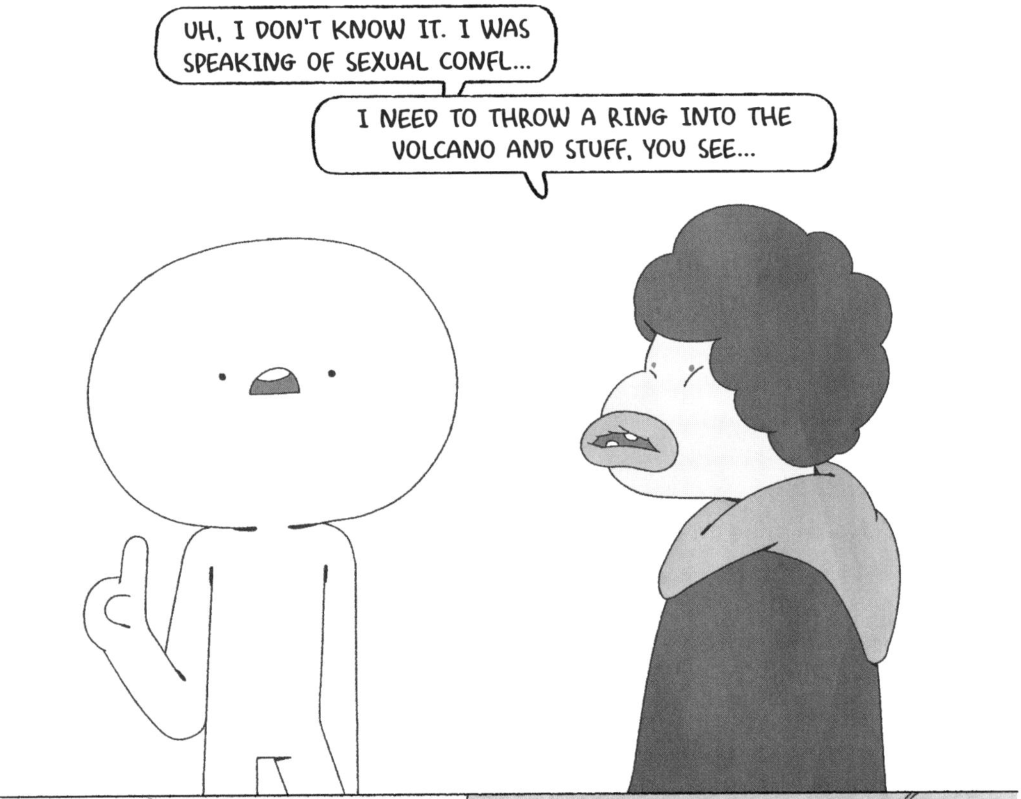
UH, I DON'T KNOW IT. I WAS SPEAKING OF SEXUAL CONFL...
I NEED TO THROW A RING INTO THE VOLCANO AND STUFF, YOU SEE...

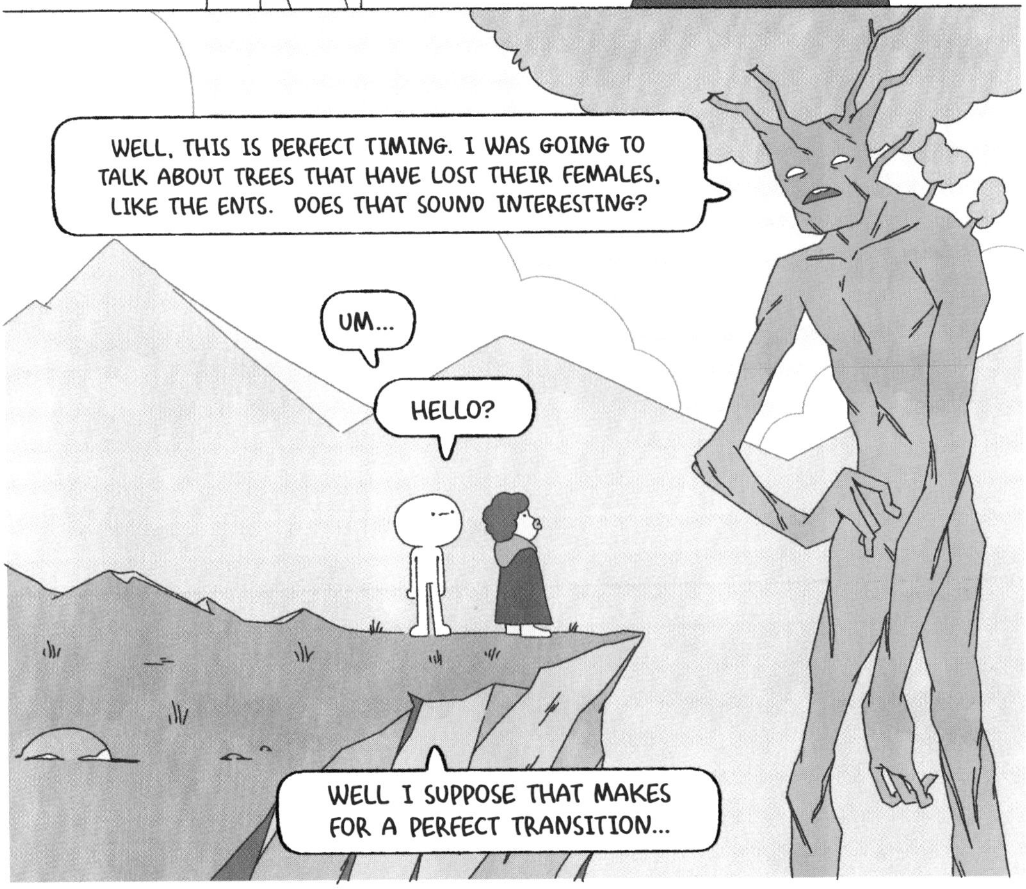
WELL, THIS IS PERFECT TIMING. I WAS GOING TO TALK ABOUT TREES THAT HAVE LOST THEIR FEMALES, LIKE THE ENTS. DOES THAT SOUND INTERESTING?
UM...
HELLO?
WELL I SUPPOSE THAT MAKES FOR A PERFECT TRANSITION...

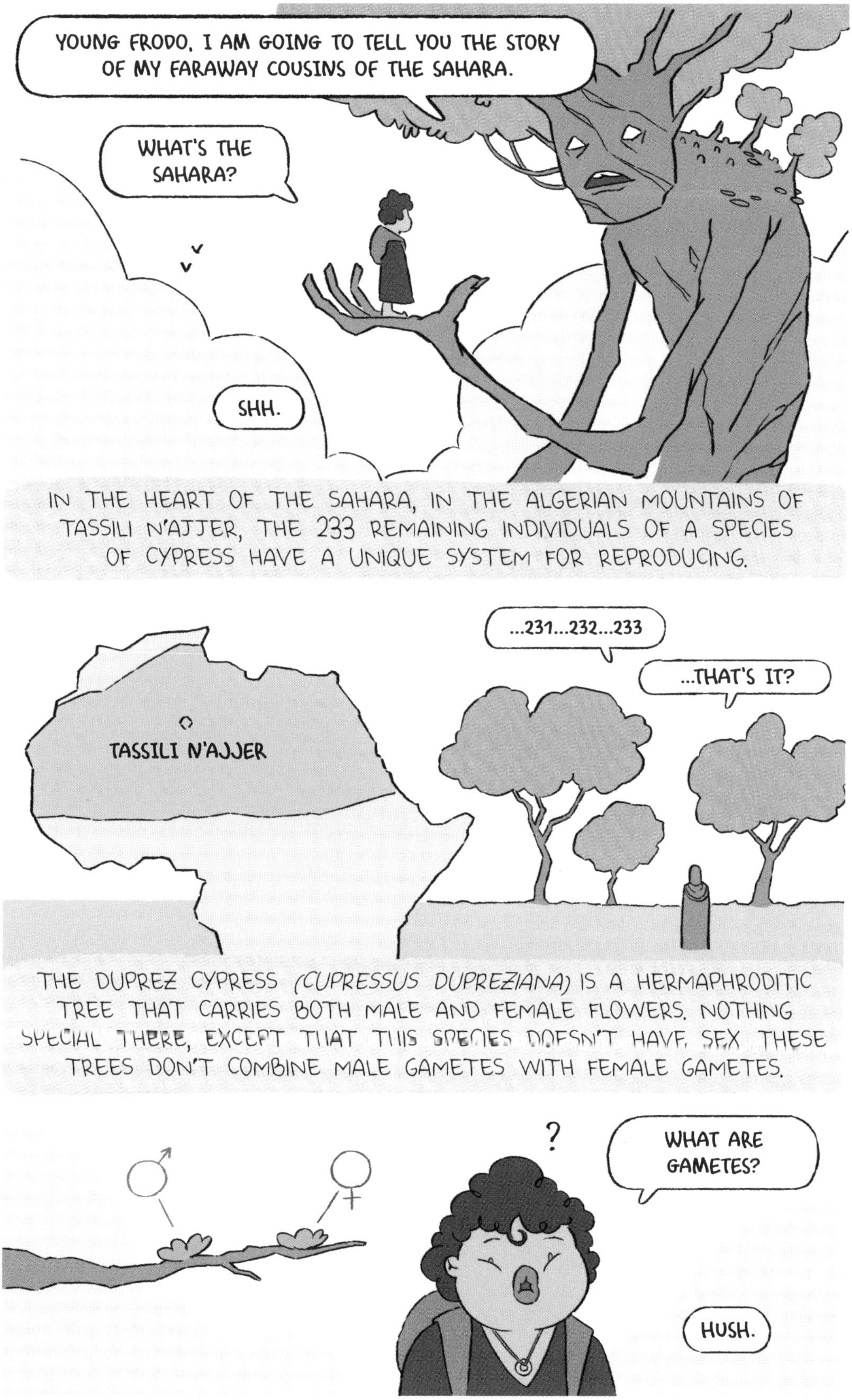
YOUNG FRODO, I AM GOING TO TELL YOU THE STORY OF MY FARAWAY COUSINS OF THE SAHARA.
WHAT'S THE SAHARA?
SHH.
IN THE HEART OF THE SAHARA, IN THE ALGERIAN MOUNTAINS OF TASSILI N'AJJER, THE 233 REMAINING INDIVIDUALS OF A SPECIES OF CYPRESS HAVE A UNIQUE SYSTEM FOR REPRODUCING.
TASSILI N'AJJER
...231...232...233
...THAT'S IT?
THE DUPREZ CYPRESS *(CUPRESSUS DUPREZIANA)* IS A HERMAPHRODITIC TREE THAT CARRIES BOTH MALE AND FEMALE FLOWERS. NOTHING SPECIAL THERE, EXCEPT THAT THIS SPECIES DOESN'T HAVE SEX. THESE TREES DON'T COMBINE MALE GAMETES WITH FEMALE GAMETES.
?
WHAT ARE GAMETES?
HUSH.

NORMALLY, POLLEN HAS ONLY HALF OF THE GENES OF THE PLANT THAT PRODUCED IT. THE FEMALE EGG CELL WILL THEN PROVIDE THE OTHER HALF.

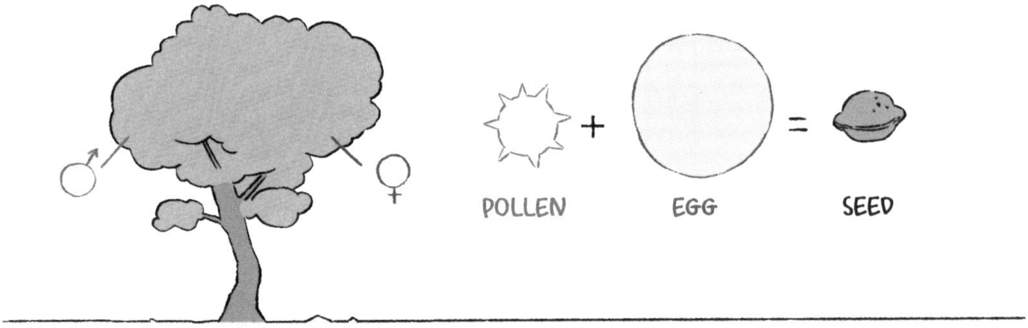
POLLEN
EGG
SEED

BUT THAT'S NOT THE CASE HERE:

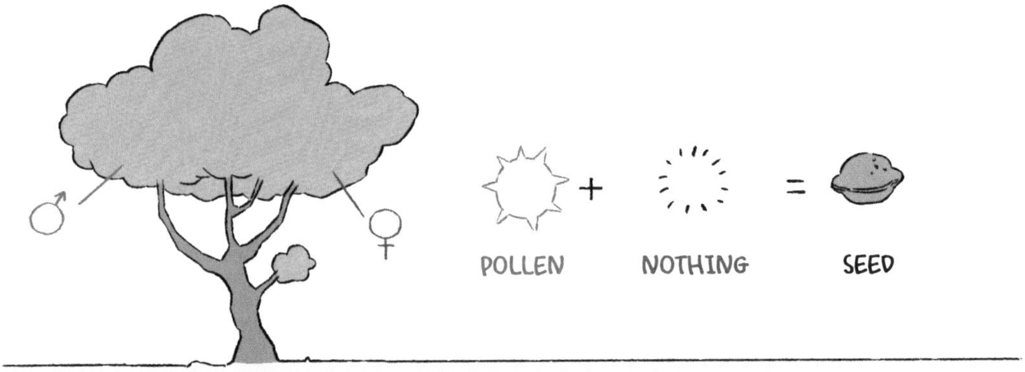
POLLEN
NOTHING
SEED

IT WILL SETTLE IN THE FEMALE FLOWERS, REMOVE ANY FEMALE CELLS THAT GET IN ITS WAY, AND DEVELOP AS AN EMBRYO.

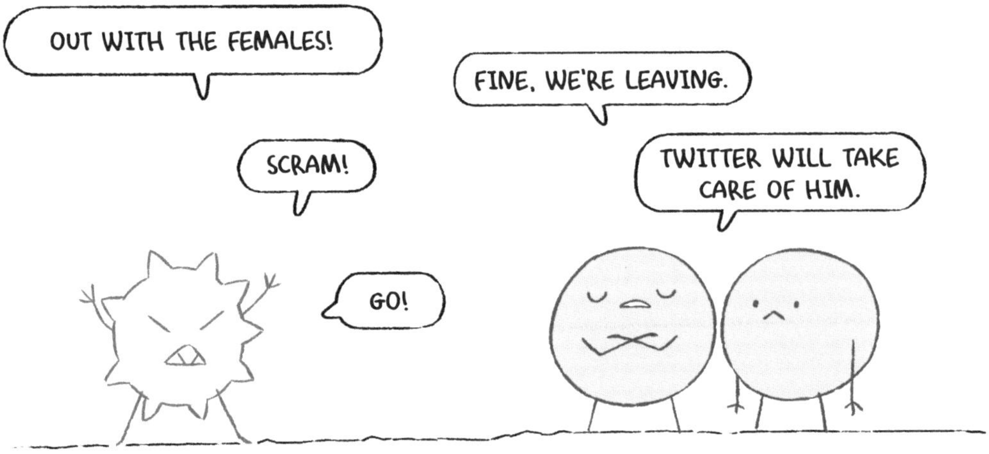
OUT WITH THE FEMALES!
FINE, WE'RE LEAVING.
SCRAM!
TWITTER WILL TAKE CARE OF HIM.
GO!

POLLEN FROM THE MALE CONTAINS THE FULL PACKAGE OF GENES FROM THE PARENT PLANT!

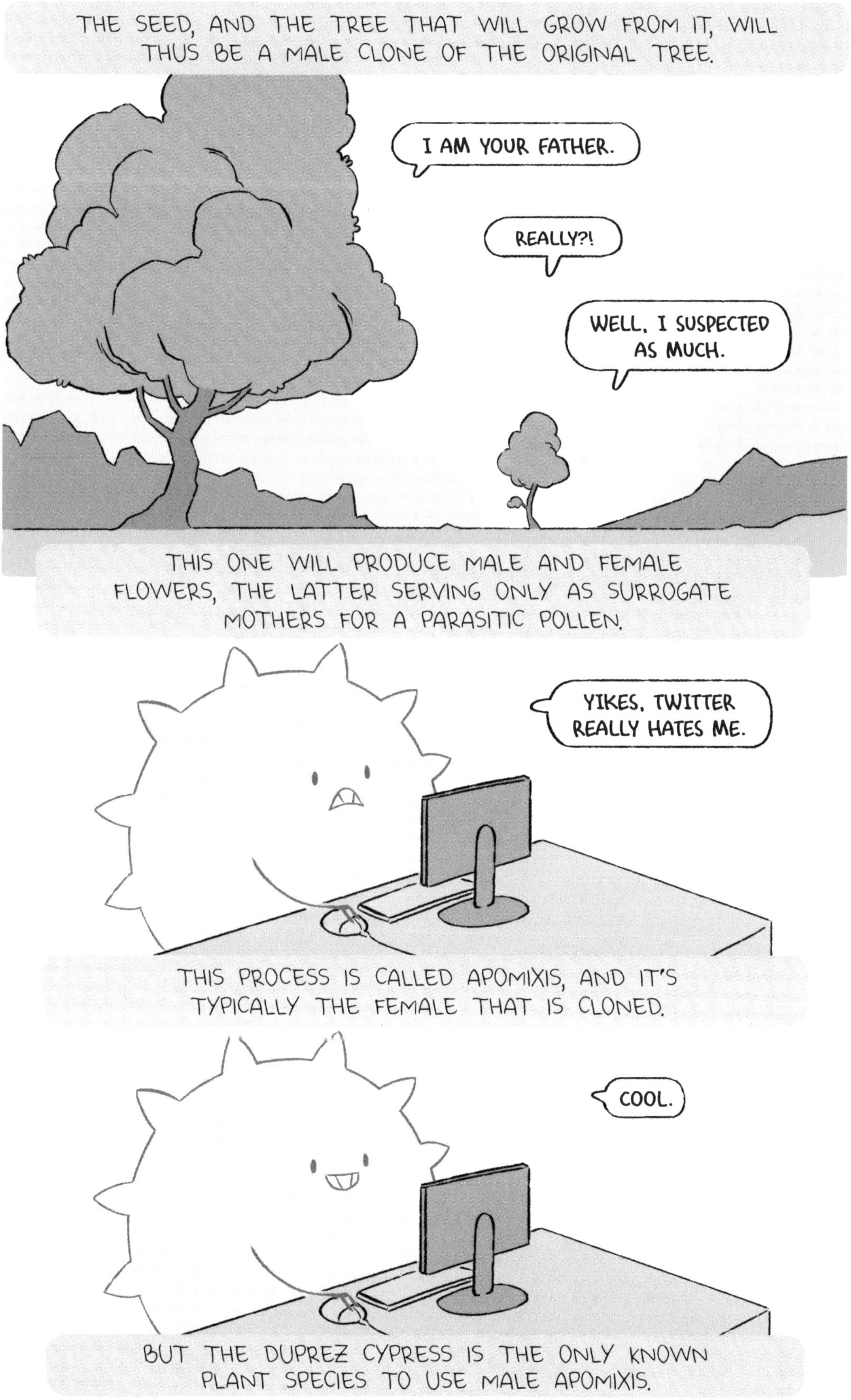
THE SEED, AND THE TREE THAT WILL GROW FROM IT, WILL THUS BE A MALE CLONE OF THE ORIGINAL TREE.
I AM YOUR FATHER.
REALLY?!
WELL, I SUSPECTED AS MUCH.
THIS ONE WILL PRODUCE MALE AND FEMALE FLOWERS, THE LATTER SERVING ONLY AS SURROGATE MOTHERS FOR A PARASITIC POLLEN.
YIKES, TWITTER REALLY HATES ME.
THIS PROCESS IS CALLED APOMIXIS, AND IT'S TYPICALLY THE FEMALE THAT IS CLONED.
COOL.
BUT THE DUPREZ CYPRESS IS THE ONLY KNOWN PLANT SPECIES TO USE MALE APOMIXIS.

THE SPECIMENS THAT EXIST TODAY ARE FOR THE MOST PART VERY OLD, UNIQUELY MALE, AS OLD AS THE FIRST HUMANS, AND LIVING IN A REMOTE AREA.

THESE DESCENDANTS OF THE VERY OLDEST PATRIARCHY IN THE WORLD ARE A BIT LIKE THE ENTS OF OUR PLANET.

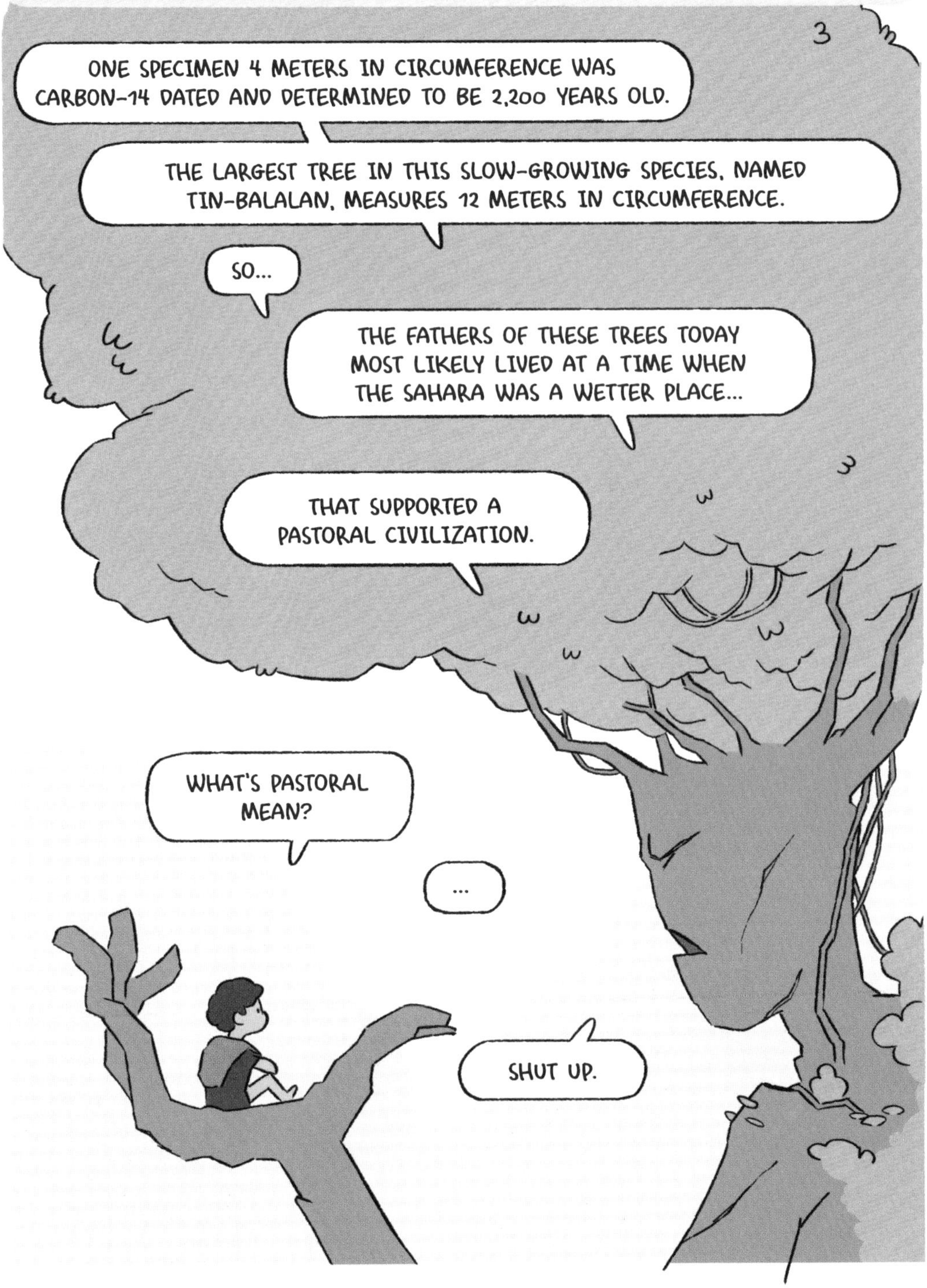
ONE SPECIMEN 4 METERS IN CIRCUMFERENCE WAS CARBON-14 DATED AND DETERMINED TO BE 2,200 YEARS OLD.
THE LARGEST TREE IN THIS SLOW-GROWING SPECIES, NAMED TIN-BALALAN, MEASURES 12 METERS IN CIRCUMFERENCE.
SO...
THE FATHERS OF THESE TREES TODAY MOST LIKELY LIVED AT A TIME WHEN THE SAHARA WAS A WETTER PLACE...
THAT SUPPORTED A PASTORAL CIVILIZATION.
WHAT'S PASTORAL MEAN?
...
SHUT UP.

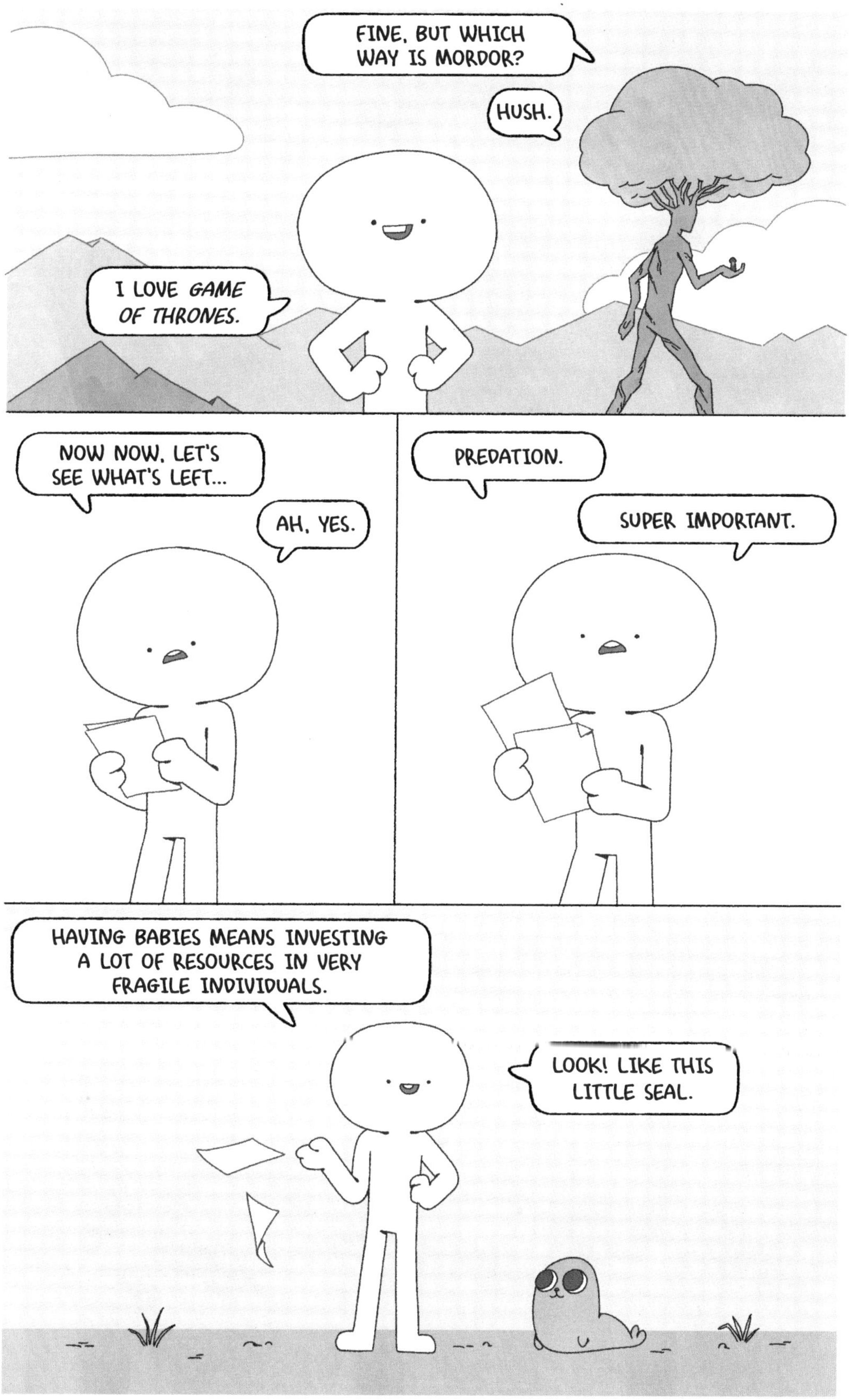
FINE, BUT WHICH WAY IS MORDOR?
HUSH.
I LOVE GAME OF THRONES.
NOW NOW, LET'S SEE WHAT'S LEFT...
AH, YES.
PREDATION.
SUPER IMPORTANT.
HAVING BABIES MEANS INVESTING A LOT OF RESOURCES IN VERY FRAGILE INDIVIDUALS.
LOOK! LIKE THIS LITTLE SEAL.

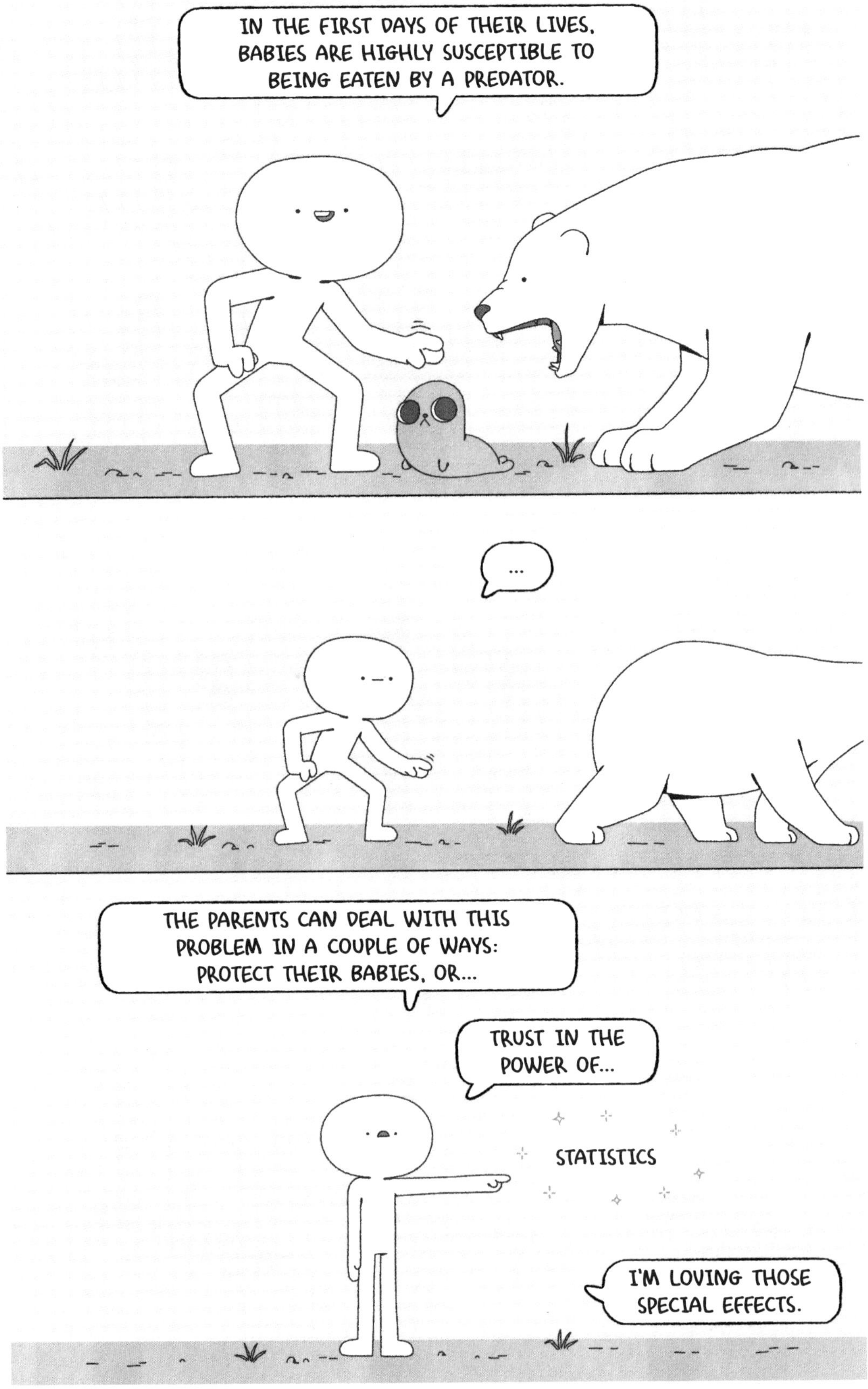
IN THE FIRST DAYS OF THEIR LIVES, BABIES ARE HIGHLY SUSCEPTIBLE TO BEING EATEN BY A PREDATOR.
...
THE PARENTS CAN DEAL WITH THIS PROBLEM IN A COUPLE OF WAYS: PROTECT THEIR BABIES, OR...
TRUST IN THE POWER OF...
STATISTICS
I'M LOVING THOSE SPECIAL EFFECTS.

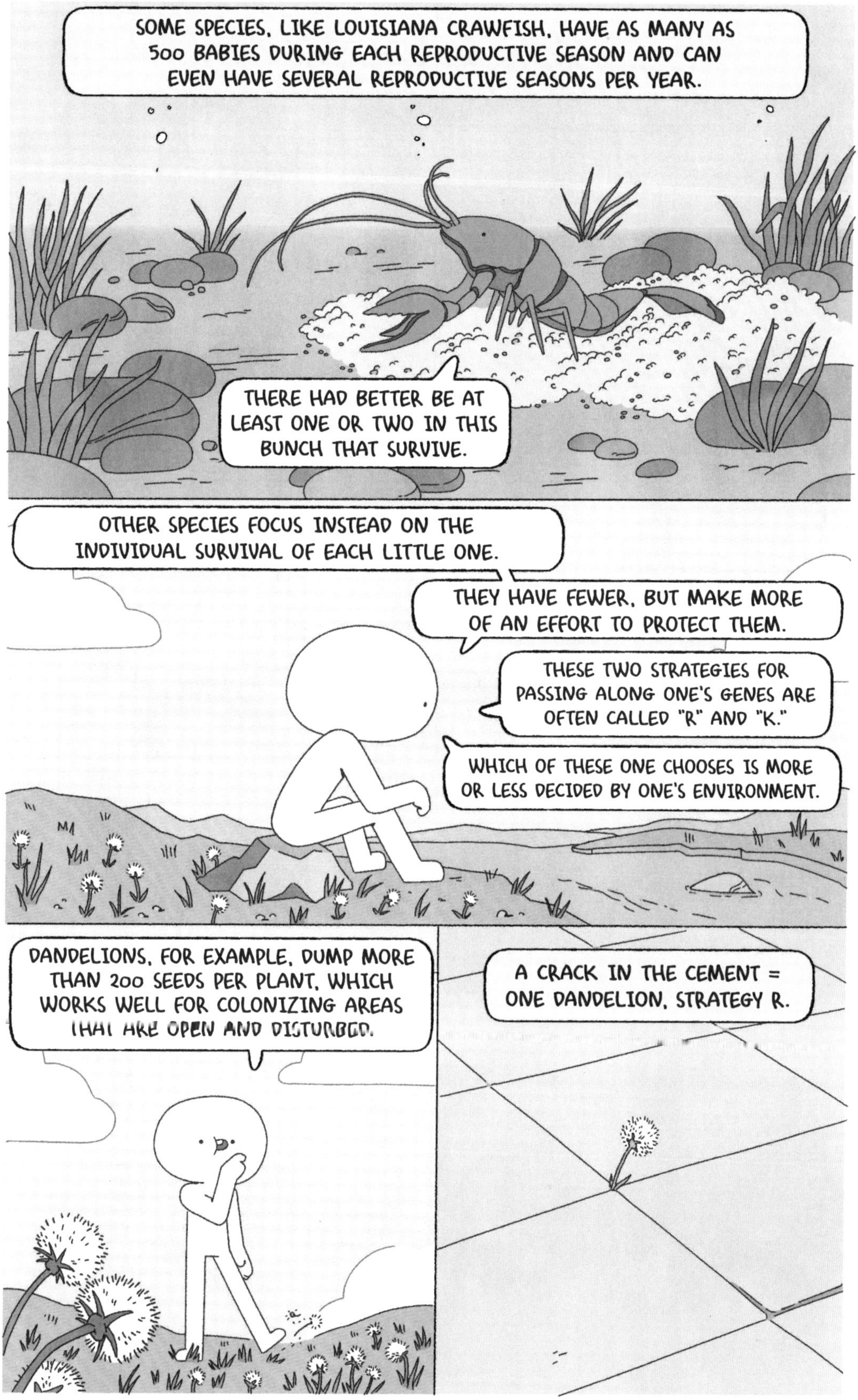
SOME SPECIES, LIKE LOUISIANA CRAWFISH, HAVE AS MANY AS 500 BABIES DURING EACH REPRODUCTIVE SEASON AND CAN EVEN HAVE SEVERAL REPRODUCTIVE SEASONS PER YEAR.
THERE HAD BETTER BE AT LEAST ONE OR TWO IN THIS BUNCH THAT SURVIVE.
OTHER SPECIES FOCUS INSTEAD ON THE INDIVIDUAL SURVIVAL OF EACH LITTLE ONE.
THEY HAVE FEWER, BUT MAKE MORE OF AN EFFORT TO PROTECT THEM.
THESE TWO STRATEGIES FOR PASSING ALONG ONE'S GENES ARE OFTEN CALLED "R" AND "K."
WHICH OF THESE ONE CHOOSES IS MORE OR LESS DECIDED BY ONE'S ENVIRONMENT.
DANDELIONS, FOR EXAMPLE, DUMP MORE THAN 200 SEEDS PER PLANT, WHICH WORKS WELL FOR COLONIZING AREAS THAT ARE OPEN AND DISTURBED.
A CRACK IN THE CEMENT = ONE DANDELION, STRATEGY R.

INVERSELY, OTHER PLANTS, SUCH AS AVOCADOS, PRODUCE ONLY A FEW SEEDS.

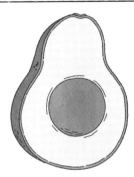

BUT EACH IS WELL PACKED INSIDE A FRUIT THAT IS RICH IN RESOURCES...

TO ATTRACT ANIMALS THAT WILL DISPERSE IT (STRATEGY K).

I GUESS THIS IS A BAD EXAMPLE, BECAUSE THE ANIMALS THAT DISPERSE AVOCADO SEEDS, GIANT SLOTHS AND CERTAIN PACHYDERMS (GOMPHOTHERES), DIED OFF THOUSANDS OF YEARS AGO. WITH THAT, THIS SPECIES LOST ITS REPRODUCTIVE PARTNER.

GIANT SLOTH

IN SHORT, DIFFERENT STRATEGIES EXIST TO FACILITATE THE TRANSFER OF GENES TO THE NEXT GENERATION.

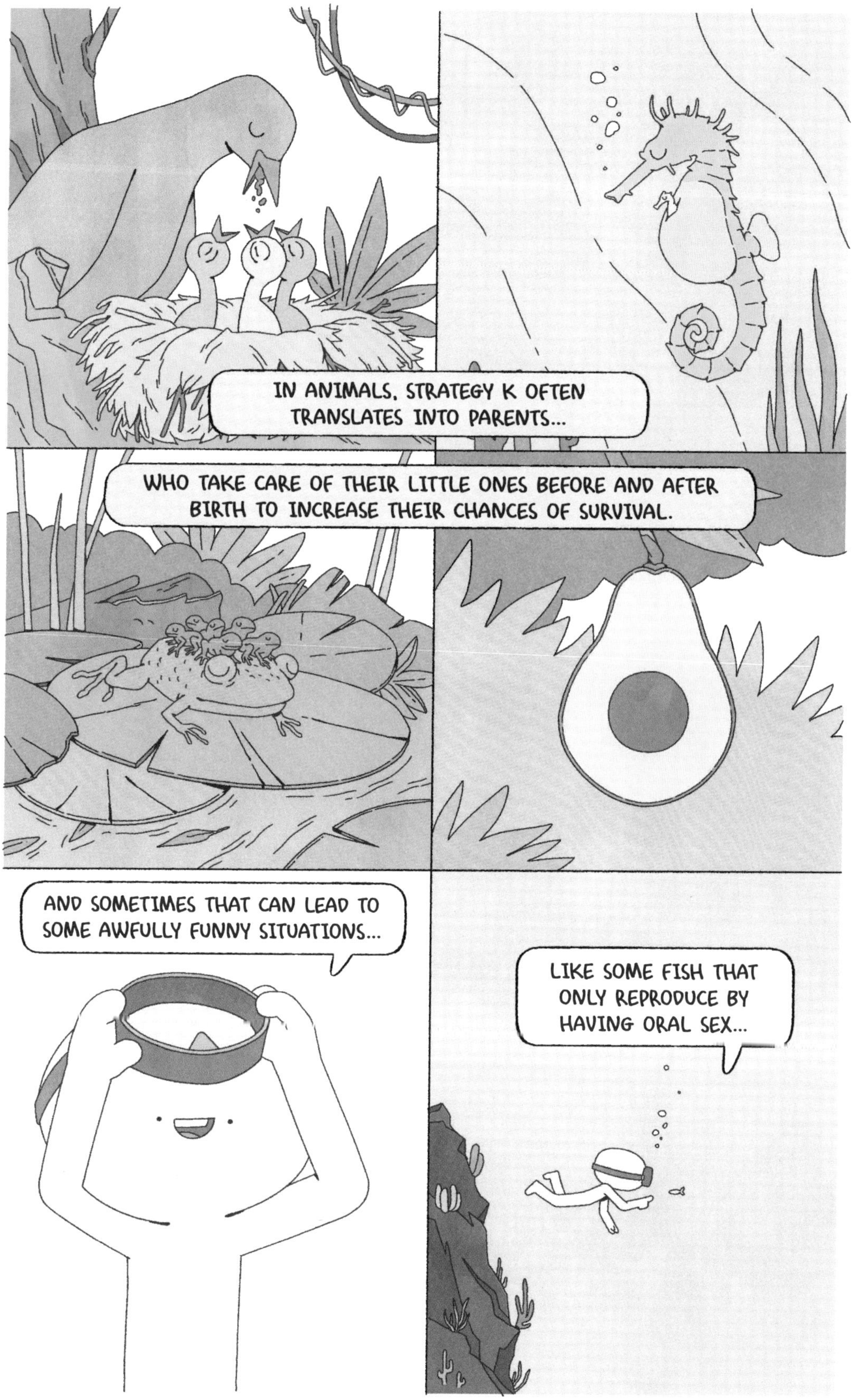
IN ANIMALS, STRATEGY K OFTEN TRANSLATES INTO PARENTS...
WHO TAKE CARE OF THEIR LITTLE ONES BEFORE AND AFTER BIRTH TO INCREASE THEIR CHANCES OF SURVIVAL.
AND SOMETIMES THAT CAN LEAD TO SOME AWFULLY FUNNY SITUATIONS...
LIKE SOME FISH THAT ONLY REPRODUCE BY HAVING ORAL SEX...

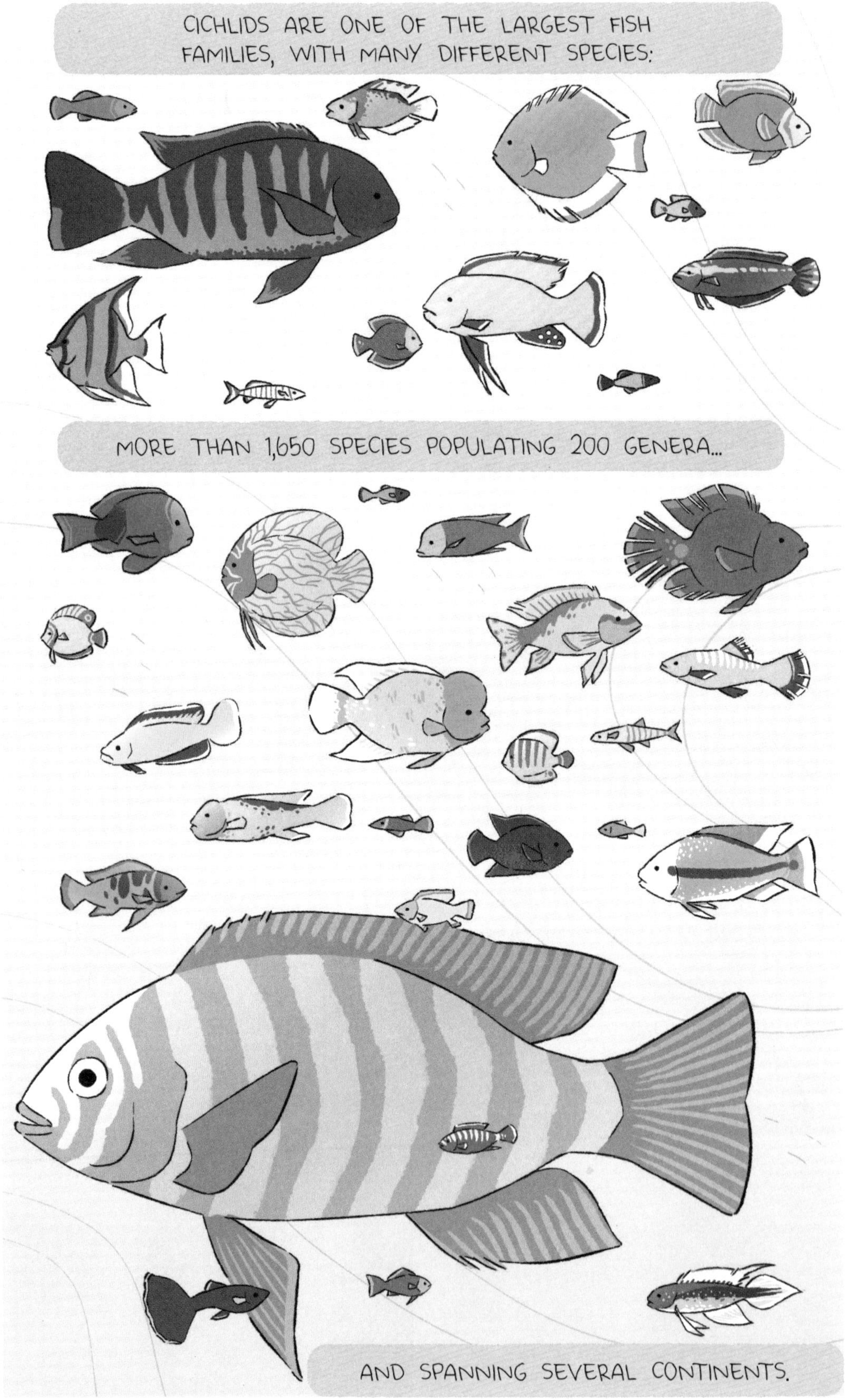
CICHLIDS ARE ONE OF THE LARGEST FISH FAMILIES, WITH MANY DIFFERENT SPECIES:
MORE THAN 1,650 SPECIES POPULATING 200 GENERA...
AND SPANNING SEVERAL CONTINENTS.

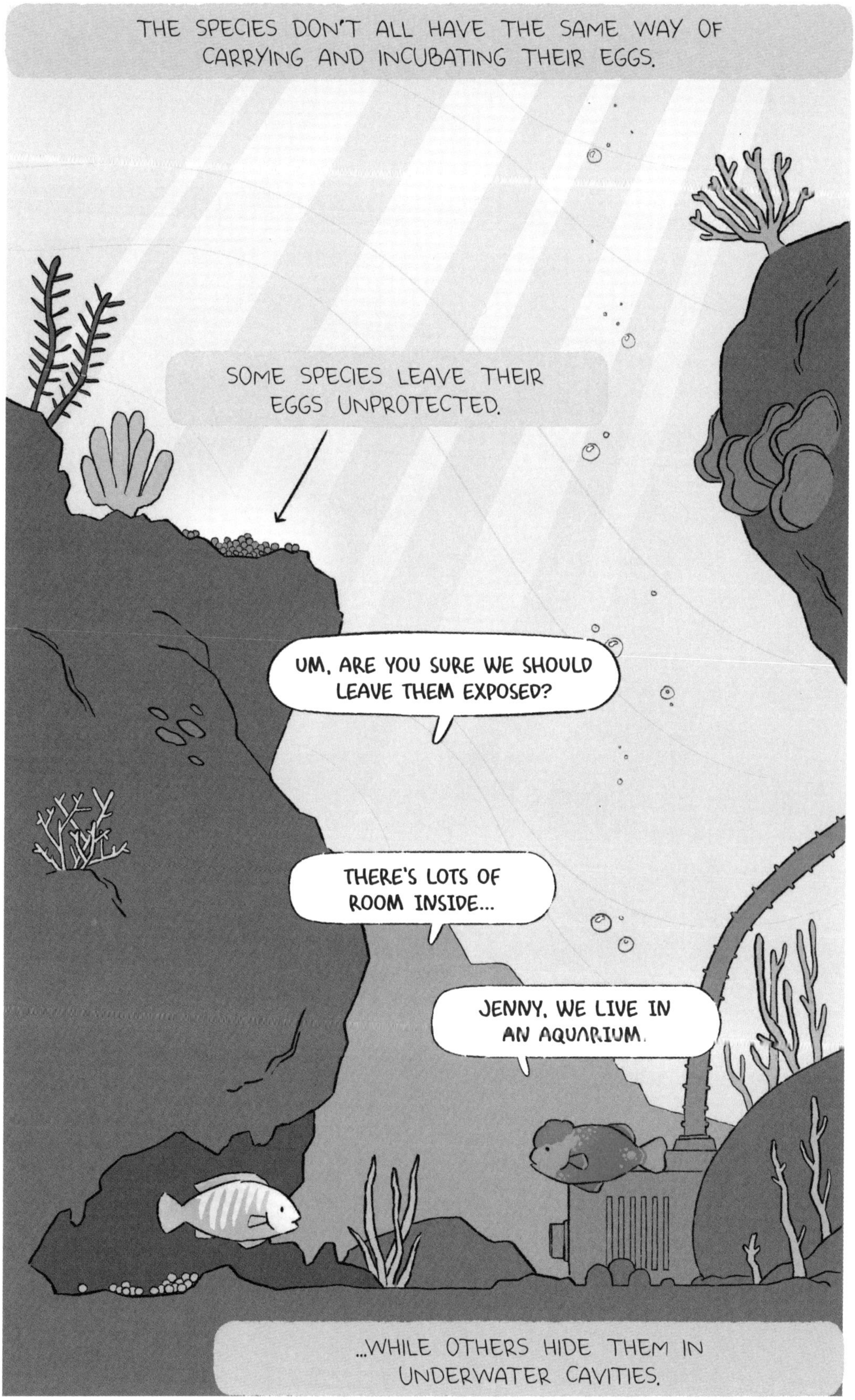
THE SPECIES DON'T ALL HAVE THE SAME WAY OF CARRYING AND INCUBATING THEIR EGGS.
SOME SPECIES LEAVE THEIR EGGS UNPROTECTED.
UM, ARE YOU SURE WE SHOULD LEAVE THEM EXPOSED?
THERE'S LOTS OF ROOM INSIDE...
JENNY, WE LIVE IN AN AQUARIUM.
...WHILE OTHERS HIDE THEM IN UNDERWATER CAVITIES.

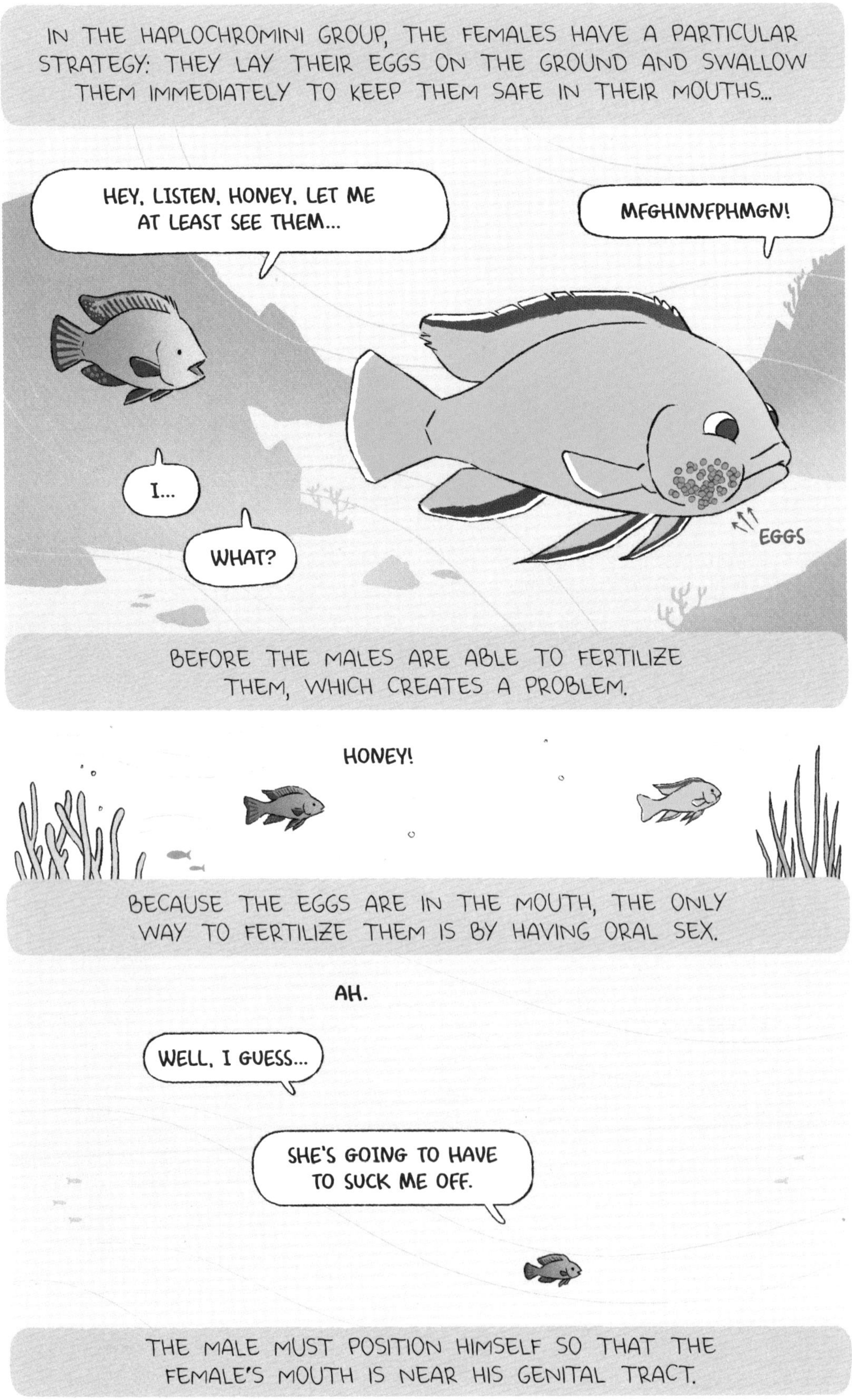
IN THE HAPLOCHROMINI GROUP, THE FEMALES HAVE A PARTICULAR STRATEGY: THEY LAY THEIR EGGS ON THE GROUND AND SWALLOW THEM IMMEDIATELY TO KEEP THEM SAFE IN THEIR MOUTHS...
HEY, LISTEN, HONEY, LET ME AT LEAST SEE THEM...
MFGHNNFPHMGN!
I...
WHAT?
EGGS
BEFORE THE MALES ARE ABLE TO FERTILIZE THEM, WHICH CREATES A PROBLEM.
HONEY!
BECAUSE THE EGGS ARE IN THE MOUTH, THE ONLY WAY TO FERTILIZE THEM IS BY HAVING ORAL SEX.
AH.
WELL, I GUESS...
SHE'S GOING TO HAVE TO SUCK ME OFF.
THE MALE MUST POSITION HIMSELF SO THAT THE FEMALE'S MOUTH IS NEAR HIS GENITAL TRACT.

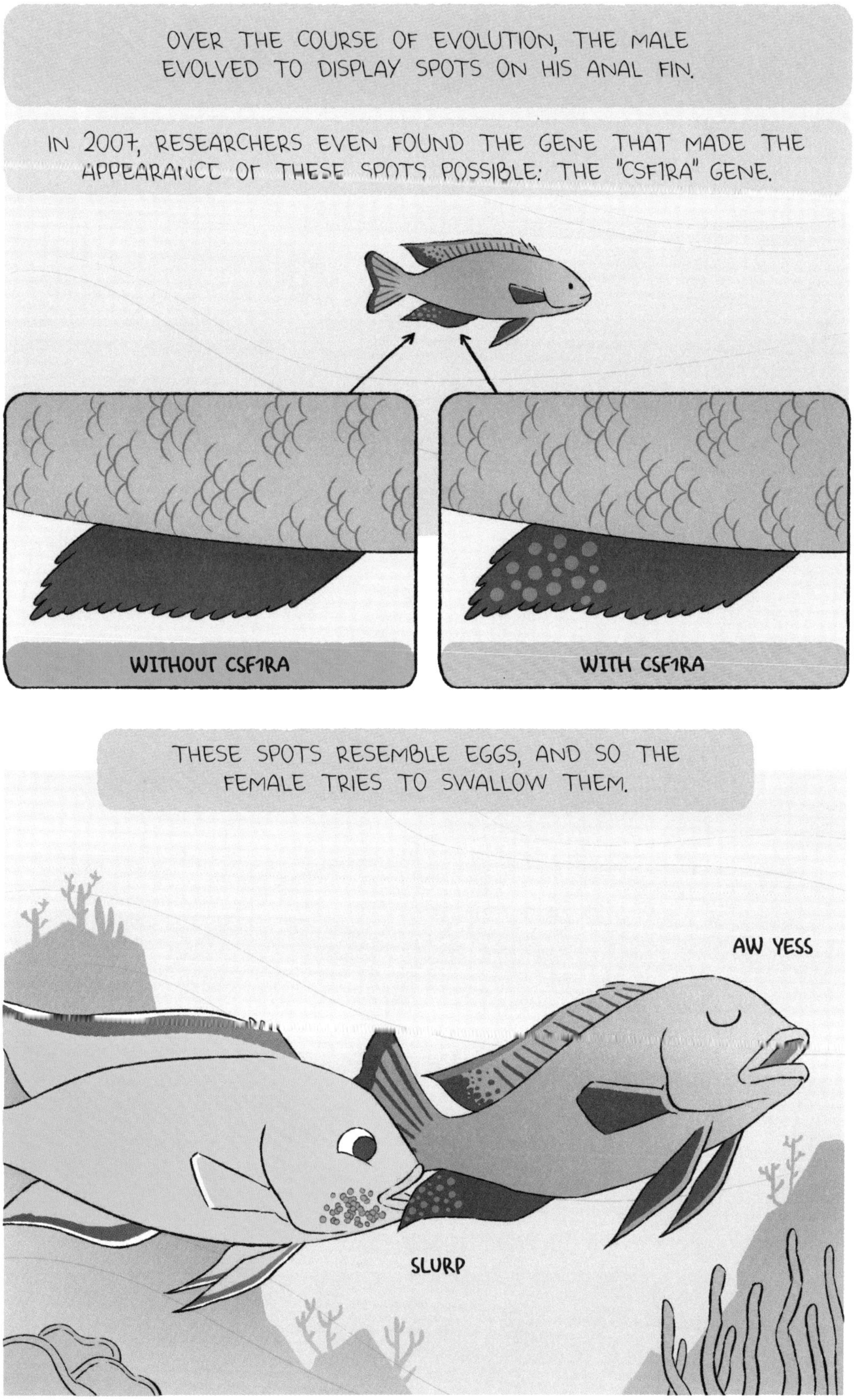
OVER THE COURSE OF EVOLUTION, THE MALE EVOLVED TO DISPLAY SPOTS ON HIS ANAL FIN.
IN 2007, RESEARCHERS EVEN FOUND THE GENE THAT MADE THE APPEARANCE OF THESE SPOTS POSSIBLE: THE "CSF1RA" GENE.
WITHOUT CSF1RA
WITH CSF1RA
THESE SPOTS RESEMBLE EGGS, AND SO THE FEMALE TRIES TO SWALLOW THEM.
AW YESS
SLURP

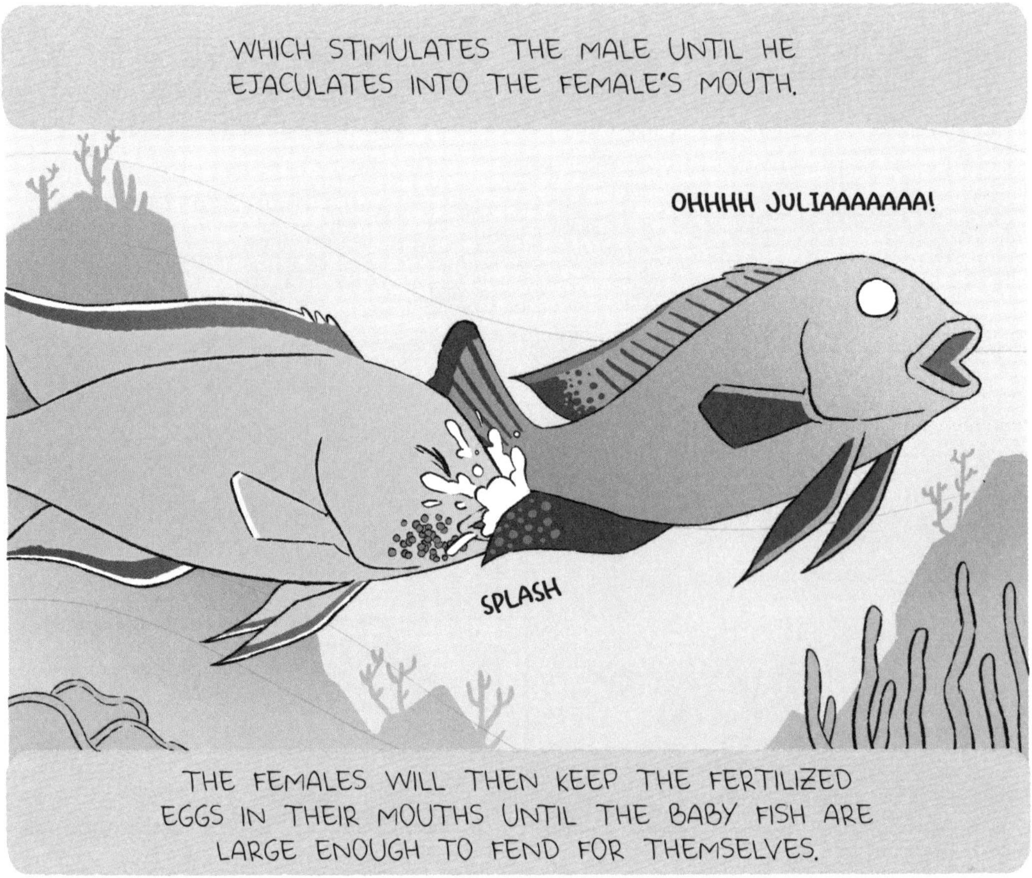
WHICH STIMULATES THE MALE UNTIL HE EJACULATES INTO THE FEMALE'S MOUTH.
OHHHH JULIAAAAAAA!
SPLASH
THE FEMALES WILL THEN KEEP THE FERTILIZED EGGS IN THEIR MOUTHS UNTIL THE BABY FISH ARE LARGE ENOUGH TO FEND FOR THEMSELVES.

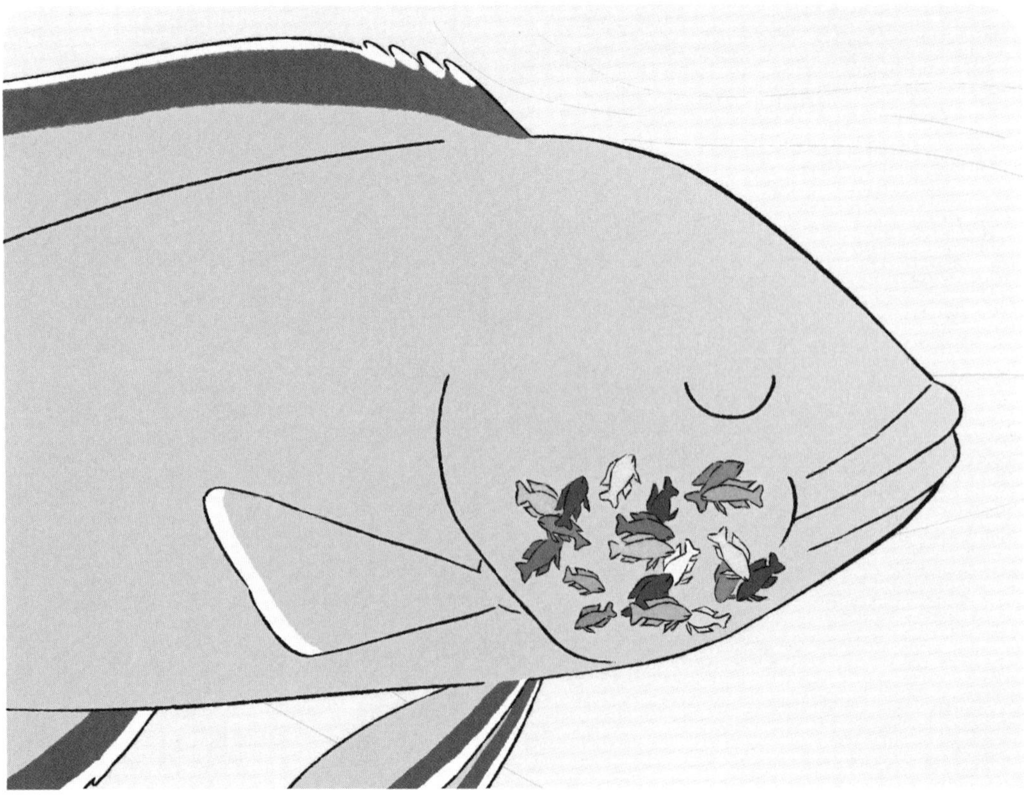

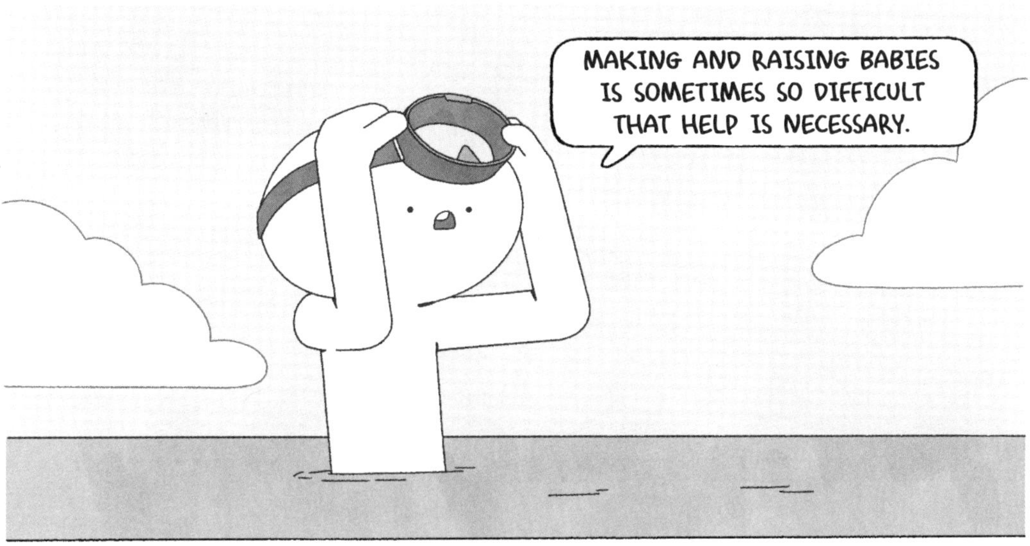
MAKING AND RAISING BABIES IS SOMETIMES SO DIFFICULT THAT HELP IS NECESSARY.

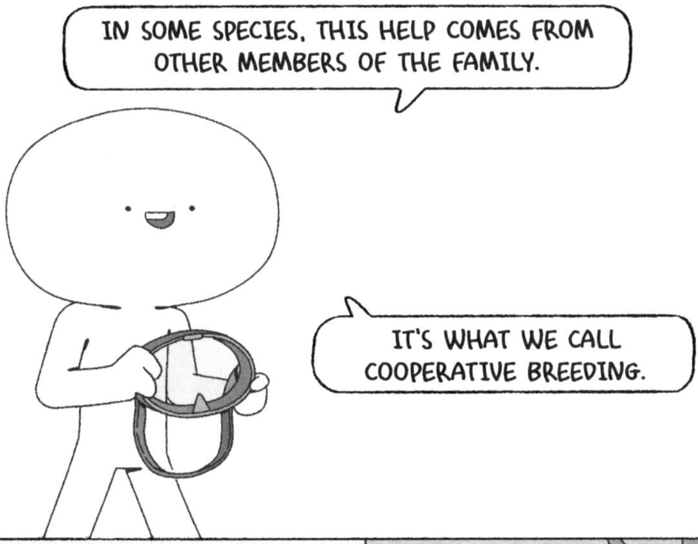
IN SOME SPECIES, THIS HELP COMES FROM OTHER MEMBERS OF THE FAMILY.
IT'S WHAT WE CALL COOPERATIVE BREEDING.

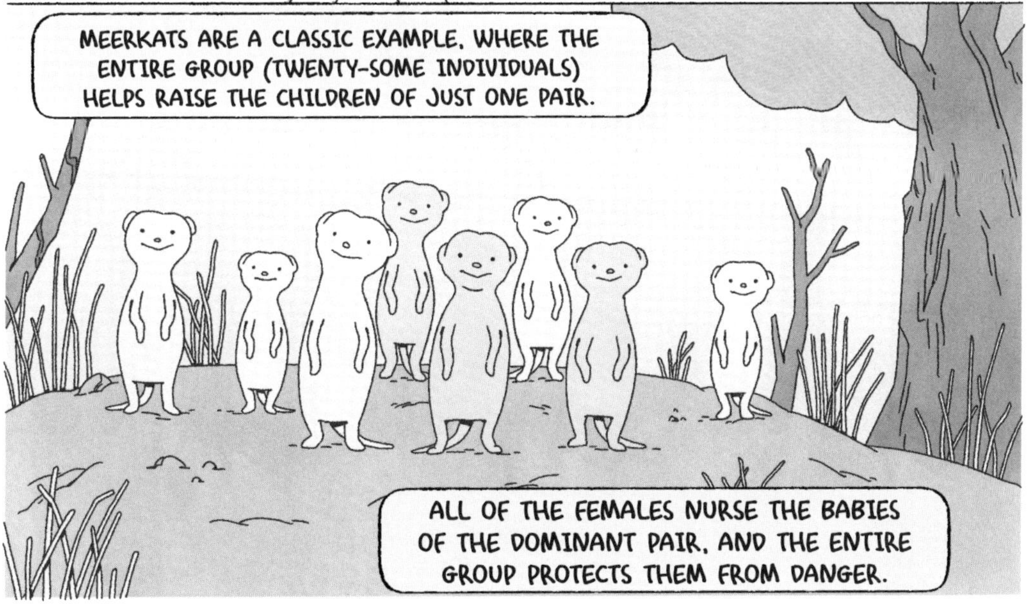
MEERKATS ARE A CLASSIC EXAMPLE, WHERE THE ENTIRE GROUP (TWENTY-SOME INDIVIDUALS) HELPS RAISE THE CHILDREN OF JUST ONE PAIR.
ALL OF THE FEMALES NURSE THE BABIES OF THE DOMINANT PAIR, AND THE ENTIRE GROUP PROTECTS THEM FROM DANGER.

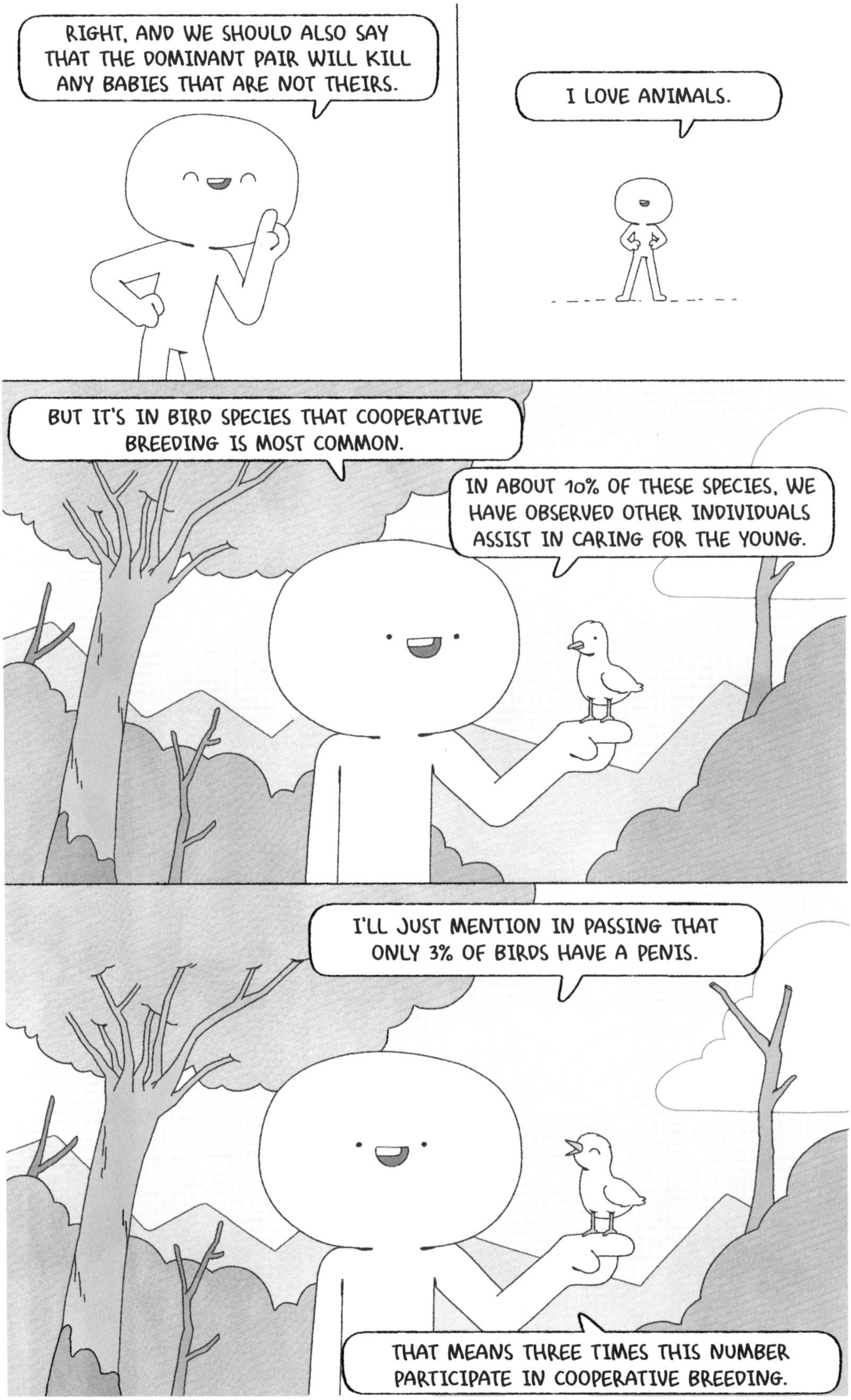
RIGHT, AND WE SHOULD ALSO SAY THAT THE DOMINANT PAIR WILL KILL ANY BABIES THAT ARE NOT THEIRS.
I LOVE ANIMALS.
BUT IT'S IN BIRD SPECIES THAT COOPERATIVE BREEDING IS MOST COMMON.
IN ABOUT 10% OF THESE SPECIES, WE HAVE OBSERVED OTHER INDIVIDUALS ASSIST IN CARING FOR THE YOUNG.
I'LL JUST MENTION IN PASSING THAT ONLY 3% OF BIRDS HAVE A PENIS.
THAT MEANS THREE TIMES THIS NUMBER PARTICIPATE IN COOPERATIVE BREEDING.

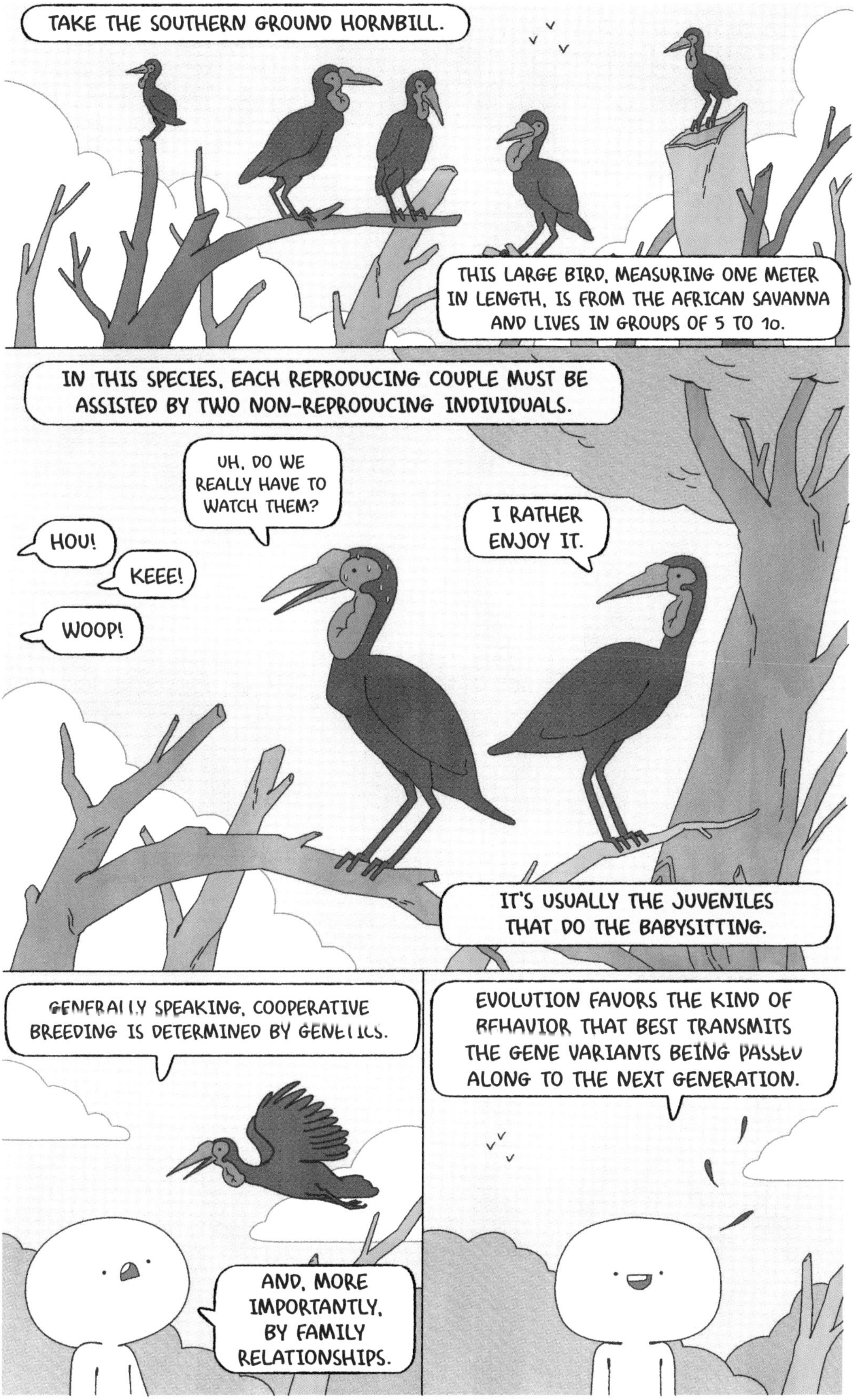
TAKE THE SOUTHERN GROUND HORNBILL.
THIS LARGE BIRD, MEASURING ONE METER IN LENGTH, IS FROM THE AFRICAN SAVANNA AND LIVES IN GROUPS OF 5 TO 10.
IN THIS SPECIES, EACH REPRODUCING COUPLE MUST BE ASSISTED BY TWO NON-REPRODUCING INDIVIDUALS.
HOU!
UH, DO WE REALLY HAVE TO WATCH THEM?
KEEE!
WOOP!
I RATHER ENJOY IT.
IT'S USUALLY THE JUVENILES THAT DO THE BABYSITTING.
GENERALLY SPEAKING, COOPERATIVE BREEDING IS DETERMINED BY GENETICS.
AND, MORE IMPORTANTLY, BY FAMILY RELATIONSHIPS.
EVOLUTION FAVORS THE KIND OF BEHAVIOR THAT BEST TRANSMITS THE GENE VARIANTS BEING PASSED ALONG TO THE NEXT GENERATION.

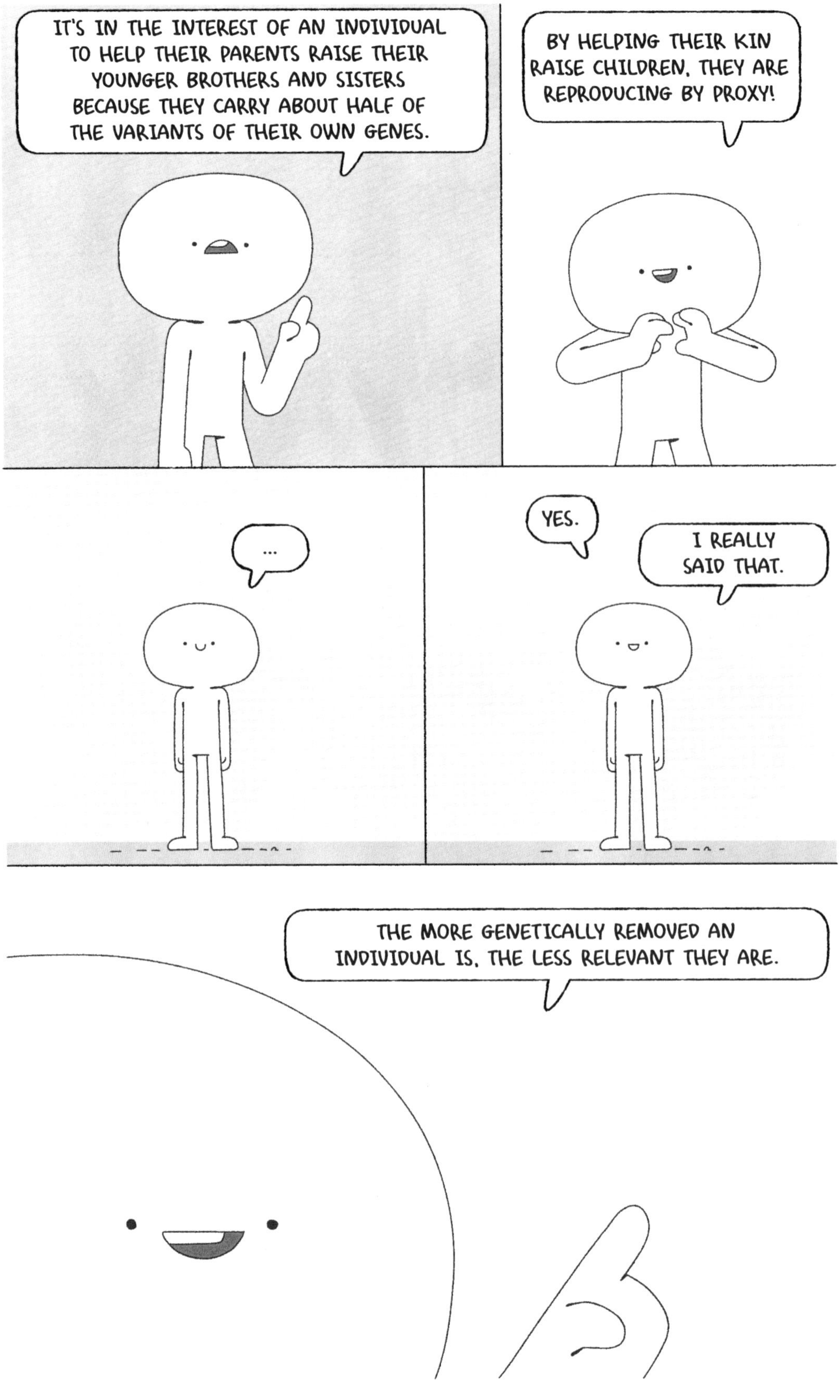
IT'S IN THE INTEREST OF AN INDIVIDUAL TO HELP THEIR PARENTS RAISE THEIR YOUNGER BROTHERS AND SISTERS BECAUSE THEY CARRY ABOUT HALF OF THE VARIANTS OF THEIR OWN GENES.
BY HELPING THEIR KIN RAISE CHILDREN, THEY ARE REPRODUCING BY PROXY!
...
YES.
I REALLY SAID THAT.
THE MORE GENETICALLY REMOVED AN INDIVIDUAL IS, THE LESS RELEVANT THEY ARE.

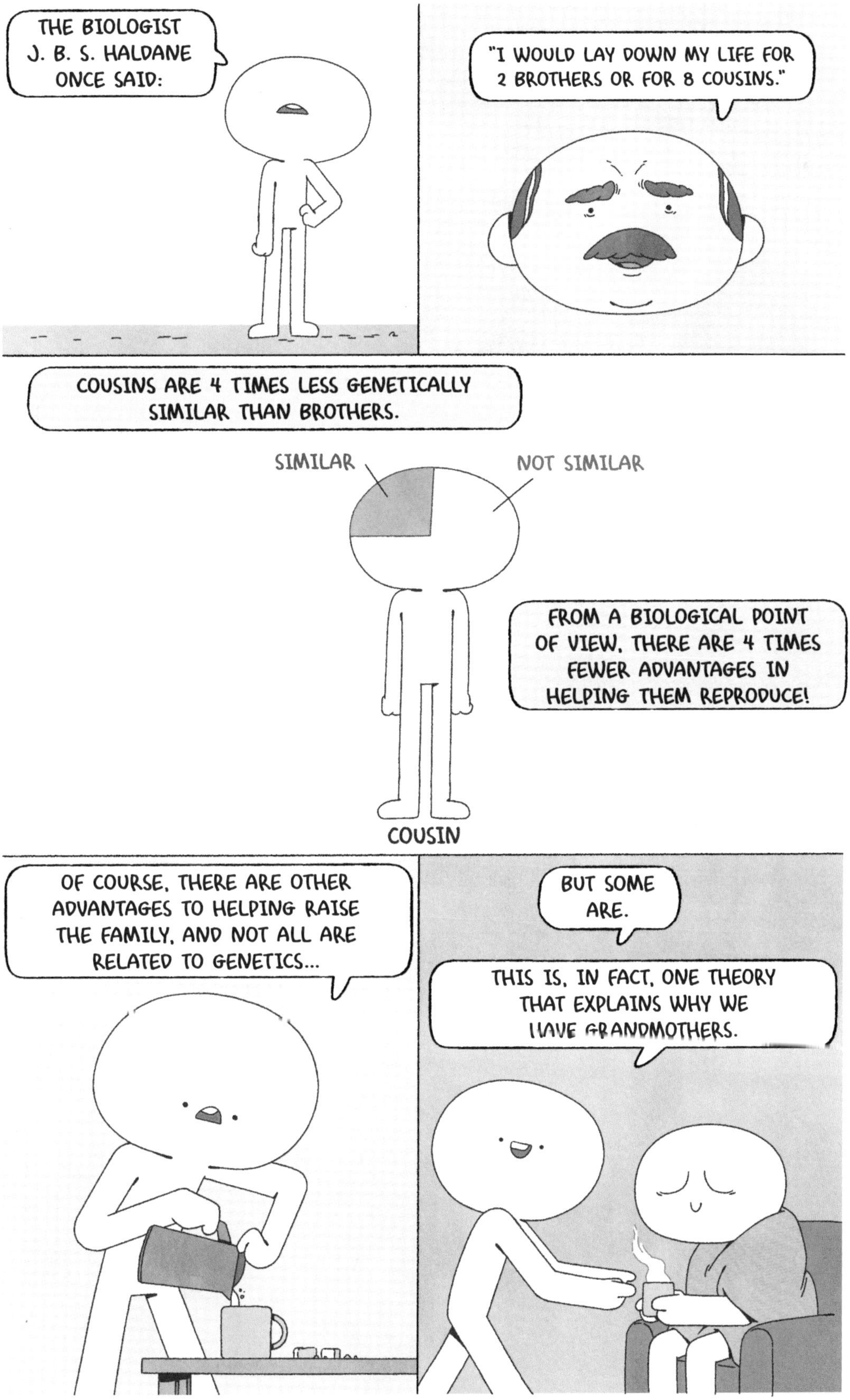
THE BIOLOGIST J. B. S. HALDANE ONCE SAID:
"I WOULD LAY DOWN MY LIFE FOR 2 BROTHERS OR FOR 8 COUSINS."
COUSINS ARE 4 TIMES LESS GENETICALLY SIMILAR THAN BROTHERS.
SIMILAR
NOT SIMILAR
FROM A BIOLOGICAL POINT OF VIEW, THERE ARE 4 TIMES FEWER ADVANTAGES IN HELPING THEM REPRODUCE!
COUSIN
OF COURSE, THERE ARE OTHER ADVANTAGES TO HELPING RAISE THE FAMILY, AND NOT ALL ARE RELATED TO GENETICS...
BUT SOME ARE.
THIS IS, IN FACT, ONE THEORY THAT EXPLAINS WHY WE HAVE GRANDMOTHERS.

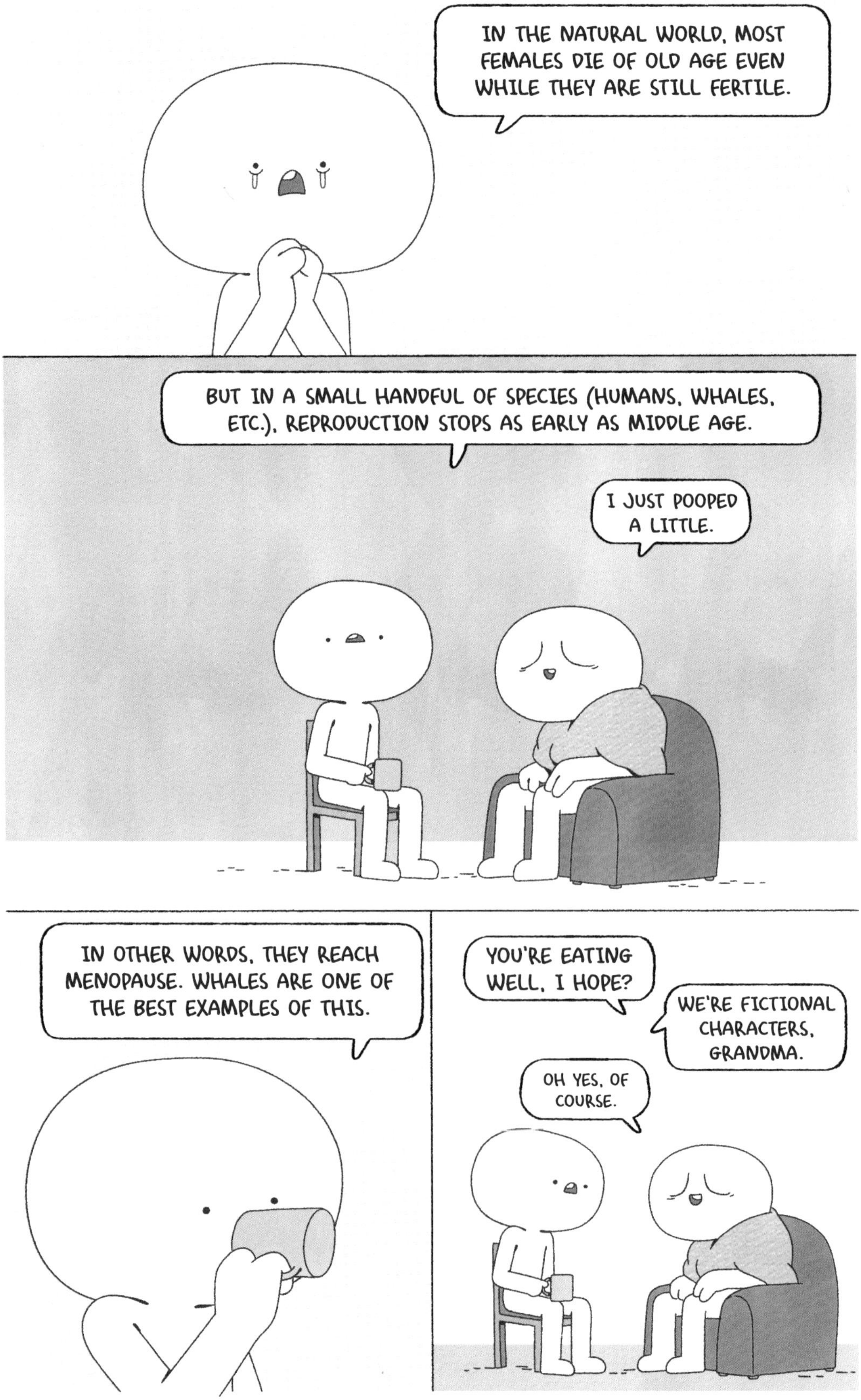
IN THE NATURAL WORLD, MOST FEMALES DIE OF OLD AGE EVEN WHILE THEY ARE STILL FERTILE.
BUT IN A SMALL HANDFUL OF SPECIES (HUMANS, WHALES, ETC.), REPRODUCTION STOPS AS EARLY AS MIDDLE AGE.
I JUST POOPED A LITTLE.
IN OTHER WORDS, THEY REACH MENOPAUSE. WHALES ARE ONE OF THE BEST EXAMPLES OF THIS.
YOU'RE EATING WELL, I HOPE?
WE'RE FICTIONAL CHARACTERS, GRANDMA.
OH YES, OF COURSE.

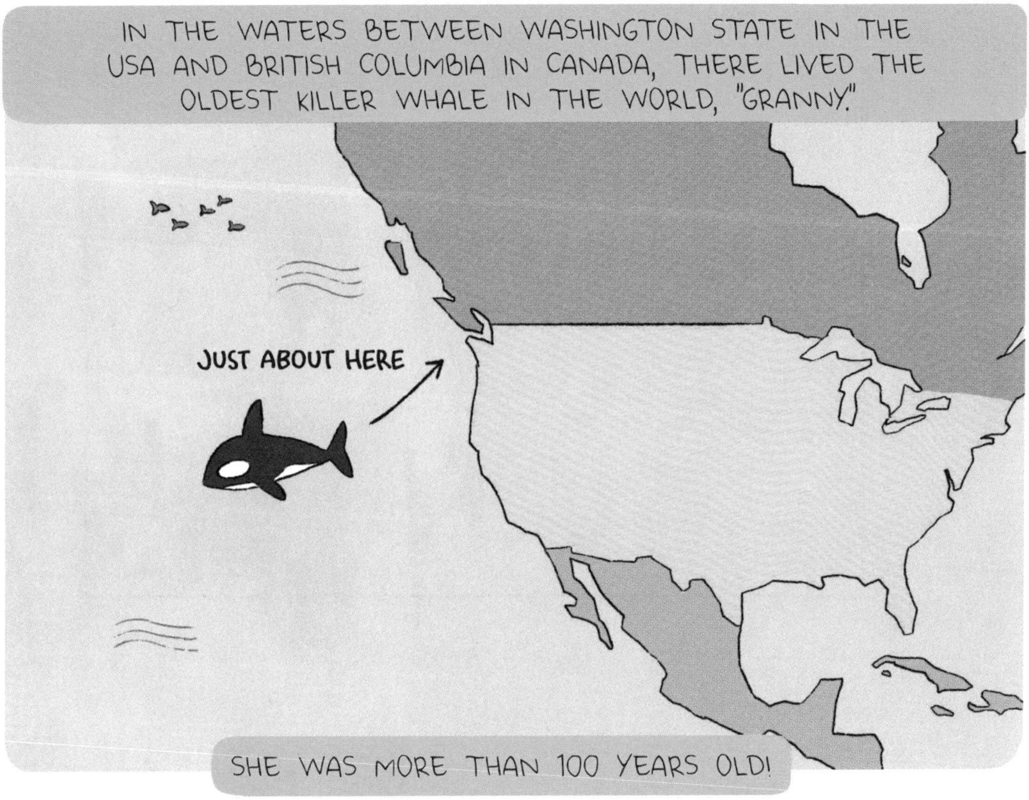
IN THE WATERS BETWEEN WASHINGTON STATE IN THE USA AND BRITISH COLUMBIA IN CANADA, THERE LIVED THE OLDEST KILLER WHALE IN THE WORLD, "GRANNY."
JUST ABOUT HERE
SHE WAS MORE THAN 100 YEARS OLD!

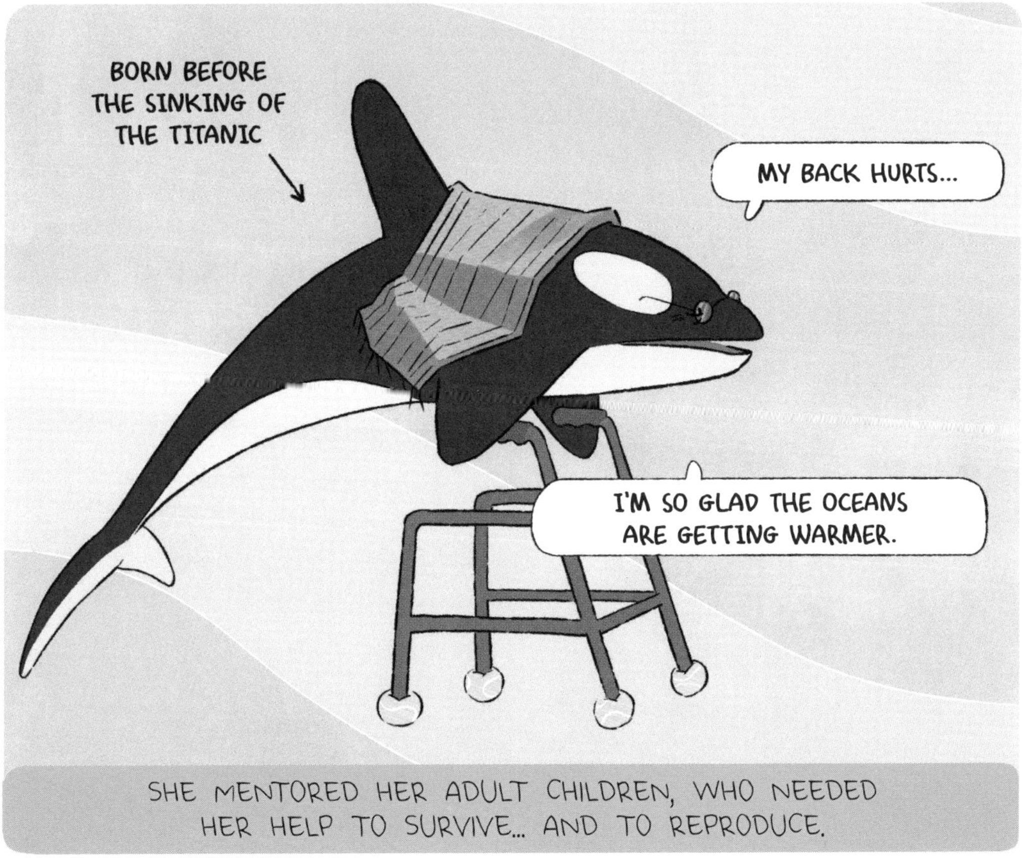
BORN BEFORE THE SINKING OF THE TITANIC
MY BACK HURTS...
I'M SO GLAD THE OCEANS ARE GETTING WARMER.
SHE MENTORED HER ADULT CHILDREN, WHO NEEDED HER HELP TO SURVIVE... AND TO REPRODUCE.

THE LIFE OF A KILLER WHALE IS NOT SO SIMPLE: THE AMOUNT OF SALMON THEY FEED ON VARIES GREATLY FROM YEAR TO YEAR.

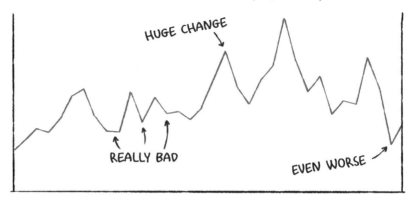

CHINOOK SALMON NUMBERS (1980-2013)

HUGE CHANGE

REALLY BAD

EVEN WORSE

IN THIS FLUCTUATING ENVIRONMENT, THE EXPERIENCE OF THE OLDER WHALES IS VERY IMPORTANT TO THEIR SURVIVAL.

WHEN THERE'S TOO MUCH OF A CHOICE I DON'T KNOW WHAT TO EAT...

THEY STOP BREEDING AROUND THE AGE OF THIRTY BUT CAN LIVE ON FOR MANY MORE DECADES. THESE ARE YEARS THEY PUT TO GOOD USE BY GUIDING YOUNG MALES, IN PARTICULAR, WHO ARE MUCH MORE DEPENDENT THAN FEMALES.

THEY ARE SO ESSENTIAL FOR THE SURVIVAL OF THEIR ADULT SONS THAT, WHEN AN OLDER MOTHER DIES, HER SONS ARE 14 TIMES MORE LIKELY TO DIE WITHIN A YEAR.

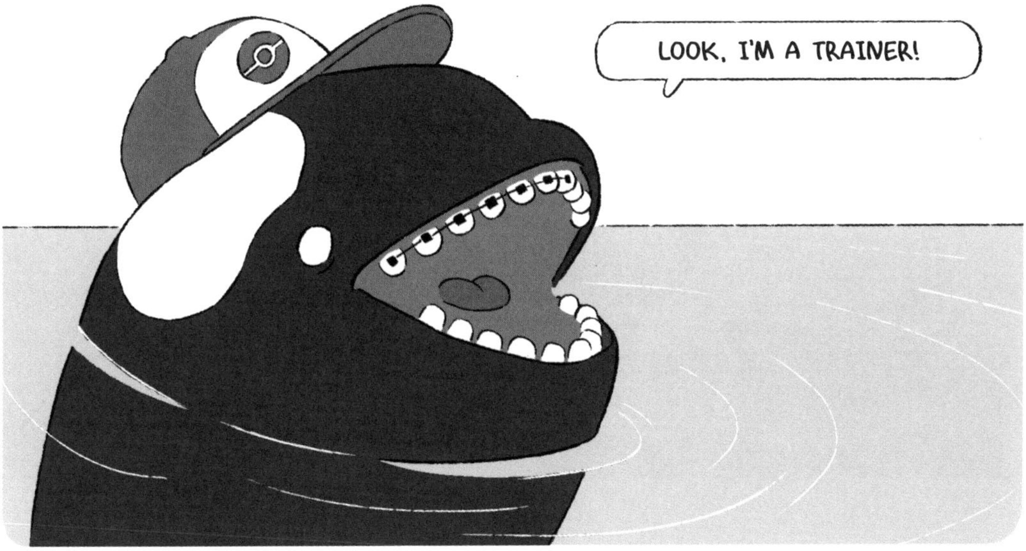
LOOK. I'M A TRAINER!

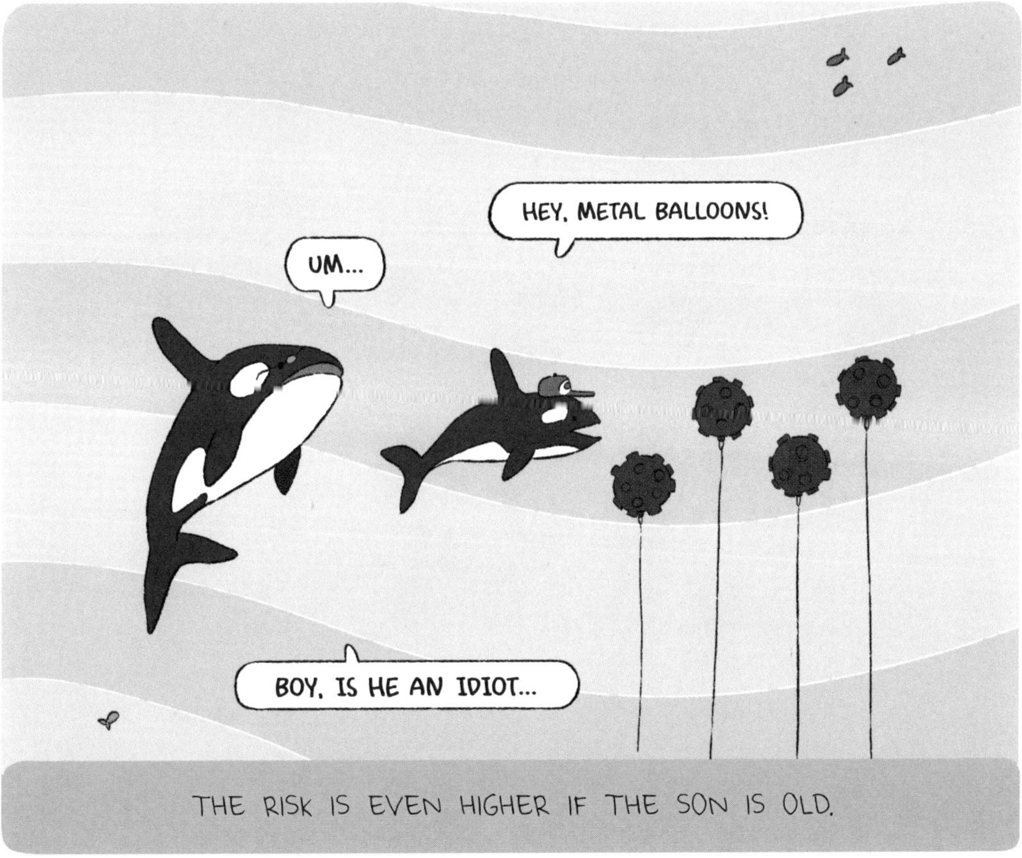
HEY. METAL BALLOONS!
UM...
BOY. IS HE AN IDIOT...

THE RISK IS EVEN HIGHER IF THE SON IS OLD.

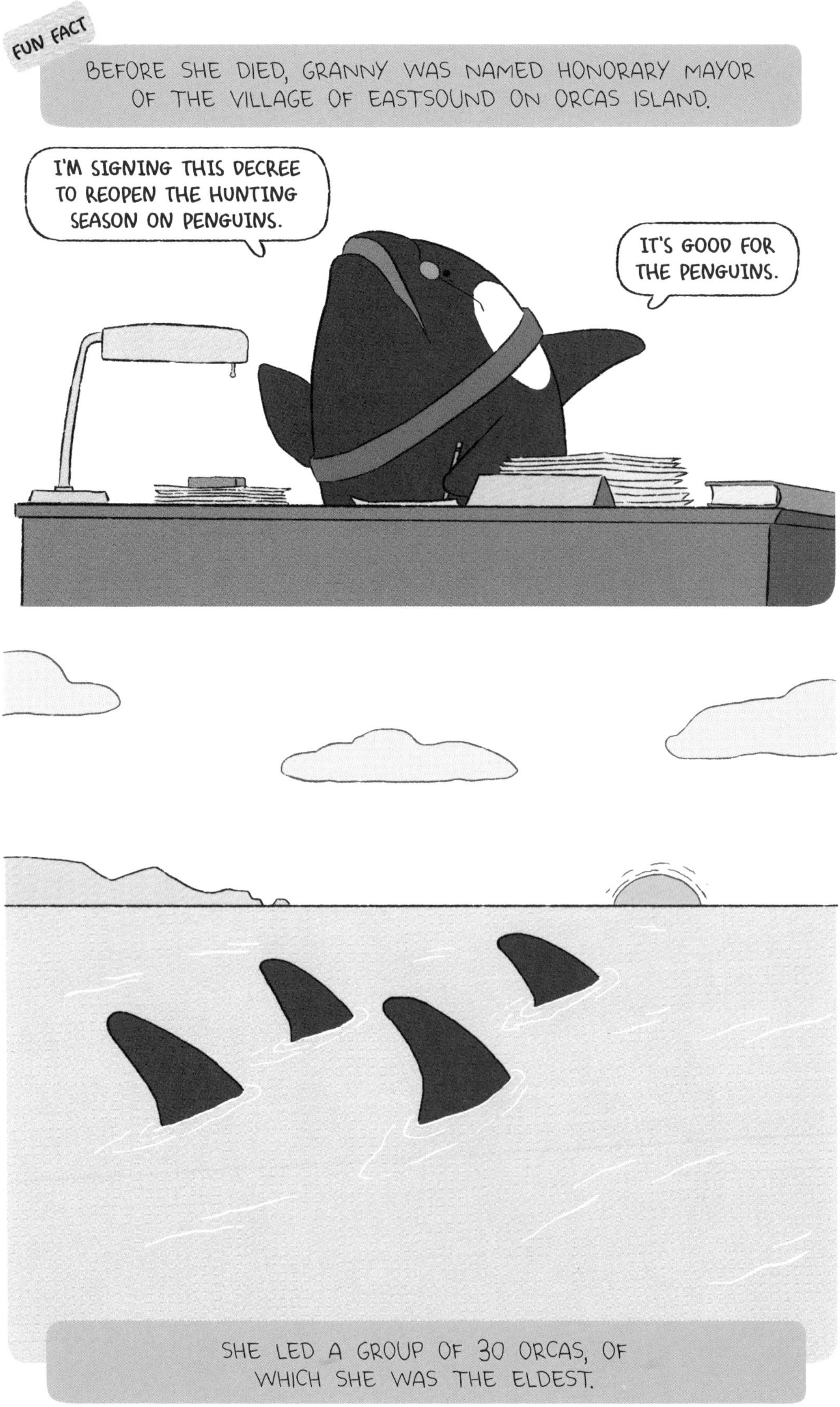
FUN FACT
BEFORE SHE DIED, GRANNY WAS NAMED HONORARY MAYOR OF THE VILLAGE OF EASTSOUND ON ORCAS ISLAND.
I'M SIGNING THIS DECREE TO REOPEN THE HUNTING SEASON ON PENGUINS.
IT'S GOOD FOR THE PENGUINS.
SHE LED A GROUP OF 30 ORCAS, OF WHICH SHE WAS THE ELDEST.

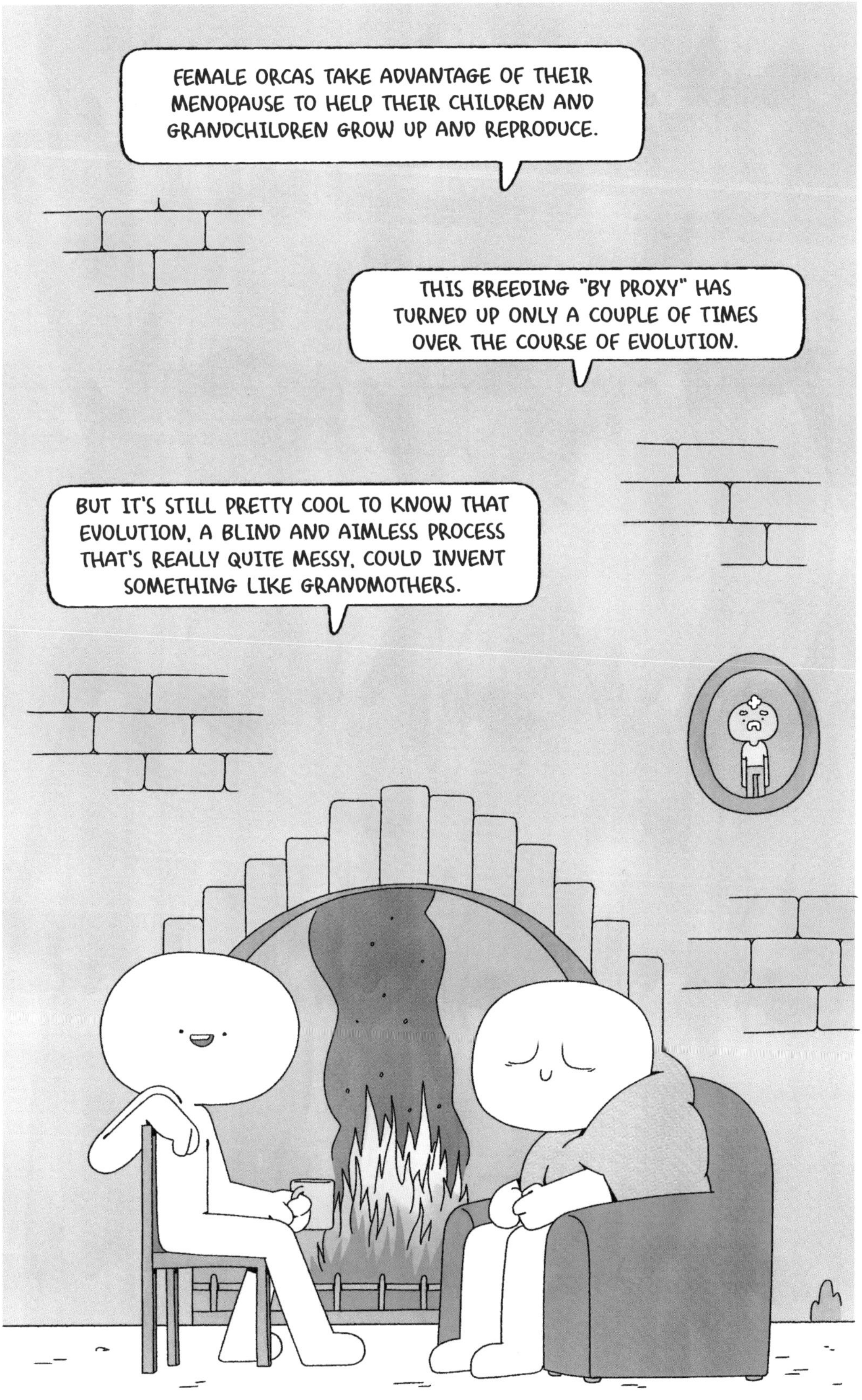
FEMALE ORCAS TAKE ADVANTAGE OF THEIR MENOPAUSE TO HELP THEIR CHILDREN AND GRANDCHILDREN GROW UP AND REPRODUCE.
THIS BREEDING "BY PROXY" HAS TURNED UP ONLY A COUPLE OF TIMES OVER THE COURSE OF EVOLUTION.
BUT IT'S STILL PRETTY COOL TO KNOW THAT EVOLUTION, A BLIND AND AIMLESS PROCESS THAT'S REALLY QUITE MESSY, COULD INVENT SOMETHING LIKE GRANDMOTHERS.

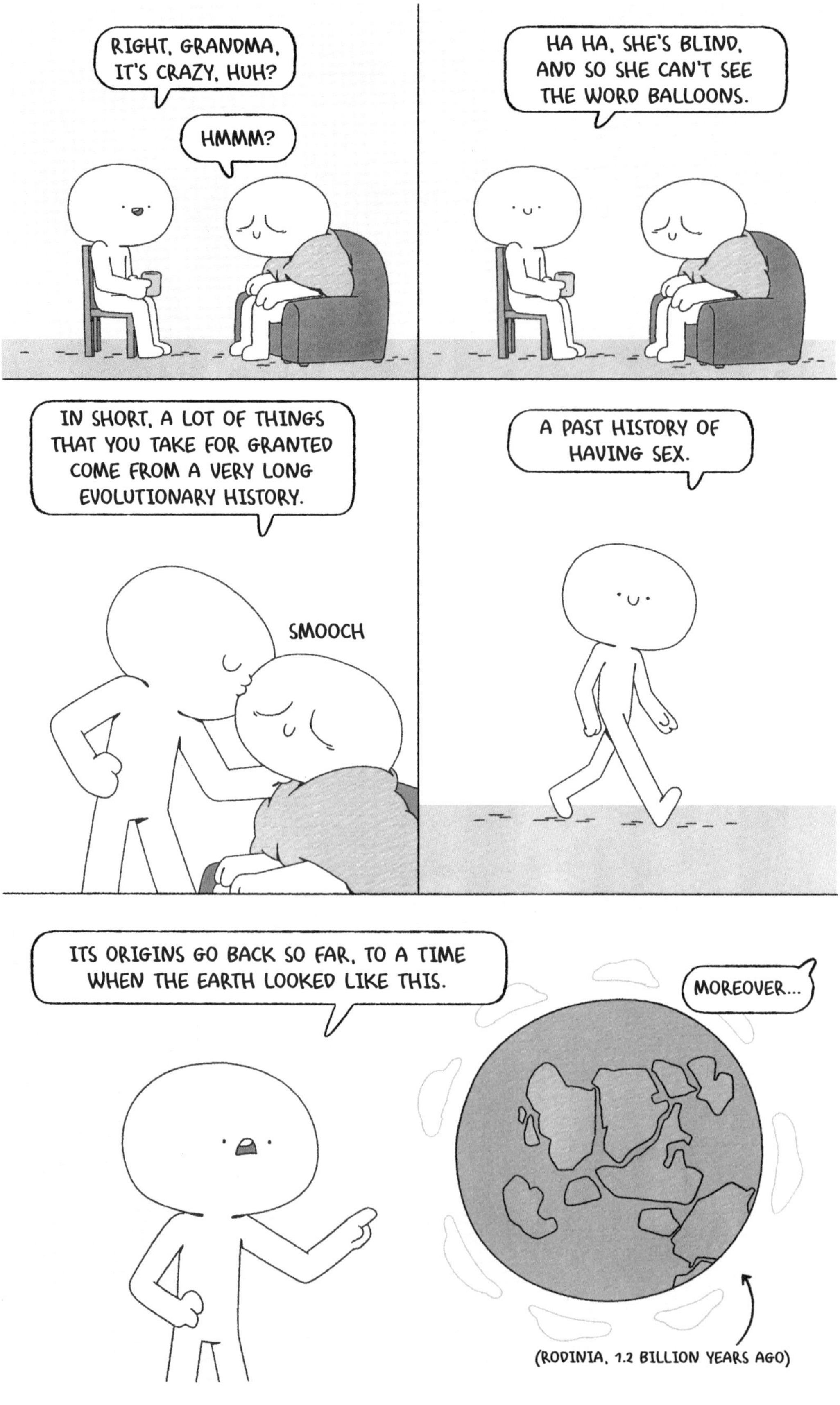
RIGHT, GRANDMA, IT'S CRAZY, HUH?
HMMM?
HA HA, SHE'S BLIND, AND SO SHE CAN'T SEE THE WORD BALLOONS.
IN SHORT, A LOT OF THINGS THAT YOU TAKE FOR GRANTED COME FROM A VERY LONG EVOLUTIONARY HISTORY.
SMOOCH
A PAST HISTORY OF HAVING SEX.
ITS ORIGINS GO BACK SO FAR, TO A TIME WHEN THE EARTH LOOKED LIKE THIS.
MOREOVER...
(RODINIA, 1.2 BILLION YEARS AGO)

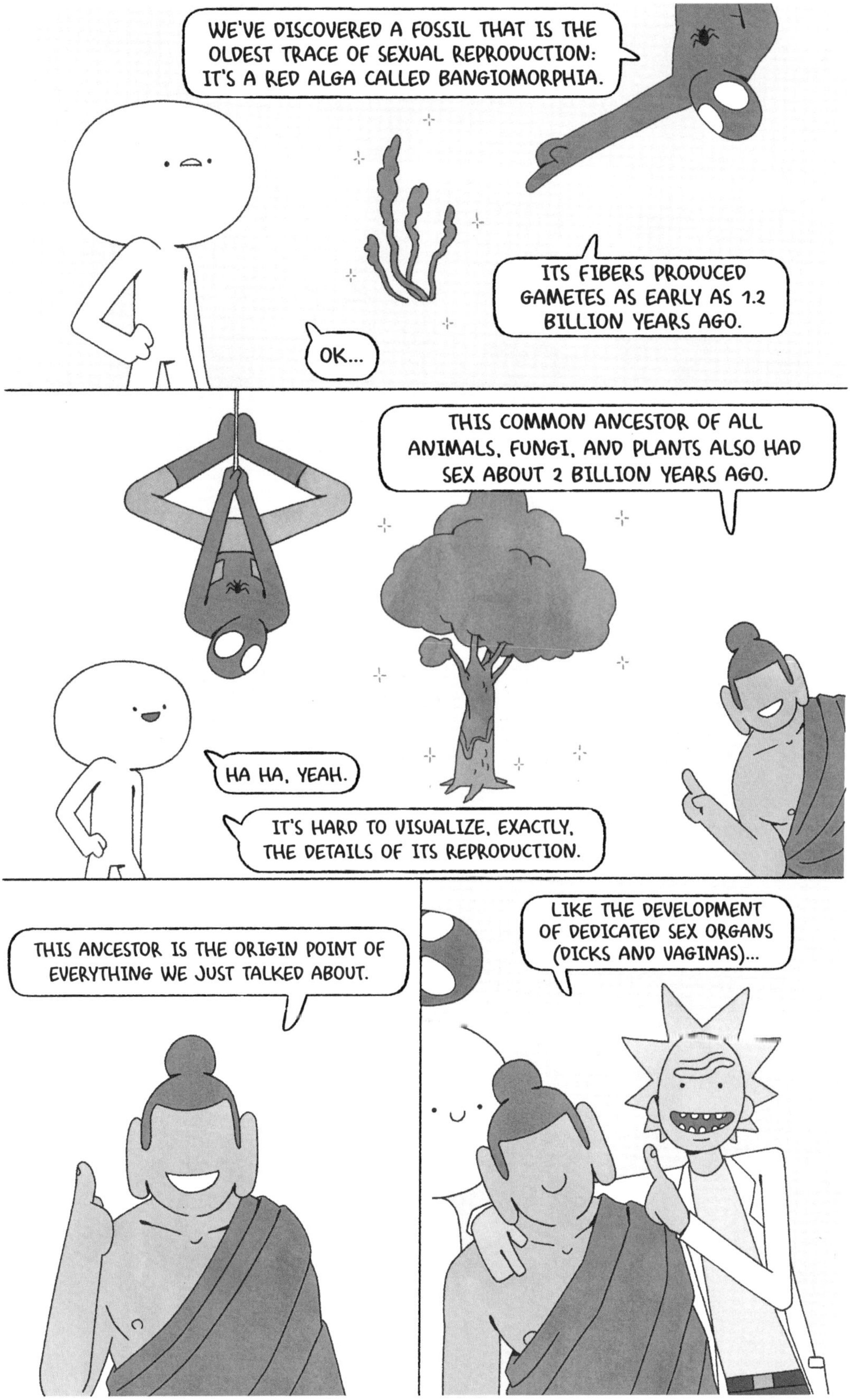
WE'VE DISCOVERED A FOSSIL THAT IS THE OLDEST TRACE OF SEXUAL REPRODUCTION: IT'S A RED ALGA CALLED BANGIOMORPHIA.
ITS FIBERS PRODUCED GAMETES AS EARLY AS 1.2 BILLION YEARS AGO.
OK...
THIS COMMON ANCESTOR OF ALL ANIMALS, FUNGI, AND PLANTS ALSO HAD SEX ABOUT 2 BILLION YEARS AGO.
HA HA, YEAH.
IT'S HARD TO VISUALIZE, EXACTLY, THE DETAILS OF ITS REPRODUCTION.
THIS ANCESTOR IS THE ORIGIN POINT OF EVERYTHING WE JUST TALKED ABOUT.
LIKE THE DEVELOPMENT OF DEDICATED SEX ORGANS (DICKS AND VAGINAS)...

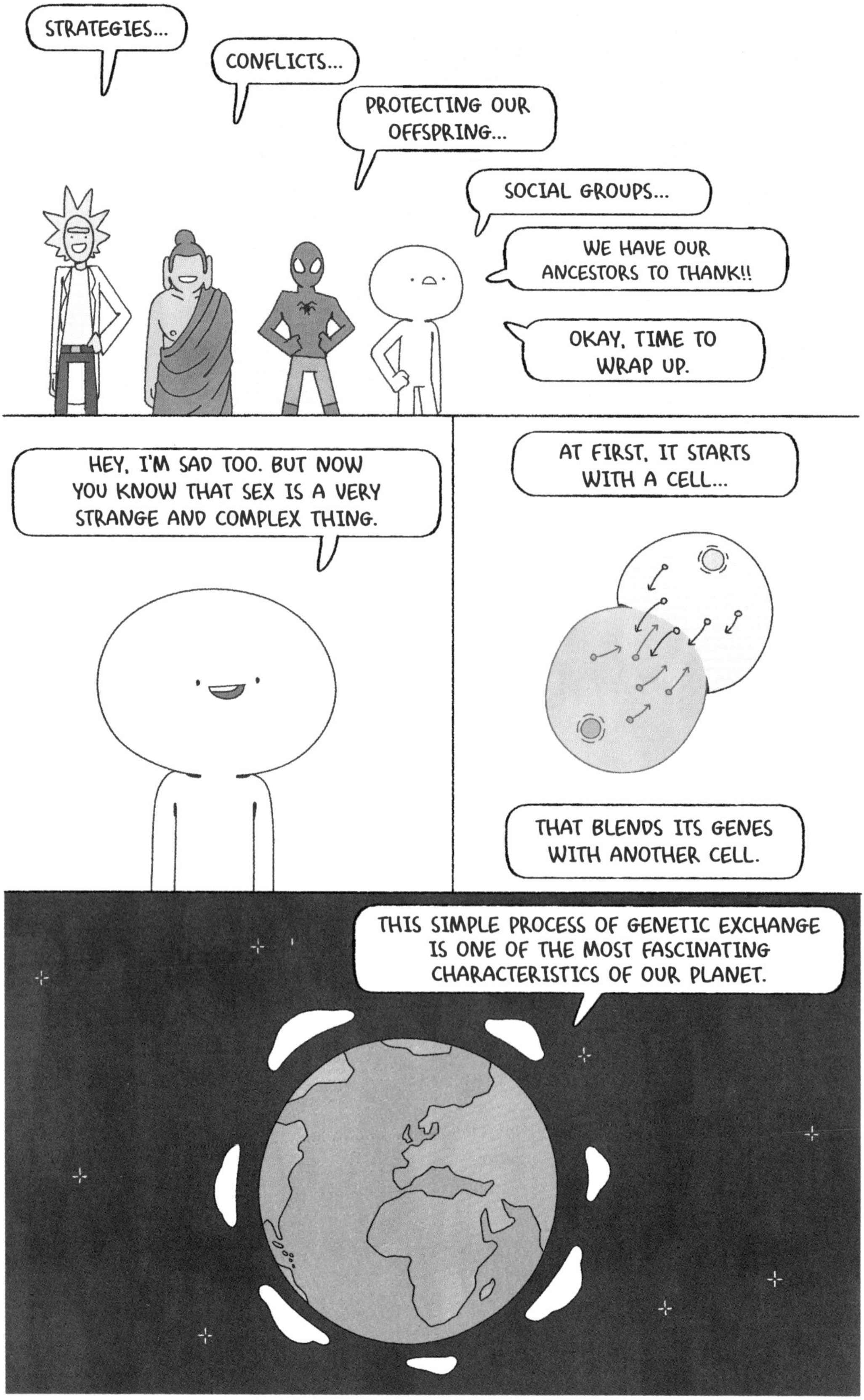
STRATEGIES...
CONFLICTS...
PROTECTING OUR OFFSPRING...
SOCIAL GROUPS...
WE HAVE OUR ANCESTORS TO THANK!!
OKAY, TIME TO WRAP UP.
HEY, I'M SAD TOO. BUT NOW YOU KNOW THAT SEX IS A VERY STRANGE AND COMPLEX THING.
AT FIRST, IT STARTS WITH A CELL...
THAT BLENDS ITS GENES WITH ANOTHER CELL.
THIS SIMPLE PROCESS OF GENETIC EXCHANGE IS ONE OF THE MOST FASCINATING CHARACTERISTICS OF OUR PLANET.

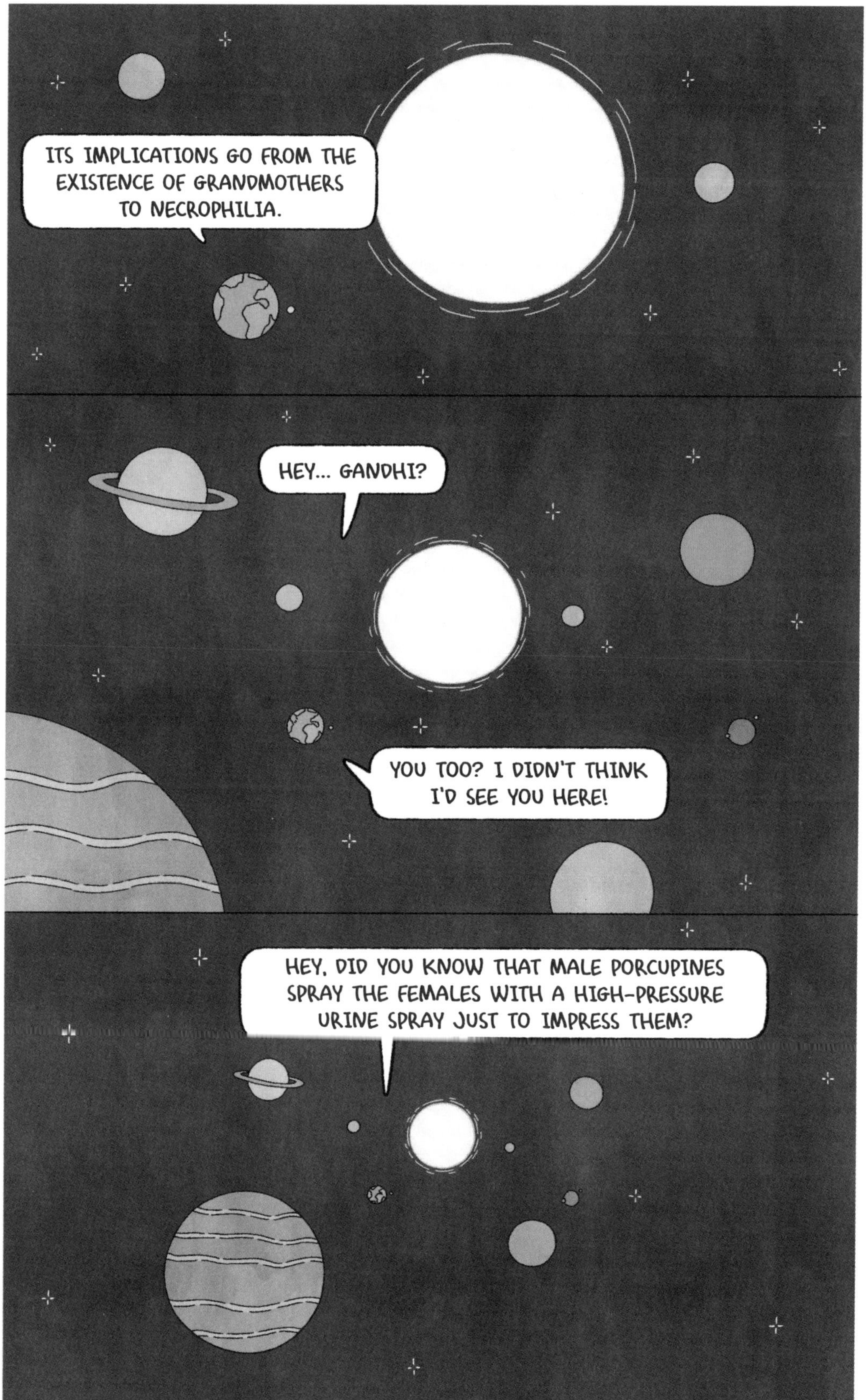
ITS IMPLICATIONS GO FROM THE EXISTENCE OF GRANDMOTHERS TO NECROPHILIA.
HEY... GANDHI?
YOU TOO? I DIDN'T THINK I'D SEE YOU HERE!
HEY, DID YOU KNOW THAT MALE PORCUPINES SPRAY THE FEMALES WITH A HIGH-PRESSURE URINE SPRAY JUST TO IMPRESS THEM?

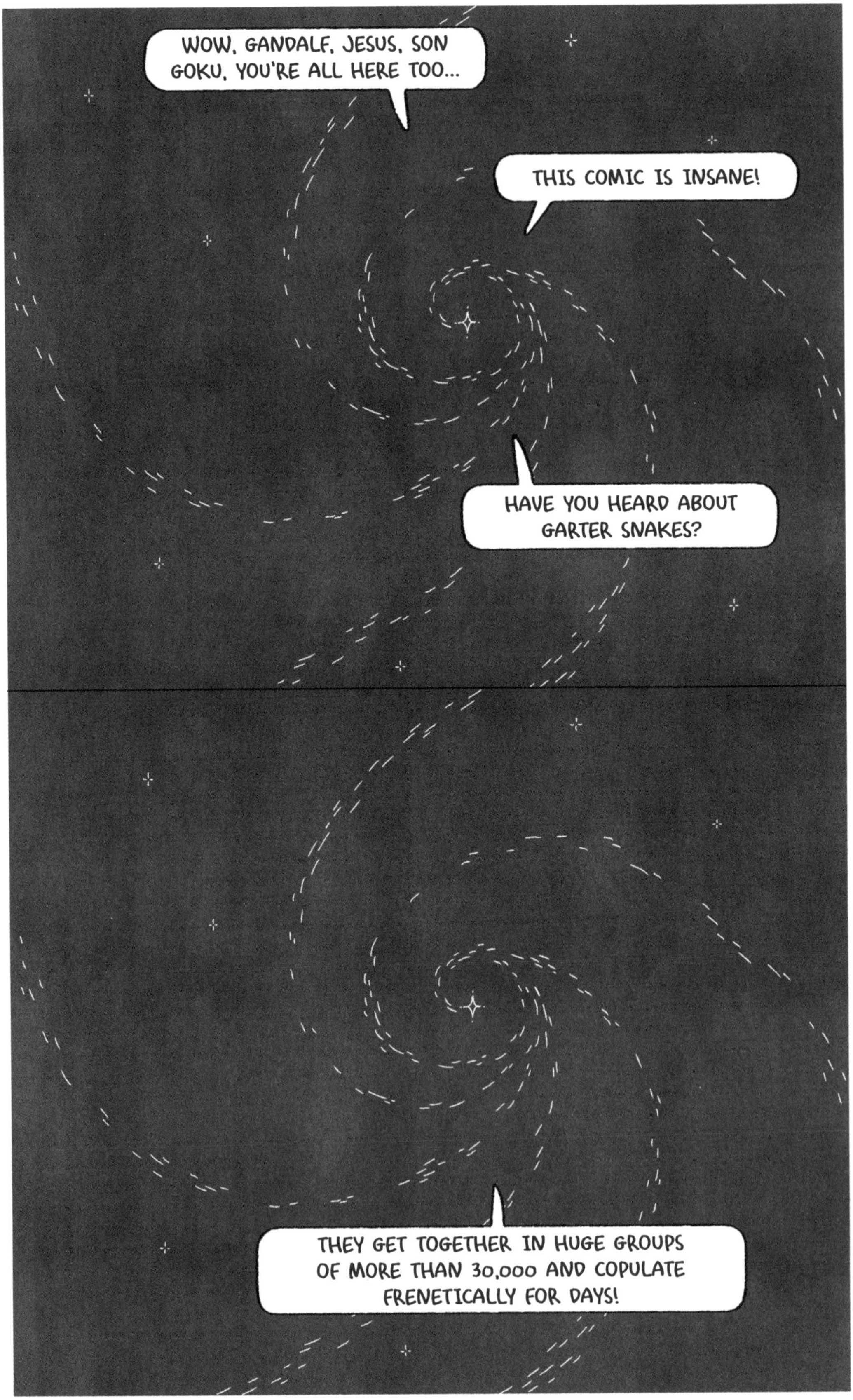
WOW, GANDALF, JESUS, SON GOKU, YOU'RE ALL HERE TOO...
THIS COMIC IS INSANE!
HAVE YOU HEARD ABOUT GARTER SNAKES?
THEY GET TOGETHER IN HUGE GROUPS OF MORE THAN 30,000 AND COPULATE FRENETICALLY FOR DAYS!

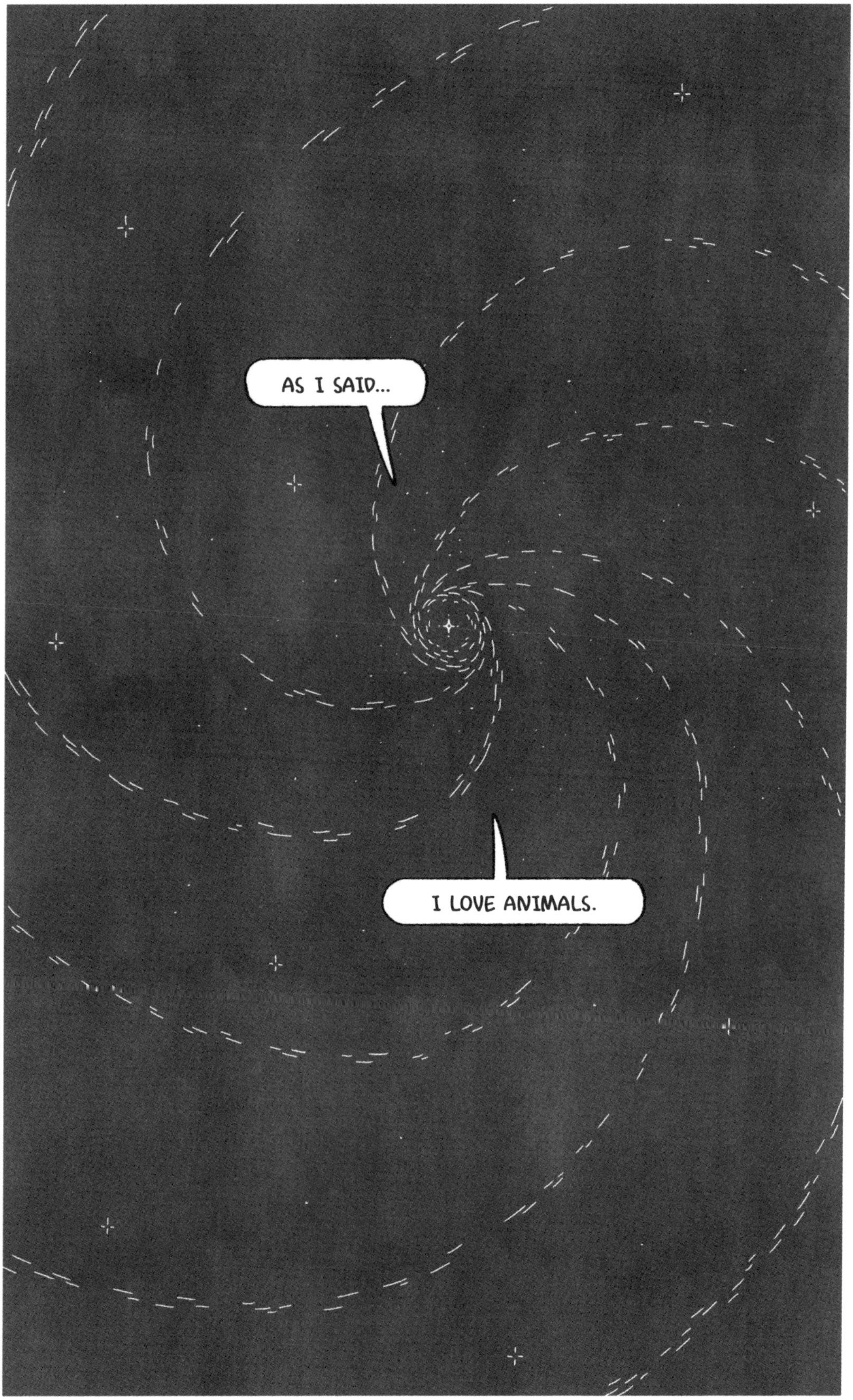
AS I SAID...
I LOVE ANIMALS.

REFERENCES

The scientific literature on the evolution of sex and the impact it has had on life-forms over time is vast, amounting to at least half of what has been published on evolutionary biology. It's impossible, then, to provide the reader with an exhaustive list of references. And because this graphic novel is already pretty huge, there's only one page left for the bibliography. So, I'm sorry to say that what follows is really just a superficial list of suggestions. But it's a start.

GENERAL READING

Nature's Nether Regions: What the Sex Lives of Bugs, Birds, and Beasts Tell Us About Evolution, Biodiversity, and Ourselves by Menno Schilthuizen (Viking, 2014).
A great popular science treatment of the complexity of sex in animals, the evolution of sex organs, and sexual behaviors.

Sex on Earth: A Celebration of Animal Reproduction by Jules Howard (Bloomsbury Sigma, 2014).
An eclectic roundup of big ideas related to reproductive biology and delivered with a light personal touch.

Dr. Tatiana's Sex Advice to All Creation: The Definitive Guide to the Evolutionary Biology of Sex by Olivia Judson (Metropolitan Books, 2002).
A sex guide for all living things and a hilarious natural history in the form of letters to and answers from the preeminent sexpert in all creation.

Other sources of inspiration:
The website birdandmoon.com posts science and nature comics that cover many aspects of animal life. Funny and informative!

Wikipedia:
Although many people are skeptical of the information we find on Wikipedia (yes, I know that includes you, but we'll keep this just between us), most of the concepts and species presented in this book have Wikipedia pages, and they are very often of excellent quality. You can also cut out the middleman and directly consult the article references.

The website treeofsex.org addresses the many factors that determine sex in the natural world: sex chromosomes, haplodiploidy, temperature, social factors, etc.

Finally, there are many things I couldn't cover in this book for lack of space. Here are some key words and phrases to use in your own research: klepton, sperm competition, conflict between mitochondrial and nuclear DNA, why bananas will disappear (and how that's linked to sex), feminization by Wolbachia bacteria, using urine to induce the Whitten effect, the Trivers-Willard hypothesis, and kakapos.... Finally, you can also visit me on the dirtybiology YouTube channel (youtube.com/c/dirtybiology), where I will inevitably end up talking about it!

xoxo, and happy reading.

ACKNOWLEDGMENTS

When you think about it, all of our ancestors had sex (and if they hadn't, we wouldn't be here to talk about it). We thank this uninterrupted chain of individuals who were kind enough to go out and get laid at least once. Thanks especially to our parents, who were very understanding when I told them that my job is to talk about animal penises on the Internet. Thanks also to our reviewers and to so many others for their support during the creation of this book. You are wonderful but far too numerous to name, so I'll just mention the people I drove nuts with this project: Valentin Manon, Marine Pierre, Camille Guile, Gwen and Cyrielle. Thank you.
—Léo

Thanks to everyone who supported and encouraged me: my family, my colleagues, my patrons (you're amazing <3), and everyone who had a hand in creating this graphic novel by offering their ideas and jokes. I hope you are all doing well. I certainly am—the weather is great here, the people are nice, and yes, I'm eating well and I try not to stay up too late, okay, me too, sending kisses. Thanks to David Douillet. Thanks to my dog Snowee and to the snakes my cats deposited under my chair while I worked on this comic. A special thanks to Pierre Kerner for the emergency assistance while Léo attempted to be the first to urinate in an unexplored part of Madagascar. Thanks to everyone for believing in your dreams and for never giving up. Close the door behind you, please.
—Colas

"To our Grandma."

Internet:
dirtybiology.com
Facebook:
dirtybiology
ayeah.prod
Twitter:
@dirtybiology
@AYEAH_Colas

Library of Congress Cataloging-in-Publication Data

Names: Grasset, Léo, 1989– author. | Grasset, Colas, 1990– illustrator.
Title: Dirty biology : the x-rated story of the science of sex / script by Léo Grasset ; art by Colas Grasset.
Other titles: Dirtybiology English
Description: University Park, Pennsylvania : Graphic Mundi, The Pennsylvania State University Press, [2021] | Includes bibliographical references.
Summary: "A graphic novel exploring the scientific details and unusual facts of sexual reproduction among various species"—Provided by publisher.
Identifiers: LCCN 2021001828 | ISBN 9780271087054 (paperback ; alk. paper)
Subjects: MESH: Reproduction—physiology | Sexuality—physiology | Biological Evolution | Graphic Novel
Classification: LCC QP251 | NLM QH 481 | DDC 612.6—dc23
LC record available at https://lccn.loc.gov/2021001828

Copyright © 2021 The Pennsylvania State University
All rights reserved
Printed in the United States of America
Published by The Pennsylvania State University Press,
University Park, PA 16802-1003

Graphic Mundi is an imprint of The Pennsylvania State University Press.

Translated by Kendra Boileau

Originally published as *DirtyBiology: La grande aventure du sexe*, by Léo and Colas Grasset
© Editions Delcourt — 2017

The Pennsylvania State University Press is a member of the Association of University Presses.

It is the policy of The Pennsylvania State University Press to use acid-free paper. Publications on uncoated stock satisfy the minimum requirements of American National Standard for Information Sciences—Permanence of Paper for Printed Library Material, ANSI Z39.48–1992.